OLD-SCHOOL
ADVANCED CALCULUS

OLD SCHOOL ADVANCED CALCULUS

BY

W. BENJAMIN FITE
Davies Professor of Mathematics
Columbia University

With A New Preface and Recommended Reading Section by Karo Maestro

THE MACMILLAN COMPANY

New York   *1938*

Blue Collar Scholar (LLC Pending)/Createspace

2018 Edition

Original Edition:

COPYRIGHT, 1938
BY THE MACMILLAN COMPANY

All rights reserved—no part of this book may be reproduced in any form without permission in writing from the publisher, except by a reviewer who wishes to quote brief passages in connection with a review written for inclusion in magazine or newspaper.

Published January, 1938

2018 Edition Produced From Public Domain Version of Original

New Preface & RR Section @2018 Karo Maestro

PREFACE

This book has been written to supply an introductory course in mathematical analysis for those who are looking forward to specializing in mathematics. It is assumed that the reader has had a preliminary course in Calculus and is familiar with the rules for differentiating and integrating the ordinary functions. Although these rules have been taken for granted, I have been at some pains to define and explain the operations of differentiation and integration as clearly as possible, since the reader cannot hope to make satisfactory progress in analysis unless he understands these fundamental operations thoroughly. If he does understand them he ought to be able to follow the rest of the book with a reasonable amount of effort. The subject, however, is a difficult one and he should not expect to master it offhand.

Many teachers will prefer to assume the real number system, which is discussed in Chapter I, and to pass directly to Chapter II. This can be done without difficulty although the argument in several places rests on theorems proved in the first chapter. Since not every one will want to proceed in this way it seemed best to provide a chapter dealing with the real number system. In the other chapters the usual topics and theorems have been given. Nothing has been said about the Lebesgue theory of integration. The reader will meet that later if he continues his study of analysis. The last two chapters are of necessity somewhat sketchy. A more complete treatment would be out of place in a book of this kind. The reader who has not had, or does not plan to take, a course in the theory of functions of a complex variable will find a study of the last chapter worth his while. The chapter on the Calculus of Variations will introduce him to an important and extensive field. It is however more special than the one on functions of a complex variable, and should for this reason be omitted if both cannot be taken.

Although I have drawn my material from many sources I have kept the number of references down to a minimum, since a too elaborate set of references in a book of this kind would be confusing to the reader. I should mention de la Vallée-Poussin's *Cours d'analyse infinitésimale* and Volume I of Courant and Hilbert's *Methoden der mathematischen Physik*. One will recognize immediately the closeness with which I have followed the latter in the chapter on Fourier series.

I gratefully acknowledge my indebtedness to Professor W. B. Carver of Cornell University who read the manuscript of the first few chapters and made constructive and helpful comments thereon. I also received many valuable suggestions from Mr. Saul Gorn who read the galley proof and worked all the exercises.

W. B. F.

NEW YORK CITY,
November, 1937.

Preface To The 2018 edition

Textbooks are a unique form of literary activity. Unlike the far more romanticized and psychosocially motivated acts of creation that produce fiction, the creation of textbooks is usually dictated by far more specific needs and motivations.

Most times, these motivations and needs are pragmatic. A professor or teacher finds there's a dearth of standard texts for his or her course on a subject that is his or her research specialty and decides to write one. For example, Saunders MacLane & Garrett Birkoff's classic *A Survey of Modern Algebra* was born in their algebra course at Harvard University-the first in the United States for undergraduates-that way.

It may be that the author feels the standard textbooks are out of date and aren't really preparing his/her students for advanced study that in turn will prepare them for the current research landscape. Walter Rudin's *Principles Of Mathematical Analysis* famously began as a set of lecture notes Rudin prepared for MIT's advanced calculus course, for which he felt there were no truly modern presentations in English of analysis on abstract metric spaces at the time. The early versions of Serge Lang's *Algebra* began similarly with Lang's attempt to modernize graduate algebra courses with the language of category theory.

Or that author has a unique "take" on the subject which they can't find in any of the standard textbooks. For example, most of John Stillwell's highly unusual textbooks attempt to give a historical presentation of their subject matter, which in turn makes them quite unorthodox in both structure and topic selection. (For the record, it also makes all of them quite wonderful and well worth checking out.)

Of course, none of these motivations are purely objective. All authors, even textbook authors (particularly those who are passionate teachers) have their own personal emotional investments which moved them to undertake the not inconsiderable task of writing a textbook.

In this particular unique case, however, the catalyst for my reproduction of this archaic work is as simple as it is unusual: It was born from a complaint.

Or, to less politely but more accurately describe the incident in the language of Brooklyn where I was born and nurtured in my early years: The republication of this book is a product of bitching.

I should clarify that last remark somewhat. *Bitching*, in the street jargon of my old stomping grounds of the brownstones of Cypress Hills and Prospect Park, refers to a very particular kind of complaining. It's a complaining borne of frustration with what you see as a completely pointless system which you know in all likelihood will be completely indifferent to your objections and criticisms. The hope in this kind of complaining-besides venting your frustration, of course-is that enough people who hear it will agree and in turn complain to others, who will agree with them, etc.-and if enough backlash is created, those in charge might be forced to rethink their procedures. It's the kind of harmless complaining that can only take place in a democracy like the kind my grandparents and parents grew up in (which in principle, America still is, but to what degree is at this writing is very much an open question) and which can lead to subtle but positive changes in public or private policy.

In a previous generation, it was also called griping.

The griper in this case was one Wallace Goldberg.

Wallace Goldberg is the now-retired, former long-time chairman of the mathematics department of my undergraduate alma mater, Queens College of The City University of New York. At least, that's how most of us who were mathematics majors there at that time got to know him.

In addition, I got to know him both as a private undergraduate tutor-I remember him opening his door to me after my late father would drive me to his house for paid lessons in calculus and differential equations-and later on as my teacher in undergraduate partial differential equations.

(Those were better and happier days for me-when my biggest worry was my grades with my mom and dad waiting outside for me in the car while being tutored. After my father was diagnosed with cancer-those days ended with frigid abruptness.)

It was during the aforementioned PDE course that Goldberg made this gripe.

It was when he was lecturing on the elements of Fourier series for their use in the solution of second order partial differential equations. He was discussing the fundamental approximation theorem of Fourier series for differentiable functions. He turned from the blackboard with a sour face.

"I could prove this theorem for you-if you had a real advanced calculus course. None of you have, which is why none of you understand what uniform convergence is. All of you should have an advanced calculus course at this point as mathematics and physics majors, but our department's decided it's more important to sell courses to students that don't want to work. So they split the course into 2 nonsense courses. So as a result, I can't prove this theorem for you, so just memorize it and try to understand what it means."

He was visibly annoyed, which gave me the impression this was a fight he lost when previous chairs set up the curricula in the department. Our department, like all American mathematics departments, once did have a year-long advanced calculus course that all mathematics and many physics and engineering students took. Many longtime faculty pined for it, thinking of it longingly. But about a decade before I got there, after dramatic reductions in enrollment and many complaints from mathematics majors about the inconsistency of the course content, the department split the course into 2 new courses, Math 310 (Elementary Real Analysis) and Math 202 (Advanced Calculus). The title of the latter course was really somewhat deceptive because it was really what most other colleges call Vector Calculus and its' content varied enormously depending on who was teaching it. The former course was pretty stable and it corresponded to what used to be the first semester of the old year long AC course.

When I decided to republish this old jewel by Fite, I couldn't help but think of Dr. Goldberg's gripe and what I've since learned about the fascinating history of this most important and protean of courses that has, for the most part in American universities, gone the way of the dodo. His gripe was one I've become sympathetic to over the years-and wondering if perhaps its' time to reconsider its place in the undergraduate mathematics major.

****

Trying to determine where and when American universities began offering advanced calculus courses is a bit like trying to definitively answer the question of where and when the hamburger first was served in America. To answer the question in any reliable empirical manner would probably require years of well-funded historical research, excavating the course offerings of the oldest university mathematics departments in America going back to at least the 1890's.

That's assuming, of course that the entirety of those records still exist in some form.

What *can* be said definitively without much hard research is that the "advanced calculus" course as has commonly come be to be understood is a completely American invention. While mathematical analysis courses have been offered in European universities practically since the time of Euler-and the nature and content of those courses changed as analysis itself evolved from Newton to Dedekind and Cantor-they were not what we would think of as advanced calculus courses. They were either practical, manipulative courses in applied calculus or at top universities, rigorous courses in real or complex variables.

(And to go into more detail here would require a comparative and comprehensive study of the history of mathematics in the European and American university systems, which of course would require several volumes in and of itself. For the reader who is fascinated and determined to begin such a study, we heartily recommend beginning with the wondrous classic by Kline (1) as well as https://en.wikipedia.org/wiki/History_of_mathematics#19th_century.)

It can be safely stated without much argument that courses in rigorous analysis at the advanced university/research level were well-established in France and Germany by the turn of the 20th century. The distribution of its' central ideas globally began at this time with the publication of seminal works such as the second edition of Jordan's *Cours d'analyse* and de la Vallée Poussin's *Cours d'analyse Infinitesmale*.

American university mathematical training at this point was considerably inferior for the most part to Europe and geared towards very practical aims of training both teachers and engineers.

Indeed, in the first 2 decades of the 20th century immediately preceding the birth of the Mathematical Association of America, which was to play a vital role in creating what we think of today as the modern mathematics undergraduate and doctoral programs, due to the influence of various "elective-based" reform movements, mathematics came to play a very reduced role in American education. Some universities were an exception-Harvard, Princeton and Yale were the first to institute required core curricula for undergraduates to counter current "democratization" trends. (For more details, see the excellent article by Tucker (2)). Beginning in the late 19th century and continuing until the beginning of World War II, many of the strongest American doctoral students in math went to study in Europe and these became the first missionaries that would bring modern mathematics-particularly rigorous analysis-to these top American programs.

Advanced calculus courses have become somewhat nostalgically mythologized as a kind of Golden Age ruined by reformers. Many who were students in this courses in the 1960's and later think the rigorous AC course was in place at every decent university from the inception, where "real" students went and today's "wussy" students couldn't hack it. Truth be told, the first advanced calculus courses were offered only at the very top programs and were extremely diverse in content and level. While some universities *did* offer what came to be thought of as an advanced calculus course, namely a rigorous undergraduate course in functions of real variables building on basic calculus, these were the outlier programs. At most universities before World War II, the course was quite literally "advanced" calculus-namely a further non-rigorous course in calculus building upon single-variable calculus with topics such as calculus of several variables, differential equations, Fourier series and partial differential equations, including a number of applications to the physical sciences. The course also varied considerably in length-at some programs, it was a one semester course, at others, it was a full year course.

The syllabus of the proto-advanced calculus course illustrates a major fact that those who aren't knowledgeable about the history of this course don't realize, one I think whose importance in understanding how the course AC evolved-and eventually dismantled-cannot be underestimated.

Namely, the course we think of as "Calculus 3" i.e. a 19th century non-rigorous presentation of multivariable calculus-*simply did not exist at most universities until the 1950's*. Most college calculus courses before that were a year-long course in single variable calculus. Functions of several variables was considered an advanced subject that the second semester of the year-long AC course was intended to cover. Indeed, it was the original purpose for the course in the mathematics curricula at most programs! Why multivariable calculus was considered a "serious" topic unlike single variable calculus was probably a result of the enormous successes of both vector analysis and partial differential equations in the second half of the 19th century, putting thermodynamics, classical mechanics and the differential geometry of curves and surfaces all on a solid mathematical footing. Therefore, a presentation of functions of several variables was considered an essential prerequisite for a detailed study of classical vector and tensor analysis for physical science majors. Eventually, this evolved into the rigorous second semester of AC that built upon the careful treatment of one variable in the first semester. (Remember, at this point, the AC course at most programs was neither necessarily a year-long nor a rigorous course.)

It was the accepted necessity of taking calculus of several variables for all physics and engineering majors in addition to mathematics majors that made the AC course such an important course in the American math curricula. It was AC's importance to the training of physical science students that also heavily influenced making applications an important part of the evolving course. The first 3 standard texts in advanced calculus authored by major mathematicians more or less prove this, based on the lecture notes of E.B.Wilson and William Osgood at Harvard University and F.Woods at MIT. The much later course authored by David Widder also followed this pattern of a jigsaw content AC course whose level of rigor varied dramatically and emphasized an introduction to multivariable calculus. The first advanced calculus courses that modern students would recognize as American university undergraduate analysis courses were probably taught by James Peirpont at Yale and Oswald Velben at Princeton in the early 1900's.

Until the early 1930's at most universities, Advanced Calculus remained the "mixed bag" course that tried to cater to the needs of practical as well as theoretically minded students. But slowly, at the top tier universities, the European trained American mathematicians began to bring a modern approach to the courses they taught. These ideas eventually began to filter down to lesser programs via their students that eventually became PhDs that taught there. Math departments with strong undergraduate programs usually offered a one semester real analysis or "functions of a real variable" course to strong seniors to compensate for the insufficient rigor in most AC courses to prepare them for graduate studies. The publication of Hardy's *A Course of Pure Mathematics* in 1908 also had a gradual but profound effect on advanced calculus courses as it made a completely rigorous approach to analysis directly accessible for the first time to students who only spoke English.

The 2 dramatic changes that made America a top research and educational nation for mathematics and the physical sciences were a) the mass migration of many prominent young European mathematicians and scientists from France and Germany fleeing fascism in the 1930's and b) the major role many played in creating a host of important applications to physics, chemistry, engineering and economics vitally important in the American war effort in the 1940's. Even more important, many of these scientists brought their best young students with them, who would later become world leaders in the field based in America. It was then that the modern perspective on calculus truly arrived to American mathematics programs-indeed, the value of mathematics and physics was impressed for decades upon American culture. At the strongest programs-and by the 1960's, *all* universities-this meant the reshaping of the advanced calculus course into the form most mathematicians of Dr. Goldberg's generation knew it as students: It became a rigorous year-long course on the foundations of calculus and its' applications for undergraduates. The first semester focused on the theory of calculus on the real line while the second semester gave a careful presentation of functions of several variables on Euclidean spaces. Both semesters also incorporated many significant applications for physics and engineering students. This is the course Dr. Goldberg was reminiscing about in his cranky tirade.

And William Fite's forgotten text is a very strong early example of the apex of this mainstay of American undergraduate mathematics courses.

****
....

I've called this reissuing of Fite's text *Old School Advanced Calculus* as an accurate description of not only the contents, but the overall approach of this excellent text. As stated elsewhere, one of Blue Collar Scholar's prime directives will be to reissue out of print quality mathematics textbooks at a low price.

And Fite's text certainly qualifies as one.

But this particular reissue has another motivation-to bring back into print a standard AC text as a fine example of what today's mathematics programs are missing by breaking this course into 2 independent courses in single-variable analysis and vector calculus. This is done in the hope that it can encourage the constantly changing academic environment to consider bringing this course back, suitably updated and modified for today's students while still preserving the considerable strengths the original year-long AC course had over the current offerings.

It's worth briefly considering why this "split" occurred in the author's opinion. Part of the reason for The Great Bifurcation was a very real ideological divide within mathematics itself that occurred globally after World War II. The rise in Europe of the absolute abstraction of pure mathematics to the exclusion of all intuition and physical applications, exemplified by the increasingly influential Bourbaki school, lead to an academic reflection of this change. As a result, in the 1970's and 1980's, analysts trained in the fanatically purist Bourbakian style who taught the advanced calculus course began to teach it with increasing levels of generalization and abstraction. They also deliberately excluded from the course applications and intuition, which the Bourbaki school and its acolytes treated with utter contempt and scorn. The adoption of such textbooks as Jean Dieudonné's *Foundations of Pure Mathematics* and Walter Rudin's *Principles Of Mathematical Analysis* as "advanced calculus done right" at many universities was symbolic of this change. This of course alienated many mathematically inclined physics and engineering majors who would take the course. These students also formed an important source of income for mathematics departments generated by the AC course. This lead to steadily increasing pressures within mathematics departments to separate out the purely abstract material from the geometric/intuitive aspects of the course-which eventually lead to a literal split. The split of the original course into a purely mathematical course in single variable analysis and a vector calculus course with varying levels of rigor was symptomatic of a more fundamental change in the undergraduate/graduate curricula. It split the required calculus/analysis/differential equations courses into an "analysis for mathematicians" track and an "analysis for physicists and engineers" track. The expansion of the standard calculus sequence from a 2 semester to a 3-4 semester sequence containing a "Calculus III" non-rigorous multivariable calculus course was an important part of this reorganization. The availability of this course effectively ended the necessity of taking the standard mathematical AC course for physical science majors. It should be noted this split happened at this time in all branches of mathematics, but it had the most profound effect on calculus and analysis courses. In fact, in some very strong programs, a further bifurcation occurred in the vector analysis course- into vector calculus for physicists and engineers, which eliminated the need for theorems and proofs altogether and a course on analysis on differentiable manifolds, which contained only the abstract theory of functions of several variables in a totally modern context! This has had several major drawbacks for all mathematics students.

The major academic result is that the splitting of the AC course into 2-3 distinct courses has resulted in a codification of the separation of pure and applied mathematics into the undergraduate majors, which has been very counterproductive for both. While this indeed began to become a problem during the last days of the AC course in American universities, it seems to have grown far worse after the separation. This inconsistency has dire consequences for either of the major clientele of this critical course. The mathematics student who ends up in a non-rigorous vector analysis course-where the instructor loathes proofs and just gives the students a bunch of theorems to memorize without proof and lots of computations that ends up being a rehash of Calculus III-is not only going to be left wondering why he bothered to take real analysis, he or she is going to have to relearn the material themselves before graduate school. It's really here that the resulting redundancy of multivariable calculus courses has done the most damage in university programs. Conversely, when serious physics or engineering majors who actually want to learn rigorous proofs with the mathematics majors attempt to take the mathematics major vector analysis instead of the applied course they're encouraged to take by their department, they find it's taught by a Bourbakian pure mathematician who thinks physical applications or geometric sketches aren't real mathematics and begins the first lecture with the sentence, "A Banach space is......" –the result is they won't be able to drop the course fast enough. Even in smaller university programs where there is a single vector analysis which is offered to both math and physical science students, the resulting course is incredibly diverse in content and rigor even within a specific university depending on who teaches it! In any case, either kind of student gets the painfully reinforced message that either aspect of mathematics is mutually exclusive to the other-which of course, couldn't be further from the truth. To me, this fracturing of the pure and applied aspects of mathematics has weakened the mathematicians produced since they only have a one dimensional view of the subject. Many of the key ideas of both classical and modern analysis had their roots in physics and geometry. While there has been some welcome reversal of this splitting over the first part of the 21st century, pure and applied analysis courses remain in separate departments when both kinds of students are really not experiencing the full power of mathematics.

An added practical problem is that this separation has been positively reinforced by perceived specialization of the courses to the needs of particular students. While having separate tracks for pure and applied mathematics and 2 resulting courses on the calculus of vector valued functions that has had the desired effect of allowing each program to tailor the course for the specific needs of either math or physical science students, which doubles the number of paying customers-yeah, this is primarily about economics, let's not kid ourselves, ok? The resulting economic realities have had some very negative unintended-at least at first-consequences. For example, many math departments have a vector calculus course that's open to both math and science majors and which differs in which approach is taken by instructor. And of course, when that becomes a regular occurrence, students that only care about bell-curve busting grades to get them into Harvard med or engineering schools and don't give a damn about actual mathematics, but want credit for a "hard" course will plan for taking it in the professor that doesn't prove anything and lets them put the entire textbook into their programmable calculators. And when it's offered, such students will pack the course-therefore creating a financial incentive for financially strapped departments to ensure the course with that teacher runs regularly and the basic analysis course that could actually teach students-even the math majors who really need the course-won't be run. (That actually happened at Queens after the old AC course was split whenever vector analysis was taught by a certain professor. No, I won't name the professor, of course.) The result is a piecemeal curricula at any except very financially profitable schools with an enormous number of electives-and further weakening of undergraduate mathematical training across the board.

The main advantage of the original AC course, as exemplified by Fite, is a unified presentation of mathematical analysis comprised of virtually all the main topics of undergraduate analysis needed by both mathematics and physical science majors, covered using a uniform terminology and level of rigor. Even if each semester was taught by a different faculty member, they were both bound by more or less the same syllabus, which limited their ability to diverge from it drastically. When the subject selection, notation and rigor level is consistent throughout like it is with books like Fine's, then a balance that benefits all involved is achieved and maintained in the entire course. Pure mathematics students get exposed to important physical and geometric applications along with mathematical rigor. Physics and engineering students get exposed to pure mathematics and the abstract minimalist deductive skills it builds in them that will be invaluable when they begin research. Fite, in particular, does a terrific job of combining a careful "epsilon-delta' presentation of calculus of one and several variables with many applications to classical physics, differential equations and geometry.

We now give a brief description of the book's contents. The book begins in Chapter I with an unusual "semi-axiomatic" presentation of the real numbers via Dedekind cuts (or in the archaic terminology Fite uses, sections) of rationals. I call the presentation "semi-axiomatic" because although most of the theorems about Dedekind cuts-which are carefully defined at the beginning-from a formal construction of the real numbers are carefully stated, almost none are proven. However, virtually all of these results are used repeatedly in the course of the book, particularly in the proofs.

Chapters II and III cover the limits and differentiability properties (derivatives, differentials, partial derivatives, etc.) of real valued functions of one and several variables. The book only discusses explicitly the case of 2 variables and leaves the generalization to n variables to the reader.

Chapter IV discusses Taylor series expansions as the prototype of local power series approximations of functions. What's noteworthy about this chapter is that local extrema and Lagrange multipliers is developed entirely in terms of term by term differentiation at a point, something I doubt you'll find in most recent calculus or real analysis texts.

Chapter V gives a careful presentation of the Riemann integral in one variable in terms of Darboux upper and lower sums. This approach makes sense as Fite wants to ensure effective computation as well as sound theory. The Darboux approach does make taking limits simpler then the direct partition construction approach. The chapter also contains many standard applications, such as solids of revolution, areas under curves and trapezoidal approximation.

Chapter VI discusses the derivation of the techniques and formulas for indefinite integrals of all the standard functions of calculus. Yes, I know most calculus books cover this extensively, but Fite does it at a higher level then you'll see it done in those books-for example, there's a careful discussion of the algebra behind the method of partial fractions. He also includes some topics which are important in analysis that you won't see, such as elliptic and Abelian integrals.

Chapter VII discusses improper integrals and their convergence conditions. It also discusses the gamma function at some length. Chapter VIII is a lengthy chapter on multiple integrals, iterated integrals, line integrals and Green's theorem, change of variables and curvilinear coordinate systems as well as a number of applications.

Chapter IX gives the general theory of infinite series. It is richer then the chapter in most analysis books, containing not only the usual theory of convergence and tests, but unusual topics such as double series, quasi-uniform convergence and a detailed discussion of Wierstrauss' example of a continuous function with no derivative at any point in its domain. Chapter X presents the theory and applications of power series,building on the general theory presented in the last chapter and the earlier chapter on Taylor expansions. Chapter XI gives a classical exposition of trigonometric series with emphasis on Fourier and orthogonal series and some of their applications to partial differential equations. Chapter XIII discusses implicit functions, the Implicit Function Theorem and gives a basic introduction to 2 x 2 and 3 x 3 determinant and their applications such as the Jacobian. Chapters XIII gives an interesting presentation of the classical differential geometry of curves and surfaces in $\mathbf{R}^2$ and $\mathbf{R}^3$ as an application of all the earlier results, covering about as much as can be covered without tensor analysis or differential forms.

The last 2 chapters give optional introductions to the calculus of variations and functions of a complex variable-really just samplers of the elements of these subjects, in the hopes students will go on and take full graduate courses in them both. I think both subjects are of incredible importance in analysis as well as physics. Sadly, basic introductions to them both given in advanced calculus is one of the most important things that was lost with the course and this gives yet another good reason to reconsider a year-long AC course in the curricula.

The style of the book is very relaxed and methodical. Each section has scant examples, but also has a large number of exercises which are mostly straightforwardly computational. So in this regard, many of the exercises act as supplemental examples for that section. There are happily many classical physics examples such as work, exact forms in differential equations, applications of Lagrange multipliers to extremum problems, potential theory and others.

A look at the contents of Fite makes one notice something rather ironic-namely, a major subject that would be unthinkable in today's undergraduate math training is missing-namely, vector analysis. This bears some explanation. While Fite is not entirely devoid of the standard topics of a vector analysis course- line and surface integrals are discussed, as are parametric curves and surfaces and classical differential geometry-all these topics are presented analytically in a mid-19th century fashion, as functions of several real variables rather than vector-valued functions. When Fite taught his course in the 1930's, vector and tensor analysis was not part of the advanced calculus course. Vector and tensor analysis was still taught as a separate year-long course intended for physics and engineering students as well as for mathematics students with a serious interest in applications. This course was presented using the late 19th century framework of Gibbs' vector algebra and Einstein/Levi-Civita's summation conventions. This is primarily because the language of linear algebra was not yet part of how this subject was taught. Without linear algebra as a foundation, these frameworks result in a lengthy and sometimes cumbersome presentation. (Although in fairness, when it's done well, the old fashioned presentation style *does* give the student a great deal of geometric and physical insight which is not always present in modern books! Which is why I believe it's still well worth studying by students.) Linear transformations and vector spaces had been developed at this point as a branch of abstract algebra unifying vectors, matrices, tensors and determinants, which had all become important tools of computation in both math and physics in the late 19th century. As such, when Fite taught this course, you would only find linear algebra presented in graduate algebra courses at the very strongest programs. These ideas would not be incorporated into the second half of the AC sequence on functions of several variables to create the modern vector analysis course until linear algebra became a standard undergraduate course in the 1960's. (Indeed, if of the incorporation of linear algebra and the resulting revisions of the presentation of multivariable calculus were the only major changes made to the AC course at that time, it might still exist in its classical form!)

When I decided to republish Fite as an example of a classical AC course, I agonized for months over whether or not to add material on modern vector analysis on Euclidean spaces to "update" the course. While a rigorous version of the "Calculus III" approach to functions of several variables such as Fite's is not incorrect, of course-it is quite limited and won't really prepare students for further work in either mathematics or the physical sciences the way the treatment did when it was written. On the other hand, Fite's treatment is rich in applications as well as being very careful and the latter is something that's been lost in the modern multivariable calculus courses.

In the end, I reluctantly decided to leave the book unchanged. I did this for a few reasons. Firstly, I felt making such radical changes to the book would defeat the entire point of republishing it. I like Fite's book quite a bit in its' original form as a very solid example of the old-fashioned AC course and wanted the untampered book to make the case for the original syllabus. The other reason is because it will be very easy for any student or teacher using the book to supplement it with vector calculus material at the desired level and style. A great deal of free vector analysis lecture notes and OpenSource textbooks are currently available online for free. And I'll be making the perfect supplemental text, Keith Miller's *Vector Analysis,* available very shortly at Blue Collar Scholar. (OoOoOoOoOoO-aren't we sneaky?) All kidding aside, it's modern, concise, very inexpensive and complements Fite perfectly. This was the best solution to me because it keeps the published texts cheap while still preserving Fite's original work. Be on the lookout for it very shortly.

One more important recommendation on the use of Fite's book as a course text or self-study text. Back when this text was written, there really was no such thing as an honors calculus course at most universities. What most mathematics departments did with their strongest incoming students was test them and give them advanced placement credits into appropriate courses. A good example was given to me when I was in honors abstract algebra by the teacher, Kenneth Kramer. When he was a very strong upper level undergraduate in math at Columbia University in the 1960's, Columbia had no honors algebra sequence. Instead, he was placed out of the undergraduate abstract algebra course into Serge Lang's graduate course. Kramer told us it was both thrilling and terrifying at the same time-like swimming in the deep part of the pool when you were a kid for the first time by yourself. (Lang taught the course out of the first edition of his classic graduate text *Algebra*.) Similarly, entering freshmen that were strong in AP calculus were usually placed directly into the first term of the advanced calculus sequence.

I bring this up because I believe there is a serious dearth of appropriate textbooks on honors calculus for strong incoming freshmen. The standard textbook, Micheal Spivak's *Calculus* , while beautifully written with a host of terrific exercises, isn't really an honors calculus course. Yes, I know, a lot of people's heads just exploded. A calculus course, to me, by definition, has many applications to the sciences and geometry-and Spivak just has too few. To me, Spivak is a text for a basic analysis text-ironically, such as for the single variable analysis course that used to be the first semester of the AC course! Don't get me wrong, it's a great book, I just don't think without supplementing it extensively with applications, that it's appropriate for the kind of course it's usually used-or *mis*used-for. There are a bare handful of books that I think *would* be appropriate. But while I was preparing Fite for republication, it hit me that this book would be absolutely perfect for a year-long honors course in calculus for strong freshman! I was seriously considering retitling the book *Modern Honors Calculus ,* but the archaic nature of the way several topics are presented would have made it awkward if not outright inappropriate. But while the course in still *Old School Advanced Calculus*, that doesn't mean it can't be used as an inexpensive honors calculus course for today's students.

And I strongly encourage professors who are fortunate enough to be teaching the next generation of mathematicians in their freshman university honors calculus courses or schoolteachers with mathematically gifted students or homeschooling parents with gifted kids dreaming of Harvard someday to take a good long look at this book. I doubt you'll regret buying it.

Ok, I've rambled on long enough. I now hereby commend this long out of print treasure to the care of all students and teachers who love mathematics, both pure and applied. I hope you'll be as impressed and inspired by it as I was-as inspiring as it must have been to those mathematics students of yesteryear who first experienced it.

Lastly, I dedicate the book to Wallace Goldberg, whose grumpy demeanor and stoic leadership carried the Queens College mathematics department through the many lean years I was there and whose gripe in that long-ago lecture on Fourier series planted the seed for this book's rebirth for a new generation.

Enjoy.

Karo Maestro

New York City

March, 2018

CONTENTS

CHAPTER I
THE SYSTEM OF REAL NUMBERS

	PAGE
Definition of Real Numbers	1
The Fundamental Operations	7
Variables and Limits	10
Criterion for the Existence of a Limit	11
Complex Numbers	13

CHAPTER II
FUNCTIONS OF ONE VARIABLE

Definition of Function	17
Definition of Continuity	18
Properties of Continuous Functions	20
The Derivative	26
The Derivative of a Composite Function	28
Geometrical Interpretation of the Derivative	29
Rolle's Theorem	31
The Mean Value Theorem	31
Derivatives of a Higher Order	33
Infinitesimals	34
Principal Part of an Infinitesimal	35
Differentials of a Higher Order	36

CHAPTER III
FUNCTIONS OF MORE THAN ONE VARIABLE

Continuity	40
Partial Derivatives	43
Total Differentials	44
An Application in Geometry	48
Invariance of the Total Differential	49
Differentiation of Equations	50
Directional Derivatives	51
Exact Differentials	54
Euler's Theorem on Homogeneous Functions	59
The Order of Differentiation	60

CONTENTS

CHAPTER IV

TAYLOR'S EXPANSION WITH THE REMAINDER. APPLICATIONS

General Remarks	64
Only One Expansion	66
Infinite Series	66
Expansion of log $(1 + x)$	67
Euler's Constant	69
Other Examples	70
Indeterminate Forms	72
The Indeterminate Form $\frac{0}{0}$	73
Applications to Geometry	75
Maxima and Minima of Functions of One Variable	76
Generalized Taylor Series	77
Maxima and Minima of Functions of Two Variables	79
Lagrange Multipliers	81
A Generalization	83

CHAPTER V

THE DEFINITE INTEGRAL

An Appeal to Geometry	88
Consecutive Methods of Subdivision	89
Darboux's Theorem	90
Integrable Functions	91
Properties of the Definite Integral	93
Volume of a Solid of Revolution	96
First Mean Value Theorem for Integrals	97
Second Mean Value Theorem for Integrals	98
The Definite Integral as a Function of Its Limits	100
Differentiation under the Integral Sign	101
The Indefinite Integral	103
Approximate Integration	105
Determination of the Error	106
Applications of the Definite Integral	109

CHAPTER VI

INDEFINITE INTEGRALS

Integration of Rational Functions	117
Unicursal Curves	121

CONTENTS

Elliptic Integrals.. 122
Numerical Computation.................................... 126

CHAPTER VII

IMPROPER AND INFINITE INTEGRALS

Improper Integrals... 133
Tests for Convergence..................................... 134
Practical Rule... 136
Infinite Integrals... 139
Integration by Parts....................................... 142
Differentiation and Integration under the Integral Sign.... 142
The Gamma Function.. 146
The Beta Function... 148
The Relation between the Beta Function and the Gamma
 Function... 148
Stirling's Formula.. 150
Stirling's Series... 153

CHAPTER VIII

DOUBLE AND TRIPLE INTEGRALS

The Double Integral....................................... 157
Simple Properties of the Double Integral.................. 158
Mean Value Theorem.. 159
Repeated Integrals.. 160
Line Integrals and Green's Theorem........................ 166
The Integral $\int P dx + Q dy$........................... 169
Simply and Multiply Connected Regions..................... 172
Change of Variables in a Double Integral.................. 174
Geometric Applications of Double Integrals................ 176
Triple Integrals.. 181
Repeated Integrals.. 182
Change of Variables in a Triple Integral.................. 185
Surface Integrals... 186
Green's Theorem in Space.................................. 187
Stokes' Theorem... 188
Potential... 195
Work.. 198
Total Differentials....................................... 200

CONTENTS

CHAPTER IX

INFINITE SERIES

Sequences	205
The Greatest Limit of a Bounded Sequence	205
Infinite Series	208
General Test for Convergence	209
Series with Positive Terms	209
Cauchy's Root Criterion	212
D'Alembert's Test	212
Application of the Greatest Limit	213
Series with Arbitrary Terms	213
Series with Variable Terms	217
The Weierstrass Test for Uniform Convergence	223
Abel's Lemma	224
Abel's Test for Uniform Convergence	225
Differentiation of an Infinite Integral	227
Test for Uniform Convergence of an Infinite Integral	230
Infinite Integrals of Infinite Series	232
Integration of an Infinite Integral	235
A Continuous Function without a Derivative	237
Quasi-uniform Convergence	239
Double Series	240
General Condition for Convergence	242
Double Series Whose Terms Are All Positive or Zero	242
Test for Convergence	243
Absolute Convergence	243
Multiplication of Simple Series	245

CHAPTER X

POWER SERIES

Definitions and Theorems	247
Connection between the Coefficients and the Radius of Convergence	249
Continuity of a Power Series	249
The Derivative of a Power Series	250
Dominant Functions	252
Substitution of One Series in Another	253
Division by a Power Series	257
Double Power Series	259

CONTENTS

CHAPTER XI

TRIGONOMETRIC SERIES AND SERIES OF ORTHOGONAL FUNCTIONS

Schwarz's Inequality	264
Definition	264
Fourier Series	267
An Application in the Theory of Heat	281
An Exponential Form for Fourier Series	284
Dirichlet Integrals	285
The Fourier Integral	288
Weierstrass' Theorem	290
Orthogonal Functions	291
A Recursion Formula	295
An Application to Potential Theory	296

CHAPTER XII

IMPLICIT FUNCTIONS. FUNCTIONAL DETERMINANTS

Existence Theorems	299
Derivatives of Implicit Functions	304
Applications of the Preceding Theory	306
Functional Determinants	308
An Important Functional Determinant	311
Dependence of Functions	313

CHAPTER XIII

APPLICATIONS TO GEOMETRY

Parametric Equations of Curves and Surfaces	319

PLANE CURVES

Envelopes of Families of Curves	321
Sufficient Conditions	324
An Application in Optics	326
Curvature	328
The Osculating Circle	329
The Evolute	331
An Example	331
Intrinsic Equation of a Curve	333

CONTENTS

SKEW CURVES

Length of an Arc.................................... 336
The Differential of Arc.............................. 338
The Osculating Plane................................ 339

SURFACES

Envelopes of One-Parameter Families of Surfaces......... 342
Developable Surfaces................................ 344
The Osculating Planes of Γ.......................... 345
Envelopes of Two-Parameter Families of Surfaces......... 345

CHAPTER XIV

CALCULUS OF VARIATIONS

The Simplest Problem................................ 350
Examples from Geometry and Mechanics................. 350
Euler's Equation.................................... 353
Minimum Surface of Revolution....................... 355
The Brachistochrone................................. 357
Two Independent Variables........................... 359
Variable End Points................................. 363
Parametric Equations................................ 365
Hamilton's Principle................................. 366
The Spherical Pendulum.............................. 367
An Application in Economics......................... 369

CHAPTER XV

FUNCTIONS OF A COMPLEX VARIABLE

The Geometric Representation of Complex Numbers....... 372
Geometric Interpretation of Addition and Multiplication... 372
Geometric Interpretation of Division.................. 374
The Roots of a Number.............................. 374
Some Elementary Functions.......................... 375
Analytic Functions.................................. 380
Condition for Differentiability....................... 381
Integration... 384
Cauchy's Integral Theorem........................... 386
Cauchy's Integral Formula........................... 386
Taylor's Theorem.................................... 389
Real Integrals...................................... 391

INDEX.. 397

ADVANCED CALCULUS

ADVANCED CALCULUS

CHAPTER I

THE SYSTEM OF REAL NUMBERS

1. This book has to do almost entirely with the theory of functions of real variables. There is a brief discussion of complex numbers and the theory of functions of a complex variable. But complex numbers are defined in terms of real numbers. So that the system of real numbers forms the basis of our whole discussion. This makes it desirable to obtain at the start as clear a notion as possible of the nature of this system.

We assume that the reader is familiar with the positive and negative integers, including zero, and with numbers of the form $\frac{a}{b}$, where a and b are integers. These constitute the *rational* numbers.

It is possible in many ways to divide all the rational numbers into two mutually exclusive classes, A and B, such that every number in class A is less than every number in B. For example, we might assign to class A all the rational numbers that do not exceed 7 and to B all those that do exceed this number.

With respect to such a classification there are three possibilities:

(a) In the first place, there may be a number in A greater than any other number in this class, as in the example just cited. In this case there is no least number in B, for if α were the greatest number in A and β the least one in B, then $\frac{\alpha + \beta}{2}$, which is greater than α and less than β, would be in neither class. But this contradicts the hypothesis that every rational number is in either A or B.

(b) In the second place, there may be a least number in B. This would be the case, for example, if we made the classification by assigning to A all the rational numbers that are less than 7 and to B all the remaining rational numbers. Here there is no greatest number in A.

(c) And finally, it may be that there is neither a greatest number in A, nor a least number in B. Suppose, for example, that we assign to A all the negative rational numbers, and also all the positive ones, including zero, whose squares do not exceed 5; and to B every other rational number.

We shall speak of a classification of the rational numbers such as we have described as a "section" and shall represent it by the symbol (A, B), or, in case there is no occasion to bring the classes A and B into prominence, by a single small letter, as α.

We shall call every such section a number. In the cases (a) and (b) this number will be taken as the greatest number in A and the least number in B respectively. The appropriateness of this designation is due to two important properties of these sections. In the first place, we can arrange these sections in a scale of relative magnitudes, and in the second place we can prescribe for them laws of operation that will be true generalizations of the four fundamental operations that we apply to rational numbers.

In the definition of a section we have assumed a definite scale of magnitude for the rational numbers. And we can bring the sections arising under (c) into this same scale in the following way: If $\alpha = (A, B)$ and $\alpha' = (A', B')$ are two such sections, we shall say that α and α' are equal in case every number in A is in A' and every number in A' is in A. If there are numbers in A that are not in A', we shall say that α is greater than α'; in symbols, $\alpha > \alpha'$. If α is a number arising under (c) and β is a rational number, we shall say that α is greater than β in case β is contained in A. If β is greater than any number of A, we shall say that β is greater than α. This means that α is greater than any number in A and less than any number not in A. But no rational number has this property. For if γ is a rational number greater than any number in A, it must be in B. But by hypothesis B has no least number, and therefore there are numbers in B that are less than γ.

We shall call the new numbers arising under (c) *irrational numbers*, and the totality of rational and irrational numbers *real numbers*.

If we change the signs of every number in A and B, we get two new classes which we shall designate by $-A$ and $-B$ respectively. Then $(-B, -A)$ is a section, or number. If $\alpha = (A, B)$ we shall denote $(-B, -A)$ by $-\alpha$. Obviously $-(-\alpha) = \alpha$. If α is a rational number different from zero, one of the numbers α and $-\alpha$ is positive, that is, greater than zero. This is also true of every irrational number. For if α is irrational and less than zero, zero is greater than any number in A, and is therefore in B. But B has no least number and hence contains numbers less than zero, or negative numbers. This means that there are positive numbers in $-B$, and hence that $-\alpha > 0$.

DEFINITION. That one of the numbers α and $-\alpha$ which is positive is called the *absolute value* of α and is denoted by the symbol $|\alpha|$.

2. Preliminary theorems. We shall need the following theorems:

THEOREM 1. *If $\alpha = (A, B)$ is any real number, there are rational numbers a and b in A and B respectively such that $b - a = \epsilon$, where ϵ is any rational number.*

Select any rational number a_1 in A and form the sequence

$$a_1, \quad a_1 + \epsilon, \quad a_1 + 2\epsilon, \quad \cdots, \quad a_1 + n\epsilon, \quad \cdots.$$

Then by the axiom of Archimedes [1] from a certain point on the terms of the sequence will exceed a given number in B, and will therefore be contained in B. Let $a_1 + n\epsilon = b$

---

[1] See, for example, Young, *Fundamental Concepts of Algebra and Geometry*, pp. 148, 149.

be the first of these terms that are in B. It is obviously rational. Then $a = a_1 + (n-1)\epsilon$ is a rational number in A, and $b - a = \epsilon$.

If α and α' are any two rational numbers such that $\alpha < \alpha'$, there is a rational number β that is greater than α and less than α'. For example, $\beta = \dfrac{\alpha + \alpha'}{2}$. It follows that there is an unlimited number of rational numbers between α and α'. We show that this is also true if α and α' are any distinct real numbers.

THEOREM 2. *If α and α' are any distinct real numbers such that $\alpha < \alpha'$, there are rational numbers β such that $\alpha < \beta < \alpha'$.*

In view of what has just been said, we can assume that at least one of the given numbers is irrational. If both of them are irrational, the theorem follows from the definition of the term "greater than." If one of them, as $\alpha = (A, B)$, is irrational while α' is rational, then α' belongs to class B. But there is no least number in this class, and every number in this class is greater than α. Hence the theorem. If α is rational and α' irrational, the argument is similar. It follows that there is an infinity of rational numbers between α and α'.

EXERCISE. Show that in any interval there is only a limited number of rational numbers $\dfrac{p}{q}$, where p and q are integers and $|q| < M$, any positive number.

THEOREM 3. *If α and α' are distinct real numbers such that $\alpha < \alpha'$, there is an infinity of irrational numbers β such that $\alpha < \beta < \alpha'$.*

We know from Theorem 2 that there are two rational numbers p and q such that

$$\alpha < p < q < \alpha'.$$

We shall assume at first that p and q are positive. Select any rational number m between p^2 and q^2. If there are two integers a and b such that $\dfrac{a^2}{b^2} = m$, that is, if m is the square of a rational number, let r be a prime integer not a divisor of b such that $m + \dfrac{1}{r}$ is between p^2 and q^2. If there were two relatively prime integers a' and b' such that $\dfrac{a'^2}{b'^2} = m + \dfrac{1}{r}$, we should have

$$\frac{a'^2}{b'^2} - \frac{a^2}{b^2} = \frac{a'^2 b^2 - a^2 b'^2}{b^2 b'^2} = \frac{1}{r};$$

or

$$r(a'^2 b^2 - a^2 b'^2) = b^2 b'^2.$$

This last equation requires that r be a divisor of b', and therefore of a'. But a' and b' are relatively prime. Hence m and $m + \dfrac{1}{r}$ are rational numbers between p^2 and q^2, one of which is not the square of a rational number. We designate this one by m and form a section by assigning to class C all negative rational numbers and those positive rational numbers (including zero) whose squares are less than m, and to D all the remaining rational numbers. The number $\gamma = (C, D)$ is an irrational number between p and q and therefore between α and α'. In the light of this discussion the case in which p and q are not both positive will present no difficulty. And if there is one irrational number between α and α', there is an infinity of such numbers.

THEOREM 4. *If we divide the set of all real numbers into two mutually exclusive classes A and B such that every number in A is less than every number in B, there is a number m, which may be rational or irrational, such that every real number less than m is in A and every real number greater than m is in B.*

Let A' be the class of all the rational numbers in A and B' the class of all rational numbers in B. If there is a greatest number m in A', this number m will also be the greatest number in A. For if there were in A a number $\beta > m$, there would be rational numbers between m and β

and therefore in A and A'. But this is a contradiction. We can show in a similar way that if there is a least number m in B', this m will also be the least number in B. In these two cases the number m satisfies the conditions of the theorem and is rational.

If there is neither a greatest number in A' nor a least number in B', the number $\alpha = (A', B')$ is irrational. It belongs either to A or to B. If it belongs to A it is the greatest number of this class, since every number greater than α is also greater than an infinity of rational numbers of B'. But these latter belong to B. Hence every number greater than α belongs to class B, and α is the greatest number of A. If α belongs to B, it can be shown in a similar way that it is the least number in B.

3. Definition. A set of real numbers is said to have an *upper bound* if there is a number M that is greater than any number of the set; and to have a *lower bound* if there is a number M' that is less than any member of the set. If the numbers M and M' both exist, we say that the set is *bounded*.

The set of all positive numbers, for example, has a lower bound but no upper bound; while the set of all negative numbers has an upper bound but no lower bound. The set of numbers whose squares do not exceed 3 is bounded.

THEOREM 5. *If a set of real numbers has an upper bound, there is a number λ such that:*

(a) *No member of the set is greater than λ.*

(b) *If ϵ is an arbitrary positive number, there is a member of the set that is greater than $\lambda - \epsilon$.*

We can make a classification of all real numbers by assigning to one class all numbers that are exceeded by members of the set, and to the other class all other numbers. The conclusion then follows immediately from Theorem 4. This number, when it exists, is called *the upper limit* of the set.

For example, if the set is composed of all numbers of the form $1 + \dfrac{2^n - 1}{2^n}$, where n is any positive integer, $\lambda = 2$.

If the set has a lower bound, we can see in like manner that there is a number λ' such that:

(a) No member of the set is less than λ'.

(b) If ϵ is an arbitrary positive number, there is a member of the set that is less than $\lambda' + \epsilon$.

This number, when it exists, is called *the lower limit* of the set. If the set is composed of all numbers of the form $\frac{1}{2^n}$, where n is a positive integer, $\lambda' = 0$.

4. Addition. We are now in a position to define the first of the fundamental operations. Let $\alpha = (A, B)$ and $\alpha' = (A', B')$ be any two real numbers. We select any two rational numbers a and a' that are less than α and α' respectively, and any two rational numbers b and b' that are greater than α and α' respectively. For every such choice of these four numbers we shall have

$$a + a' < b + b'.$$

If we assign to A_1 all the rational numbers that do not exceed all numbers of the form $a + a'$, and to B_1 all the remaining rational numbers, the number $\alpha_1 = (A_1, B_1)$ is by definition the *sum* of the numbers α and α'.

If α and α' are not both rational, there is no greatest number of the form $a + a'$, nor least one of the form $b + b'$; and α_1 is greater than any number of the first of these forms and less than any number of the second form. Moreover it is the only number that has these two properties. For if α_2 were another such number, we could find two rational numbers r_1 and r_2 ($r_2 > r_1$) that lie between α and α' (Theorem 2). But we know from Theorem 1 that we can select the four numbers a, a', b, and b' in such a way that

$$(b + b') - (a + a') < r_2 - r_1.$$

And this is a contradiction.

We have based this definition upon the ordinary definition of the sum of two rational numbers. We leave it to the reader to verify that the two definitions are equivalent, and that when α and α' are both rational, the operation of addition as thus defined has the following properties:

$$\alpha + \alpha' = \alpha' + \alpha,$$
$$(\alpha + \alpha') + \alpha'' = \alpha + (\alpha' + \alpha''),$$
$$\alpha + 0 = \alpha,$$
$$\alpha + (-\alpha) = 0,$$
$$|\alpha + \alpha'| \leq |\alpha| + |\alpha'|.$$

5. Subtraction. If α and α' are any real numbers, there is one, and only one, real number β such that

$$\alpha = \beta + \alpha'.$$

In the first place, there is one such number; namely, $\alpha + (-\alpha')$, since $\alpha + (-\alpha') + \alpha' = \alpha$. Moreover, there is only one such number. For if

$$\alpha = \beta + \alpha',$$
$$\alpha = \beta_1 + \alpha',$$

then

$$\beta + \alpha' = \beta_1 + \alpha',$$
$$\beta + \alpha' + (-\alpha') = \beta_1 + \alpha' + (-\alpha'),$$
$$\beta + 0 = \beta_1 + 0,$$
$$\beta = \beta_1.$$

This number β is called the *difference* between α and α'. It is formed by adding $-\alpha'$ to α. This operation is called *subtraction*. It is the inverse of addition.

6. Multiplication. We suppose in the first place that α and α' are two positive real numbers. If a and a' are two positive rational numbers such that $a < \alpha$ and $a' < \alpha'$; and b and b' are two rational numbers such that $b > \alpha$ and $b' > \alpha'$, we form a section by assigning to A_1 all negative rational numbers, together with zero, and all positive rational numbers that do not exceed all products of the form aa', and to B_1 all the remaining rational numbers. The number $\alpha_1 = (A_1, B_1)$ shall be by definition the *product* of the positive numbers α and α'.

§7] THE SYSTEM OF REAL NUMBERS

This product is greater than every number of the form aa' and less than every number of the form bb', and it is the only number with these two properties.

If one or both of the factors are negative and neither is zero, the product shall be the number obtained by forming the product of the absolute values of the factors and then applying the usual rule of signs. If one of the factors is zero, we define the product as zero. The reader can verify that multiplication as thus defined has the following properties:

$$\alpha\alpha' = \alpha'\alpha,$$
$$(\alpha\alpha')\alpha'' = \alpha(\alpha'\alpha''),$$
$$\alpha(\alpha' + \alpha'') = \alpha\alpha' + \alpha\alpha'',$$
$$\alpha \cdot 1 = \alpha,$$
$$|\alpha\alpha'| = |\alpha| \cdot |\alpha'|.$$

7. The reciprocal of a real number. If α is any rational number different from zero, there is a unique rational number α' such that $\alpha\alpha' = 1$. Either of these numbers is called the *reciprocal* of the other. We proceed to show that every irrational number has a unique reciprocal in this sense.

Suppose in the first place that $\alpha = (A, B)$ is a positive irrational number. We form a section by assigning to A' all negative numbers, zero, and the reciprocals of all the numbers in B; and to B' all the remaining rational numbers. Then $\alpha' = (A', B')$ is such that $\alpha\alpha' = 1$. For if ϵ is an arbitrarily small positive rational number, there is a rational number a in A and a rational number b in B such that $b - a = \epsilon$ (Theorem 1). Hence $1 - \frac{a}{b} = \frac{\epsilon}{b}$. Since every b is greater than any a, we can make $\frac{\epsilon}{b}$ as small as we please by properly selecting ϵ, and therefore $\frac{a}{b}$ as near to 1 as we please. If for any b we put $a' = \frac{1}{b}$, then for any a we have aa' less than 1. This is equivalent to saying that $\alpha\alpha' = 1$.

Moreover α' is the only number that has this property. For if
$$\alpha\alpha_1' = 1,$$
then
$$\alpha'\alpha\alpha_1' = \alpha'$$
and
$$\alpha_1' = \alpha'.$$

This unique number α' is called the *reciprocal* of the positive number α and is represented by the symbol $\dfrac{1}{\alpha}$.

If α is a negative irrational number, we define its reciprocal as the negative of the reciprocal of the absolute value of α.

8. Division. If α is any real number and β is any real number except zero, the operation of finding a number γ such that $\alpha = \beta\gamma$ is called the operation of *dividing α by β*, and γ is called the *quotient* obtained by this operation.

Since
$$\alpha = \beta\left(\alpha \cdot \frac{1}{\beta}\right),$$
$\alpha \cdot \dfrac{1}{\beta}$ is a value of γ that satisfies the equation $\alpha = \beta\gamma$. Moreover it is the only solution, since if
$$\alpha = \beta\gamma = \beta\gamma',$$
then
$$\frac{1}{\beta} \cdot \beta\gamma = \frac{1}{\beta} \cdot \beta\gamma',$$
or
$$\gamma = \gamma'.$$

9. Variables and limits. If we use a letter to represent any one of a given set of numbers, we shall refer to it as a *variable*. For example, the numbers represented by the letter x may be the speeds with which a given body is moving at different times. If the variable x represents a succession of numbers in such a way that, for an arbitrary positive number ϵ, the difference between x and a number a becomes, and remains, less than ϵ in absolute value, we shall say that *x approaches the limit a;* or, in symbols, $x \to a$.

A variable may not have a limit. For example, $\sin x$ as x increases beyond any assigned value. On the other hand, $5 + \dfrac{1}{x}$ approaches the limit 5 under the same circumstances. But a variable cannot have two limits. For if these limits were a and b ($b > a$) and we take $\epsilon < \dfrac{b-a}{2}$, we could not have at the same time $|x - a| < \epsilon$ and $|x - b| < \epsilon$, since these inequalities imply that

$$a - \epsilon < x < a + \epsilon$$

and

$$b - \epsilon < x < b + \epsilon.$$

But $a + 2\epsilon < b$, or $a + \epsilon < b - \epsilon$, and x cannot at the same time be less than $a + \epsilon$ and greater than $b - \epsilon$.

10. Criterion for the existence of a limit. The following theorem gives a general criterion for determining whether a variable has a limit, or not:

THEOREM 6. *In order that the variable x approach a limit, it is necessary and sufficient that for an arbitrary positive number ϵ all the values of the variable from a given one on be such that the difference between any two of them is less than ϵ in absolute value.*

In the first place, the condition is necessary, since if x has a limit a, all the values of x from a certain one on, as x' and x'', differ from a by less than $\dfrac{\epsilon}{2}$ in absolute value. That is,

$$|x' - a| < \frac{\epsilon}{2},$$

$$|x'' - a| < \frac{\epsilon}{2}.$$

Hence

$$|x' - x''| < 2 \cdot \frac{\epsilon}{2} = \epsilon.$$

To prove that the condition is sufficient we consider in the first place the possibility that all the values of x from a certain one on are neither all greater than, nor all less than, some number a. Since the differences in the values of x approach zero and there are always some of these values less than a and some greater than a, the difference between a and x approaches zero, and a is the limit of x.

Now by hypothesis for any positive ϵ there is a value x' of x such that any succeeding value is less than $x' + \epsilon$. If all the values of x from a certain one on are greater than some number, we form a section by assigning to class A all real numbers that are exceeded by all the values of x from a certain one on, and to B all the remaining real numbers. We know from what has just been said that there will be numbers in class B. We know also from Theorem 4 that there is a number a such that every number less than a is in A and every number greater than a is in B. Then all the values of x from a certain one on are between $a - \epsilon$ and $a + \epsilon$. That is, x approaches the limit a. The same conclusion can be reached by a similar argument in case all the values of x from a certain one on are less than some number.

This criterion is perfectly general, whereas the one described in the following theorem is more restricted in its application, although easier to apply and therefore more useful.

THEOREM 7. *If a variable never increases or never decreases, a necessary and sufficient condition that it approach a limit is that it remain less than a given number in absolute value.*

The necessity of the condition is obvious. To prove its sufficiency we suppose first that the variable always increases. We then form a section by assigning to class A all real numbers that are exceeded by values of the variable, and to class B all remaining real numbers. The fact that all the values of the variable are less than a given number insures the existence of numbers in B. We know from Theorem 4 that there is a number a such that $a - \epsilon$ is in class A and $a + \epsilon$ in class B, when ϵ is any positive number. It follows from this that a is the limit of the variable.

Variables such as those described in this theorem are said to be *monotonic*.

EXERCISES

1. Prove that if $x \to a$ and $y \to a$, while z is always between x and y, then $z \to a$.

2. Prove that if $x \to a$ and $y \to b$, then $x + y \to a + b$, $x - y \to a - b$, and $xy \to ab$. Also if $b \neq 0$, then $\dfrac{x}{y} \to \dfrac{a}{b}$.

11. Complex numbers. We define a complex number as an ordered pair (a, b) of real numbers subject to certain laws of combination which we shall define and to the condition that $(a, 0)$ shall be the real number a. We shall later find it convenient to introduce another symbol for these numbers.

Addition. The *sum* of the two numbers (a, b) and (c, d) is defined as the number $(a + c, b + d)$. This is a generalization of the definition of the sum of two real numbers, since $a + c = (a, 0) + (c, 0) = (a + c, 0) = a + c$.

Multiplication. The *product* of the two numbers (a, b) and (c, d) is defined as the number $(ac - bd, ad + bc)$. The reason for this definition is not as obvious as the reason for the definition given of addition. It will become clear later. It is immediately obvious, however, that this is a generalization of the definition of the product of two real numbers, inasmuch as $ac = (a, 0) \cdot (c, 0) = (ac, 0) = ac$.

Subtraction. Since the product of the real number a by minus one is $-a$ we shall say that the product of the complex number $\alpha = (a, b)$ by minus one is $-\alpha$. But this product is $(-a, -b)$. We agree then to say that $-\alpha = (-a, -b)$. We then define the *difference* of $\alpha = (a, b)$ and $\beta = (c, d)$, or $\alpha - \beta$, as the sum $\alpha - \beta$ of $\alpha = (a, b)$ and $-\beta = (-c, -d)$, or $(a - c, b - d)$.

The operations of addition, subtraction, and multiplication as thus defined obey the associative, commutative, and distributive laws to which the corresponding operations with real numbers are subject. Thus, if $\alpha = (a, b)$,

$\beta = (c, d)$, and $\gamma = (e, f)$, then

$\alpha + \beta = \beta + \alpha$ Commutative law for addition

$\alpha \cdot \beta = \beta \cdot \alpha$ Commutative law for multiplication

$(\alpha + \beta) + \gamma = \alpha + (\beta + \gamma)$ Associative law for addition

$(\alpha \cdot \beta)\gamma = \alpha(\beta \cdot \gamma)$ Associative law for multiplication

$\alpha(\beta + \gamma) = \alpha\beta + \alpha\gamma$ Distributive law for multiplication.

It follows from the definition of multiplication that $(0, 1) \cdot (0, 1) = (-1, 0) = -1$. Hence $(0, 1)$ is a square root of -1. It is conventionally represented by the letter [1] i. That is, $i = (0, 1) = \sqrt{-1}$. Moreover $(0, b) = (b, 0)(0, 1) = bi$. It follows that $(a, b) = (a, 0) + (0, b) = a + bi$. We shall take advantage of this fact and discard the symbol (a, b) in favor of the more suggestive symbol $a + bi$. We call a the *real* part and bi the *imaginary* part of $a + bi$. A number is *real* if its imaginary part is zero. If its real part is zero, the number is said to be a *pure imaginary*.

Definition of equality. The two numbers $a + bi$ and $c + di$ are said to be *equal* if $a = c$ and $b = d$. Otherwise they are unequal. If therefore we know that two complex numbers are equal, we know also that their real parts are equal and that their imaginary parts are equal.

The number $a - bi$ is said to be the *conjugate* of $\alpha = a + bi$ and is represented by the symbol $\bar{\alpha}$. A real number is its own conjugate and a pure imaginary is minus its conjugate. The product $\alpha\bar{\alpha} = (a + bi) \cdot (a - bi) = a^2 + abi - abi - b^2i^2 = a^2 + b^2$, since $i^2 = -1$.

12. Division. If $\beta = c + di \neq 0$ and

$$\alpha = \frac{c}{c^2 + d^2} - \frac{di}{c^2 + d^2} = \frac{\bar{\beta}}{c^2 + d^2},$$

then

$$\alpha\beta = \frac{\beta\bar{\beta}}{c^2 + d^2} = \frac{c^2 + d^2}{c^2 + d^2} = 1.$$

[1] This notation is due to Euler (1777).

That is, $\dfrac{\bar{\beta}}{c^2+d^2}$ is the reciprocal of β and we represent it by the symbol $\dfrac{1}{\beta}$. Thus, $\dfrac{1}{\beta} = \dfrac{\bar{\beta}}{c^2+d^2}$.

By the *quotient* of any number α divided by β we mean the number γ such that $\alpha = \beta\gamma$. (See §8.) Since, as we have just seen, $\dfrac{\beta\bar{\beta}}{c^2+d^2} = 1$, it follows that $\gamma = \dfrac{\alpha\bar{\beta}}{c^2+d^2}$ is one solution of the equation $\alpha = \beta\gamma$. Moreover it is the only solution. For if $\beta\gamma = \beta\gamma_1$, then $\bar{\beta}\beta\gamma = \bar{\beta}\beta\gamma_1$ and $\gamma = \gamma_1$. We are assuming that $\beta \neq 0$, since if $\beta = 0$ there is no solution of the equation in case $\alpha \neq 0$, and if $\alpha = 0$ every number is a solution. If α and β are real numbers, say, a and c respectively, we have $\dfrac{\alpha}{\beta} = \dfrac{ac}{c^2} = \dfrac{a}{c}$. Thus division as here defined is a true generalization of division as applied to real numbers.

13. Inasmuch as the four fundamental operations can be applied to the entities $a + bi$ in the same way as to real numbers, it seems appropriate to call them numbers, and indeed *complex numbers*, since it takes two real numbers to define one of them.

There *is* this important difference between these new numbers and real numbers: If a and b are two distinct real numbers, one of them is greater than the other one; but we cannot say this of two distinct complex numbers.

In applying the fundamental operations to complex numbers we proceed then exactly as if we were dealing with real numbers. We can often simplify the result at any stage by replacing i^2 by -1 wherever it occurs. Thus

$$(a+bi)\cdot(c+di) = ac + bci + adi + bdi^2$$
$$= (ac-bd) + (ad+bc)i.$$

The reader will see from this the reason for the definition of multiplication that has been given.

14. Complex numbers play an important rôle in almost all branches of mathematical analysis. They are also met

with frequently in theoretical physics where it might seem that they have no application inasmuch as the numbers dealt with are what we have called real numbers. There are, in fact, many problems in mathematics that seem to depend on real numbers only which can be readily solved by recourse to complex numbers.

For example, a cubic equation with real coefficients and three real irrational roots cannot be solved in terms of real radicals only.[1]

As another example consider the Euler formula

$$e^{ix} = \cos x + i \sin x,$$

where x is real. We shall later (§208) define the symbol e^{ix} and shall also give a proof of this formula. We cite it here to show how it can be used to express $\sin nx$ and $\cos nx$ (n a positive integer) in terms of powers of $\sin x$ and $\cos x$. We have

$$e^{inx} = \cos nx + i \sin nx$$

and also

$$e^{inx} = (e^{ix})^n = (\cos x + i \sin x)^n.$$

Hence

$$\cos nx + i \sin nx = (\cos x + i \sin x)^n.$$

From this last equation we get the desired relations immediately by equating the real parts, and also the imaginary parts, of the two sides.

The use of complex integration to evaluate real definite integrals, as explained in the last chapter, affords another illustration of the point in question.

15. We might form still other classes of numbers by considering in a similar way ordered triples, or in general, ordered n-tuples of real numbers. But it is impossible to define the fundamental operations on these numbers in such a way that all the rules of ordinary algebra shall apply.[2] These numbers are called *hypercomplex numbers*.

[1] Weber and Wellstein, *Encyclopädie der Elementar-Mathematik*, I, ed. 3, p. 364.
[2] Hankel, *Theorie der complexen Zahlensysteme*, p. 107.

CHAPTER II

FUNCTIONS OF ONE VARIABLE

16. Definition. A variable y is said to be a *function* of a variable x if the number represented by it depends upon the number represented by x.

Thus, if $y = x + 5$, or $y = \sin x$, or $y = \log x$, y is a function of x. It may be that the nature of the dependence of y upon x is different for different values of x. For example, the dependence may be such that

$$y = 0 \quad \text{when} \quad x = 0$$

and

$$y = \sin \frac{1}{x} \quad \text{when} \quad x \neq 0.$$

Or we may have

$$y = 1 + x \quad \text{when} \quad x \leq 0$$

and

$$y = 1 - x \quad \text{when} \quad 0 \leq x.$$

The graph of this function is shown in Fig. 1. The preceding function will be discussed in the following section.

The function may show the dependence of y upon x only for values of x within a certain range; as, for example, $a \leq x \leq b$. In this case we say that the function is defined in the interval (a, b).

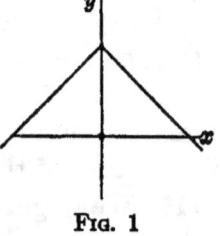

Fig. 1

If $y = ax^m$, y is a function of x in the sense here considered, provided that $m \neq 0$. If $m = 0$, $y = a$. That is, y is a constant. In order not to be under the necessity of resorting to a cumbersome circumlocution in this and similar cases, we shall say that y is a function of x even when we know that it may be a constant. Moreover we shall say that x is a function of x. It is not essential to the definition that the dependence of y upon x be expressible by means of mathematical symbols. For example, y may represent the number of prime numbers that do not exceed x. We use the symbol $f(x)$ to denote a function of x; and at times we shall also use other letters than f for this purpose.

17. Definition of continuity. The function $f(x)$ is said to be *continuous* at $x = a$ if the limit of $f(a + h)$ as h approaches zero is $f(a)$; or, in symbols,

$$\lim_{h \to 0} f(a + h) = f(a).$$

This definition implies in the first place that the function is defined throughout a neighborhood of the point $x = a$. In the second place it implies that the limit exists and is the same however h approaches zero. A consideration of the following function will illustrate the importance of this last remark:

$$y = \frac{e^{1/x}}{1 + e^{1/x}} \quad \text{when} \quad x \neq 0,$$
$$y = 0 \quad \text{when} \quad x = 0.$$

If x approaches zero through positive values, y approaches 1. But if x approaches zero through negative values, y approaches zero.[1] The function therefore has a break, or *discontinuity*, at $x = 0$, as shown in Fig. 2.

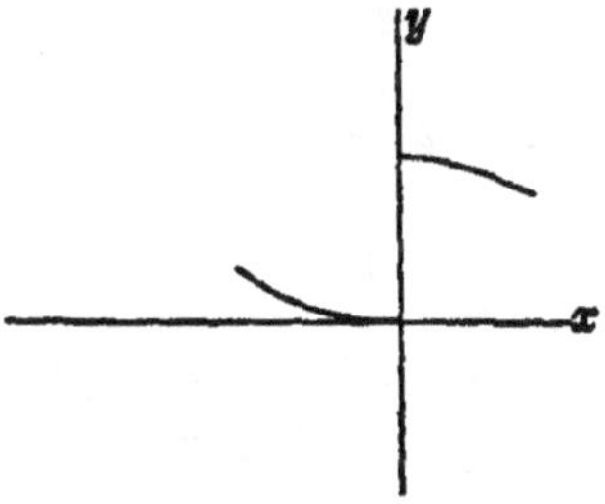

We can bring out the significance of the condition

$$\lim_{x \to a} f(x) = f(a)$$

more fully by stating the definition as follows: *The function $f(x)$ is continuous at $x = a$ if for every positive number ϵ there is a positive number η such that*

$$|f(a + h) - f(a)| < \epsilon$$

whenever $|h| < \eta$.

Fig. 2

---

[1] In this case take $a = 0$ and $a + h = x$, or $h = x$.

This is equivalent to the condition that for every value of x inside the interval $(a - \eta, a + \eta)$ the difference between $f(a)$ and $f(x)$ is less than ϵ in absolute value. If the function is defined only for the interval (a, b) and we are considering the question of continuity at the point $x = a$, we take into account only the values of x in the interval $(a, a + \eta)$. A similar remark applies to the point $x = b$. It should be clearly understood that the value we assign to η depends upon the value we have previously assigned to ϵ. The gist of the matter is that, whatever positive value we assign to ϵ, it shall be possible then to assign a suitable positive value to η.

A point at which $f(x)$ is not continuous is called a point of *discontinuity*. There are different kinds of discontinuities.

(a) It may be that as $x \to a$ from one or both sides the function increases in absolute value without limit and has no definite value at $x = a$; as for example the function $\frac{1}{x}$ at $x = 0$. In this case the function is said to be *infinite* at $x = a$ and to be discontinuous at this point. The point $x = a$ would still be a point of discontinuity if $f(x)$ had a definite value for this value of x; as, for example, the function $f(x) = \frac{1}{x}$ when $x \neq 0$ and $f(x) = 1$ when $x = 0$.

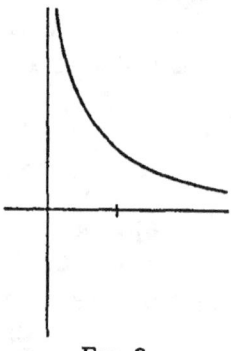

Fig. 3

(b) It may be that $f(x)$ approaches different finite limits according as x approaches a through values greater than a or through values less than a. We have already had an example of this kind of discontinuity. It is called a discontinuity of the *first kind*. A function that is continuous in (a, b) or has no other discontinuities than a finite number of the first kind is said to be *sectionally continuous* in (a, b). If its derivative also has this property the function is said to be *sectionally smooth*.

(c) It may be that the function $f(x)$ does not approach any limit as x approaches a. For example, the function $\sin \frac{1}{x}$ oscillates between 1 and -1 as x approaches zero.

(d) It is further possible for $f(x)$ to approach a definite limit as x approaches a, but that this limit is not the value of the function at $x = a$. Consider, for example, the infinite series

$$x^2 + \frac{x^2}{1+x^2} + \frac{x^2}{(1+x^2)^2} + \cdots + \frac{x^2}{(1+x^2)^2} + \cdots.$$

When $x \neq 0$ this series is a geometric progression whose ratio is $\frac{1}{1+x^2}$ and therefore less than 1 in absolute value. The series accordingly converges to

$$\frac{x^2}{1 - \frac{1}{1+x^2}} = 1 + x^2.$$

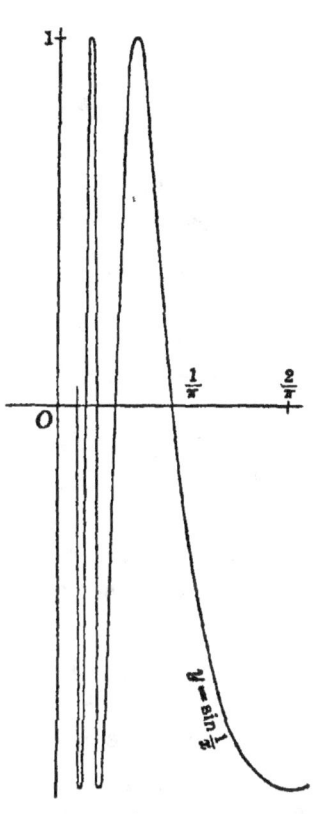

Fig. 4

That is, the series represents the function $1 + x^2$ when $x \neq 0$. This approaches 1 as x approaches zero. But the function represented by the series is obviously zero when $x = 0$.

If a function of x is continuous for every value of x in the interval (a, b), we say that it is continuous in (a, b).

18. Properties of continuous functions. When we speak of the interval (a, b) we may, or may not, include the end points. If we do we say that the interval is *closed*. If we do not we say that it is *open*. It may be that one of the end points is included and the other one not. In this case we shall say that the interval is closed at one end and open at the other.

THEOREM 1. *If $f(x)$ is continuous in the closed interval (a, b) and ϵ is an arbitrary positive number, we can divide (a, b) into partial intervals such that the difference between the values of $f(x)$ at any two points in the same partial interval will be less than ϵ in absolute value.*

If the subdivision described in the theorem were not possible, it would not be possible for at least one of the intervals (a, c) and (c, b), where $c = \dfrac{a+b}{2}$. If it is not possible in (a, c) we take $a = a_1$ and $c = b_1$. If it is possible in (a, c) we take $c = a_1$ and $b = b_1$. Then the interval (a_1, b_1) can be treated just as (a, b) has been, and we can continue this process indefinitely. This will give rise to two unlimited sequences of end points, $a, a_1, \cdots, a_n, \cdots$ and $b, b_1, \cdots, b_n, \cdots$. For every n we have $a_{n-1} \leqq a_n < b$, and $a < b_n \leqq b_{n-1}$ ($a_0 = a$, $b_0 = b$). Let λ be the upper limit of the set of a's and λ' the lower limit of the set of b's (see Theorem 7, Chapter I). Now the length of each interval is one-half the length of the preceding one, and therefore $b_n - a_n \to 0$ as n increases indefinitely. Hence $\lambda = \lambda'$. Moreover in each one of these partial intervals there are two points x_n' and x_n'' such that $|f(x_n') - f(x_n'')| \geqq \epsilon$. It is clear that λ is either within the interval (a, b) or at one of the end points. In either case $f(x)$ is continuous at $x = \lambda$, since the interval of continuity is closed. There is therefore a positive number h such that

$$|f(x) - f(\lambda)| < \frac{\epsilon}{2}$$

for all values of x within (a, b) such that $\lambda - h < x < \lambda + h$. For any two such values of x, as x_1 and x_2, we have therefore

$$|f(x_1) - f(\lambda)| < \frac{\epsilon}{2},$$

$$|f(\lambda) - f(x_2)| < \frac{\epsilon}{2},$$

and hence
$$|f(x_1) - f(x_2)| < \epsilon.$$

On the other hand, a_n is, for sufficiently large values of n, in the interval $(\lambda - h, \lambda)$ and b_n is in the interval $(\lambda, \lambda + h)$. Then for certain two points x_1 and x_2 in (a_n, b_n) we should have the contradictory relations
$$|f(x_1) - f(x_2)| \geqq \epsilon,$$
$$|f(x_1) - f(x_2)| < \epsilon.$$

The importance of the restriction that the function be continuous in the closed interval may be seen from a consideration of the function $\sin \dfrac{1}{x}$. This is continuous in the open interval $0 < x \leqq 1$, and yet such a subdivision of this interval as is described in the theorem is obviously impossible. (See Fig. 4.)

If the distance between two points, x' and x'', of the interval (a, b) is less than the length of any of these partial intervals, these two points must either be in the same partial interval or in adjacent partial intervals. In the former case
$$|f(x') - f(x'')| < \epsilon,$$

In the latter, if x' is in the partial interval (x_{n-1}, x_n) and x'' in (x_n, x_{n+1}), then
$$|f(x') - f(x_n)| < \epsilon,$$
and
$$|f(x_n) - f(x'')| < \epsilon.$$
Hence
$$|f(x') - f(x'')| < 2\epsilon.$$

This proves the theorem:

THEOREM 2. *If $f(x)$ is continuous in the closed interval (a, b) and ϵ is an arbitrary positive number, there is a positive number η such that the difference between the values of $f(x)$ at any two points whose distance apart does not exceed η is less than ϵ in absolute value.*[1]

---

[1] In the proof we had $|f(x') - f(x'')| < 2\epsilon$; but since ϵ is an arbitrary positive number it is immaterial whether we say ϵ or 2ϵ.

DEFINITION. If for a given interval (a, b) and an arbitrary positive ϵ there is a positive number η such that

$$|f(x') - f(x'')| < \epsilon$$

for any two points, x' and x'', of (a, b) for which $|x' - x''| < \eta$ we say that $f(x)$ is *uniformly continuous in* (a, b).

We have just seen that a function that is continuous in a closed interval is uniformly continuous there.

The reader should be sure that the distinction between continuity and uniform continuity is clear in his mind. He should observe that while we speak of a function being continuous or discontinuous at a point, it would mean nothing to say that a function is uniformly continuous, or non-uniformly continuous, at a point. When we speak of uniform continuity we always have reference to an interval. A consideration of the function $\sin \dfrac{1}{x}$ in the interval $0 < x \leqq a$ should convince him of the difference between the two notions.

THEOREM 3. *If $f(x)$ and $\varphi(x)$ are continuous at a point, then $f(x) \pm \varphi(x)$, $f(x) \cdot \varphi(x)$, and $\dfrac{f(x)}{\varphi(x)}$ are continuous there, provided that in the case of the quotient the divisor is not zero at the point.*

The proof is simple and is left to the reader.

THEOREM 4. *If $f(x)$ is continuous at $x = a$ and $\varphi(u)$ is continuous at $u = f(a)$, then $\varphi[f(x)]$ is a continuous function of x at $x = a$.*

For, as $x \to a$, $\lim \varphi[f(x)] = \varphi[\lim f(x)] = \varphi[f(a)]$.

THEOREM 5. *If $f(x)$ is continuous at $x = a$ and $f(a) \neq 0$, then $f(x)$ has the same sign as $f(a)$ within a sufficiently small interval around $x = a$.*

For by virtue of the continuity of $f(x)$ at $x = a$ there is an $\eta > 0$ such that

$$|f(x) - f(a)| < |f(a)|$$

whenever $|x - a| < \eta$. This means that $f(a) + f(x) - f(a)$, which is the same as $f(x)$, has the same sign as $f(a)$ for the values of x under consideration.

THEOREM 6. *If $f(x)$ is continuous in the closed interval (a, b) there is at least one value of x in this interval for which $f(x) = N$, where N is any number between $f(a)$ and $f(b)$.*

In the first place, if $f(a)$ is negative and $f(b)$ is positive, let λ be the upper limit of the bounded set of values of x in (a, b) for which $f(x)$ is negative. If $f(\lambda)$ were negative, we know from Theorem 5 that there would be values of x in (a, b) greater than λ for which $f(x) < 0$. But this is impossible. If $f(\lambda)$ were positive, $f(x)$ would be positive within a sufficiently small interval around λ, and λ would not be the upper limit of the set described. Hence $f(\lambda) = 0$. The same conclusion would follow if $f(a)$ were positive and $f(b)$ negative. If now $f(a)$ and $f(b)$ have the same sign, let N be any number between them, and consider the continuous function $\varphi(x) = f(x) - N$. It is clear that $\varphi(a)$ and $\varphi(b)$ have opposite signs. Hence there is a value λ of x in (a, b) for which $\varphi(\lambda) = 0$, or $f(\lambda) = N$.

DEFINITION. If the function $f(x)$ is bounded in the interval (a, b), the difference between its upper and lower limits is called its *oscillation* in the interval.

THEOREM 7. *If $f(x)$ is continuous in the closed interval (a, b) it is bounded in this interval.*

For, having chosen a positive number ϵ, we can by Theorem 1 subdivide (a, b) into a finite number of partial intervals such that the oscillation of $f(x)$ in any partial interval is less than ϵ. This having been done, if x is anywhere in the first partial interval (a, x_1), we have

$$|f(x) - f(a)| < \epsilon,$$

and hence

$$|f(x)| < |f(a)| + \epsilon.$$

This includes the statement

$$|f(x_1)| < |f(a)| + \epsilon.$$

If x is in the second partial interval (x_1, x_2), we have likewise
$$|f(x) - f(x_1)| < \epsilon,$$
and hence
$$|f(x)| < |f(x_1)| + \epsilon < |f(a)| + 2\epsilon.$$
Continuing in this way we see that, if there are n partial intervals in all, for every value of x in the interval (a, b)
$$|f(x)| < |f(a)| + n\epsilon.$$

Theorem 8. *If $f(x)$ is continuous in the closed interval (a, b) it has an upper limit and a lower limit in the interval and there is at least one value of x in the interval for which $f(x)$ equals its upper limit and at least one value for which $f(x)$ equals its lower limit.*

It follows from Theorem 7 that the function is bounded in (a, b). Let M be the upper limit of the set of values assumed by $f(x)$ in (a, b). The existence of this upper limit follows from Theorem 5, Chapter I. Since M is the upper limit of the function in (a, b), it is its upper limit in one or both of the intervals (a, c) and (c, b), where $c = \dfrac{a+b}{2}$.

If M is the upper limit of $f(x)$ in (a, c) we take $a = a_1$ and $c = b_1$. If M is not the upper limit of $f(x)$ in (a, c), we take $c = a_1$ and $b = b_1$. Then, M being the upper limit of $f(x)$ in (a_1, b_1), we can treat this interval in the same way as the original one and obtain an interval (a_2, b_2) within which M is the upper limit of $f(x)$. By continuing in this way we obtain an unlimited sequence of partial intervals (a_1, b_1), (a_2, b_2), $\cdots$, (a_n, b_n), $\cdots$, each of which after the first one is one-half the preceding one and contained within it. Moreover $a_{n-1} \leq a_n < b$ and $a < b_n \leq b_{n-1}$, for $n > 1$. The two sequences $a_1, a_2, \cdots, a_n, \cdots$ and $b_1, b_2, \cdots, b_n, \cdots$ have a common limit λ.

If we prove $f(\lambda) = M$, our theorem will be proved as far as concerns the upper limit of $f(x)$. Now $f(\lambda) \leq M$; and if $f(\lambda) < M$, that is, if $f(\lambda) = M - h$, where h is positive,

there is a positive number η such that when $\lambda - \eta < x < \lambda + \eta$ we have
$$|f(x) - f(\lambda)| < \frac{h}{2}.$$
This would make the maximum value of $f(x)$ in the interval $(\lambda - \eta, \lambda + \eta)$ less than $M - \frac{h}{2}$. But for a sufficiently large value of η we have a_n in the interval $(\lambda - \eta, \lambda)$ and b_n in the interval $(\lambda, \lambda + \eta)$, and M is the maximum value of $f(x)$ in (a_n, b_n). This gives us a contradiction and the supposition that $f(\lambda) = M - h$ is untenable. A similar argument applies to the lower limit of $f(x)$.

EXERCISES

Point out the discontinuities of the following functions:

1. $\dfrac{1}{x-2}$.   2. $\dfrac{5x}{x^2-4}$.   3. $\dfrac{x^2+2}{x^2-x-6}$.   4. $\tan x$.   5. $\cot x$.
6. $\sec x$.   7. $\csc x$.

Have the following functions any discontinuities?

8. $\tan 4x$.   9. $\sin(x+3)$.   10. $\dfrac{\sin x}{2-\cos x}$.   11. $\sin \dfrac{1}{x}$.

12. $f(x) = \sin \dfrac{1}{x}$ for $x \neq 0$ and $f(0) = a \neq 0$.   13. $\dfrac{3+\cos x}{1+\sin x}$.

Show that each of the following functions is continuous for every finite value of x:

14. $2 + 3x$.   15. $x^2 - 2x + 5$.   16. $\dfrac{1}{2+x^2}$.   17. $\dfrac{x^3+1}{x^2+2}$.

18. $\dfrac{x^3+1}{x^4+1}$.

19. Show that if e^x is continuous for $x = 0$ it is continuous for every finite value of x.

20. Show that $f(x)$ is discontinuous at $x = 0$ if
$$f(x) = 3^{1/x} \quad \text{when} \quad x \neq 0,$$
and
$$f(x) = 2 \quad \text{when} \quad x = 0.$$

21. Given $y = \dfrac{xe^{1/x}}{1+e^{1/x}}$ when $x \neq 0$, and $y = 0$ when $x = 0$. Plot the function for values of x near to zero.

19. The derivative. Suppose that $f(x)$ is defined in the interval (a, b) and that x_1 is on the interior of this interval.

§19] FUNCTIONS OF ONE VARIABLE

If
$$\frac{f(x_1 + h) - f(x_1)}{h}$$

approaches a unique limit as h approaches zero, regardless of the manner of this approach, this limit is called the *derivative of $f(x)$* at $x = x_1$. We shall denote the derivative of $f(x)$ at x by the symbol $f'(x)$, or the symbol $\frac{df(x)}{dx}$.

Theorem 9. *If $f(x)$ has a derivative at $x = x_1$, it is continuous at this point.*

For
$$\lim_{h \to 0} \frac{f(x_1 + h) - f(x_1)}{h} = f'(x_1),$$

and therefore
$$\lim_{h \to 0} [f(x_1 + h) - f(x_1)] = \lim_{h \to 0} [h f'(x_1)] = 0.$$

However a function may be continuous at a point and have no derivative at this point. For example, the function $f(x)$ defined as follows:

$$f(x) = x \sin \frac{1}{x} \quad \text{when} \quad x \neq 0,$$
$$f(x) = 0 \quad \text{when} \quad x = 0.$$

In this case
$$\frac{f(0 + h) - f(0)}{h} = \frac{h \sin \frac{1}{h} - 0}{h} = \sin \frac{1}{h}.$$

Now $\sin \frac{1}{h}$ oscillates between 1 and -1 as h approaches zero, and therefore approaches no limit. On the other hand it is easy to show that $f(x)$ is continuous at $x = 0$. This particular function is continuous for every finite value of x, and has a derivative for all these values, except one. We shall later (§133) describe a function that is continuous throughout an interval and has no derivative anywhere in the interval. The functions with which the reader will have to deal are for the most part those that have derivatives at their points of continuity, and he is familiar with the rules for finding these derivatives in most cases.

If a function has a derivative at a point, or at every point of an interval, we say that it is *differentiable* at the point, or over the interval, respectively.

20. Derivative of a composite function. If we have $y = f(u)$ and $u = \varphi(x)$, these equations indirectly give us y as a function of x. The natural way to get the derivative of y with respect to x is to express y in terms of x by eliminating u and then to differentiate the resulting function. But it is often impracticable to eliminate u. In this case we can proceed in a way indicated by the following discussion:

Suppose that $f(u)$ is a differentiable function of u and $\varphi(x)$ a differentiable function of x. If we give the increment Δx to x, u will receive an increment which we shall call Δu, and this increment in u will cause an increment in y which we shall call Δy. Then

$$\frac{\Delta y}{\Delta x} = \frac{f(u + \Delta u) - f(u)}{\Delta u} \cdot \frac{\Delta u}{\Delta x}.$$

Since $\varphi(x)$ is a differentiable function of x and therefore continuous, $\Delta u \to 0$ as $\Delta x \to 0$. Then

$$\lim_{\Delta x \to 0} \frac{\Delta y}{\Delta x} = \lim_{\Delta u \to 0} \frac{f(u + \Delta u) - f(u)}{\Delta u} \cdot \lim_{\Delta x \to 0} \frac{\Delta u}{\Delta x};$$

or

$$\frac{dy}{dx} = \frac{dy}{du} \cdot \frac{du}{dx}. \tag{1}$$

We have assumed that there is a positive number η such that $\Delta u \neq 0$ when $0 < |\Delta x| < \eta$. There is however another possibility. It may be that for any positive η there is a Δx different from zero and less than η in absolute value for which [1] $\Delta u = 0$. If this is the case, then since $\dfrac{du}{dx}$ exists it must be zero. Hence if we let $\Delta x \to 0$ through values for which $\Delta u \neq 0$, it follows that $\dfrac{\Delta y}{\Delta x} \to 0$. Moreover if $\Delta u = 0$ for a given value of Δx the corresponding value of $\dfrac{\Delta y}{\Delta x}$ is also zero. Hence $\lim\limits_{\Delta x \to 0} \dfrac{\Delta y}{\Delta x} = 0$ and formula (1) holds in this case also.

If we assume that $f(u)$ has a continuous derivative we can use the mean value theorem (see §23) to give another proof of (1). For $\Delta y = f(u + \Delta u) - f(u) = f'(u + \theta \Delta u)\Delta u$, and therefore $\dfrac{dy}{dx} = \lim \dfrac{\Delta y}{\Delta x} = \lim f'(u + \Delta u) \dfrac{\Delta u}{\Delta x} = f'(u) \dfrac{du}{dx}$. (See §34.)

21. Geometrical interpretation of the derivative. If we have given a function $f(x)$ which is continuous in the interval (a, b) we associate with it the locus of the equation $y = f(x)$. Now $\dfrac{f(x_1 + h) - f(x_1)}{h}$ is the slope of the chord connecting the two points on the curve $(x_1, f(x_1))$ and $(x_1 + h, f(x_1 + h))$; whereas the tangent to the curve at the point on the curve for which $x = x_1$ is by definition the line through this point whose slope is the limiting value of the slope of this chord as h approaches zero. But the limiting value of this slope is the derivative $f'(x_1)$. Hence the derivative of the function $f(x)$ at any point x is the slope of the tangent to the curve $y = f(x)$ at the point on the curve whose abscissa is this value of x.

Consider now the space curve whose equations are (see §176)
$$x = f(t), \qquad y = \varphi(t), \qquad z = \psi(t),$$
where $f(t)$, $\varphi(t)$, and $\psi(t)$ are continuous functions of t in

---

[1] An example of a situation of this kind is furnished by the function $u = \varphi(x) = x^2 \sin \dfrac{1}{x}$ when $x \neq 0$ and $\varphi(x) = 0$ when $x = 0$, in case we are considering the derivative of y with respect to x for $x = 0$.

the interval (t_1, t_2). If t_0 and $t_0 + h$ both lie between t_1 and t_2, the equations of the chord connecting the points (x_0, y_0, z_0) and $(x_0 + h_1, y_0 + h_2, z_0 + h_3)$, where
$$x_0 = f(t_0), \quad y_0 = \varphi(t_0), \quad z_0 = \psi(t_0), \quad x_0 + h_1 = f(t_0 + h),$$
$$y_0 + h_2 = \varphi(t_0 + h), \quad \text{and} \quad z_0 + h_3 = \psi(t_0 + h)$$
are
$$\frac{x - x_0}{f(t_0 + h) - f(t_0)} = \frac{y - y_0}{\varphi(t_0 + h) - \varphi(t_0)} = \frac{z - z_0}{\psi(t_0 + h) - \psi(t_0)},$$
or
$$\frac{x - x_0}{\dfrac{f(t_0 + h) - f(t_0)}{h}} = \frac{y - y_0}{\dfrac{\varphi(t_0 + h) - \varphi(t_0)}{h}} = \frac{z - z_0}{\dfrac{\psi(t_0 + h) - \psi(t_0)}{h}}.$$

The denominators are proportional to the direction cosines of the chord. If these approach limits not all zero as $h \to 0$, the line through the point (x_0, y_0, z_0) with direction cosines proportional to these limits is in the limiting position of the chord and is therefore tangent to the curve. Now these denominators do approach the respective limits $f'(t_0)$, $\varphi'(t_0)$, and $\psi'(t_0)$; and therefore the line

$$\frac{x - x_0}{f'(t_0)} = \frac{y - y_0}{\varphi'(t_0)} = \frac{z - z_0}{\psi'(t_0)}$$

is tangent to the curve at (x_0, y_0, z_0), provided that the derivatives in the denominators are not all zero. The plane through the point $t = t_0$ and perpendicular to the tangent to the curve at this point is called a *normal plane* to the curve.

EXERCISES

Find the tangent line and the normal plane to each of the following curves at the designated point:

1. $y^2 = x$, $z^2 = 1 - x$ $\quad (1, 1, 0)$.
2. $x = 2 \cos t$, $y = \sin t$, $z = 3t$ $\quad (t = 2)$.
3. $xyz = 1$, $y^2 = x$ $\quad (1, 1, 1)$.
4. $x^2 + y^2 = 1$, $y^2 + z^2 = 1$ $\quad (3/5, 4/5, 3/5)$.
5. $x = a \sin^2 t$, $y = a \sin t \cos t$, $z = a \cos t$ $\quad (t = t_1)$.
6. Does the normal plane in Ex. 5 pass through the origin?

22. Rolle's Theorem. *Let the function $f(x)$ be continuous in the closed interval (a, b) and possess a derivative throughout the corresponding open interval. Then if $f(a) = f(b) = 0$, the equation $f'(x) = 0$ has at least one root in the open interval.*

By theorem 8 $f(x)$ has a maximum M and a minimum m in (a, b). If M and m are both zero, $f(x)$ is identically zero in the interval, as is its derivative, since the derivative of a constant is zero. If $M \neq 0$ let ξ be a point in the interval for which $f(\xi) = M$. Then $f'(\xi) = 0$. For if h is positive

$$\frac{f(\xi + h) - f(\xi)}{h} \leq 0$$

and

$$\frac{f(\xi - h) - f(\xi)}{-h} \geq 0.$$

Now $f'(\xi)$ is the common limit of these two expressions and must therefore equal zero. If $m \neq 0$, it can be shown in a similar way that $f'(\xi') = 0$, where ξ' is a point of the interval for which $f(\xi') = m$.

23. The mean value theorem. Consider the locus of the equation $y = f(x)$, where $f(x)$ is continuous in the closed interval (a, b) and has a derivative in the corresponding open interval. Let A and B be the points of intersection of the curve with the ordinates $x = a$ and $x = b$ respectively. It seems plausible that there

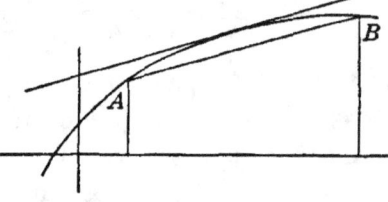

Fig. 5

is a point on the curve between A and B at which the tangent to the curve is parallel to the line AB. The analytical proof that there is such a point follows immediately from Rolle's theorem.

The equation of the chord AB is

$$y = f(a) + \frac{f(b) - f(a)}{b - a}(x - a),$$

and the difference $F(x)$ between the ordinates of the two loci at the points whose abscissa is x is

$$F(x) = f(x) - f(a) - \frac{f(b) - f(a)}{b - a}(x - a).$$

Now this function $F(x)$ satisfies the conditions of Rolle's theorem—it vanishes at $x = a$ and $x = b$, and has a derivative for every value of x between a and b. There is, therefore, some value of x, as ξ, between a and b such that

$$F'(\xi) = f'(\xi) - \frac{f(b) - f(a)}{b - a} = 0.$$

But $\dfrac{f(b) - f(a)}{b - a}$ is the slope of the chord AB. This proves the theorem:

MEAN VALUE THEOREM. *If $f(x)$ satisfies the same conditions as in Rolle's theorem, there is a value ξ of x between a and b such that*

$$f(b) - f(a) = (b - a)f'(\xi).$$

Rolle's theorem is a special case of this theorem, which in turn is a special case of the following:

GENERALIZED LAW OF THE MEAN. *If $f(x)$ and $\varphi(x)$ are continuous in the closed interval (a, b) and have derivatives in the corresponding open interval, there is a number ξ between a and b such that*

$$[\varphi(b) - \varphi(a)]f'(\xi) = [f(b) - f(a)]\varphi'(\xi).$$

Let $A = \varphi(a) - \varphi(b)$, $B = f(b) - f(a)$, and $C = f(a)\varphi(b) - f(b)\varphi(a)$. Then the function

$$\psi(x) = Af(x) + B\varphi(x) + C$$

vanishes at $x = a$ and $x = b$, and it has a derivative in the open interval (a, b). It therefore satisfies the conditions of Rolle's theorem, and there is a number ξ between a and b such that

$$\psi'(\xi) = Af'(\xi) + B\varphi'(\xi) = 0.$$

If we take account of the values of A and B we have
$$[\varphi(b) - \varphi(a)]f'(\xi) = [f(b) - f(a)]\varphi'(\xi).$$

24. It follows from the mean value theorem that if $f'(x)$ is identically zero throughout the interval (a, b), then $f(x)$ is constant in the interval—that is, has the same value for every value of x in the interval. For, if $a + h$ is any point in the interval,
$$f(a + h) - f(a) = hf'(\xi) = 0,$$
where $a < \xi < a + h$. It follows that if two functions $f(x)$ and $\varphi(x)$ have the same derivative at every point of the interval, they differ by a constant. The importance of this observation will appear when we come to the study of integration.

EXERCISES

Find the intervals of increasing and decreasing of each of the following functions: Draw the curves.

1. $y = x^2$. 2. $y = (x - 1)(x - 2)$. 3. $y = 2x^3 - 9x^2 + 12x - 1$.
4. $y = \sin x$. 5. $y = \cos x$. 6. $y = \tan x$.

7. Show that the equation $x^3 - 6x^2 + 11x - 8 = 0$ has a root between 1 and 4.

8. Show that the equation $\sin^2 x - 2 \cos x = 0$ has a root between 0 and $\frac{\pi}{2}$.

9. Does the equation $3 \sin 2x + 5 \cos 3x = 0$ have any real roots?

10. Given $y = \dfrac{xe^{1/x}}{1 + e^{1/x}}$ when $x \neq 0$, and $y = 0$ when $x = 0$. What is the derivative of y with respect to x when $x = 0$?

11. Prove that the equation $f(x) - kf'(x) = 0$ has at least one root in the interval (a, b) in case $f(x)$ and $f'(x)$ are continuous in the closed interval and $f(a) = f(b) = 0$, while $f'(a)$ and $f'(b)$ are both different from zero.

25. Derivatives of higher order. If the function $f(x)$ is differentiable in the interval (a, b), its derivative, $f'(x)$, may, or may not, be differentiable. If it is, its derivative is called the *second derivative* of $f(x)$. In a similar way we define derivatives of successively higher orders. The function $\sin x$, for example, has a derivative of every order, while the function $x^{3/2}$ has a first derivative, but none of higher order, at the origin.

If $y = f(x)$ the successive derivatives are conventionally denoted by the symbols

$$\frac{dy}{dx}, \quad \frac{d^2y}{dx^2}, \quad \cdots, \quad \frac{d^ny}{dx^n}, \quad \cdots;$$

or

$$y', \quad y'', \quad \cdots, \quad y^{(n)}, \quad \cdots;$$

or

$$f(x), \quad f'(x), \quad \cdots, \quad f^{(n)}(x), \quad \cdots.$$

26. Infinitesimals. A variable with the limit zero is called an *infinitesimal*.

A constant quantity different from zero is not an infinitesimal in this sense, although it may be very close to zero. Zero may be thought of as a variable with itself as its limit. It is therefore in conformity with convention to speak of zero as an infinitesimal.

If α and β are two infinitesimals, the ratio $\frac{\beta}{\alpha}$ may approach any finite limit, or no limit at all; it may have no upper bound, or no lower bound. For example, if $\beta = \alpha^2$, $\lim \frac{\beta}{\alpha} = 0$; if $\beta = k \sin \alpha$, where k is any number, $\lim \frac{\beta}{\alpha} = k$; if $\alpha = x^2$ and $\beta = x$, $\lim \frac{\beta}{\alpha} = \infty$; and finally if $\beta = \alpha \sin \frac{1}{\alpha}$, $\frac{\beta}{\alpha} = \sin \frac{1}{\alpha}$. But this last expression has no limit as $\alpha \to 0$.

If $\lim \frac{\beta}{\alpha} = k$, then (1) β is said to be of the *same order* as α if k is finite and different from zero; (2) β is said to be of *higher order* than α if $k = 0$; and (3) of *lower order* if $k = \pm \infty$. More particularly, we say that β is of order n as compared with α in case $\lim_{\alpha \to 0} \frac{\beta}{\alpha^n} = k \neq 0$.

The reciprocal of an infinitesimal is said to be *infinite* since it increases in absolute value beyond any preassigned limit.

EXERCISE. Show that we cannot have

$$\lim \frac{\beta}{\alpha^n} = k \neq 0$$

and

$$\lim \frac{\beta}{\alpha^m} = k' \neq 0 \quad (m \neq n)$$

as $\alpha \to 0$.

27. Principal part of an infinitesimal. If $\lim \frac{\beta}{\alpha} = k \neq 0$, we have $\frac{\beta}{\alpha} = k + \epsilon$, where ϵ is an infinitesimal; or

$$\beta = \alpha k + \alpha \epsilon.$$

That is, β is the sum of two infinitesimals, one of which is of order one as compared with α, and the other one of higher order. The part αk of β that is of order one is accordingly called the *principal part* of β. Suppose for example that $y = f(x)$, where $f(x)$ is a differentiable function. Then

$$\lim_{\Delta x \to 0} \frac{\Delta y}{\Delta x} = f'(x),$$

or

$$\frac{\Delta y}{\Delta x} = f'(x) + \epsilon,$$

where ϵ is an infinitesimal along with Δx. Hence

$$\Delta y = f'(x)\Delta x + \epsilon \Delta x,$$

and $f'(x)\Delta x$ is the principal part of Δy, if $f'(x) \neq 0$. It is called the *differential* of y and is represented by the symbol dy. In case $f'(x) = 0$ we say that $dy = 0$.

We have defined the differential of the dependent variable. In conformity with this definition we agree to call Δx the differential of x, the independent variable, and to represent it by the symbol dx. The differential of y depends upon x and Δx. The latter is perfectly arbitrary. It represents a change in the independent variable. But dy does not represent the corresponding change in y; that is represented by Δy. But since

$$y - y_1 = f'(x_1)(x - x_1)$$

is the equation of the tangent to the curve $y = f(x)$ at the point whose abscissa is x_1, we see, if we replace $x - x_1$ by its equal dx, that $y - y_1$, or the change in y as we move out from the point of contact along the tangent to the curve, is equal to $f'(x)dx$, which is the differential of y.

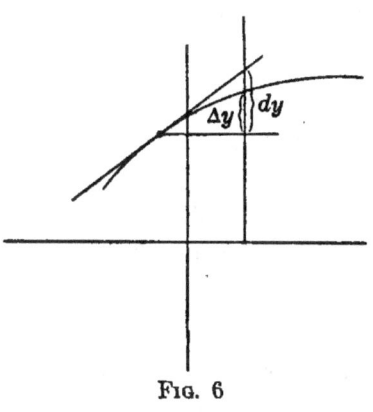

Fig. 6

By definition the differential of y is equal to the derivative of y with respect to x multiplied by the differential of x. It follows that in the symbol $\dfrac{dy}{dx}$ for the derivative we can consider the numerator as the differential of y and the denominator as the differential of x. The reader should be careful not to use the words derivative and differential interchangeably.

28. Differentials of higher order. If $y = f(x)$ has a second derivative, the differential of y' is called the *second differential* of y, and is represented by the symbol d^2y. We assume that the arbitrary value we assign to dx is the same for all values of x under consideration—that is, that dx is constant. Then

$$dy = y'dx = f'(x)dx,$$
$$d^2y = y''dx^2 = f''(x)dx^2.$$

and, in general,

$$d^{(n)}y = y^{(n)}dx^n = f^{(n)}(x)dx^n,$$

if the first n derivatives of $f(x)$ exist. In general, then,

$$\frac{d^n y}{dx^n} = f^{(n)}(x).$$

29. In case $y = f(u)$ and $u = \varphi(x)$, we have

$$dy = f'(u)du,$$

and
$$du = \varphi'(x)dx.$$
Therefore
$$dy = f'(u)\varphi'(x)dx = \frac{dy}{dx}dx$$
in view of §20.

This result is important since it shows us that when y is a composite function of x the differential of y is the same whether we take it with respect to the intermediate variable or with respect to the independent variable. The corresponding formula does not hold for differentials of higher order.

MISCELLANEOUS EXERCISES

1. Derive a simple formula for the n-th derivative of $\sin x$; of $\cos x$.
2. Show how the constants A, B, and C in §23 are determined.
3. Prove that a polynomial in x is continuous for every finite value of x.
4. If $f(x)$ and $\varphi(x)$ are relativly prime polynomials in x, is $\frac{f(x)}{\varphi(x)}$ continuous for every finite value of x, it being understood that $\varphi(x)$ is not a constant?
5. Show that $f(x)$ increases or decreases with x according as $f'(x)$ is positive or negative.
6. Write down a formula for the derivative of a determinant whose elements are differentiable functions of x.
7. Show without expanding it that the determinant
$$D = \begin{vmatrix} \sin k\theta & \sin(k\theta + \alpha) & \sin(k\theta + \beta) \\ \cos k\theta & \cos(k\theta + \alpha) & \cos(k\theta + \beta) \\ c_1 & c_2 & c_3 \end{vmatrix}$$
is independent of θ.
8. Find the equations of a tangent to the helix $x = \cos t$, $y = \sin t$, $z = t$.
9. Show that the orthogonal projection of this tangent upon the (x, y) plane is tangent to the circle $x^2 + y^2 = 1$.
10. At what angle does this curve intersect the curve $x = \sin t$, $y = \cos t$, $z = t$?
11. Find the equations of a tangent to the conical helix $x = t\cos t$, $y = t\sin t$, $z = kt$.
12. Find $\frac{d^2z}{dx^2}$ in case $z = f(x, y)$ and $y = \varphi(x)$.

13. How many real roots has the equation $a^x - x = 0$, it being assumed that $0 < a < 1$?

14. How many real roots has the equation $e^x - x = 0$?

15. Given $f(x) = x^2 \sin \frac{1}{x}$ when $x \neq 0$ and $f(x) = 0$ when $x = 0$. Find $f'(x)$ for $x = 0$ and for $x \neq 0$. Show that $f'(x)$ is discontinuous at the origin.

16. Given $f(x) = x^3 \sin \frac{1}{x}$ when $x \neq 0$ and $f(x) = 0$ when $x = 0$. Is $f(x)$ continuous at the origin? Does it have a derivative that is continuous there?

17. Prove that the derivative of a function of one variable has no discontinuities of the first kind in an interval within which it exists. (Use the mean value theorem.)

18. Prove that $f(x)$ is a constant if it is continuous and takes on only rational values.

19. What can you say if it is continuous and takes on only irrational values?

20. If $0 \leq x \leq \frac{\pi}{2}$, then $\sin x \leq x$. (See §48 for the remainder in the expansion of $\sin x$ in powers of x.) Then for all values of x we have $|\sin x| \leq |x|$. From this fact show that $\sin x$ and $\cos x$ are continuous for all values of x.

21. Then show that $\sin^2 x$ and $\cos^2 x$ are continuous for all values of x.

22. Given the function $f(x)$ defined as follows: $f(x) = 0$ when x is irrational and $f(x) = \frac{1}{q}$ when x equals a rational fraction $\frac{p}{q}$ in its lowest terms ($q > 0$). Show that $f(x)$ is continuous when x is irrational and discontinuous when x is rational.

This function is striking in that within every interval it has an infinite number of points of continuity and an infinite number of points of discontinuity. (See Chapter I, Theorems 2 and 3.)

23. $f(x) = x - [x]$, where $[x]$ denotes the greatest integer that does not exceed x. Show that $f(x)$ is continuous for non-integral values of x and discontinuous for integral values.

24. Show that $f(x)$ has a minimum in any interval containing an integer, but no maximum.

25. Prove that $\frac{a^x}{x^n} \to \infty$ as $x \to \infty$ in case $a > 1$ and $n > 0$. (Examine the derivative of the function.)

26. If $f(x)$ and $\varphi(x)$ have derivatives for $a < x < a + h$, and $\varphi'(x)$ does not vanish in this interval, we know that

$$\frac{f(a+h)-f(a)}{\varphi(a+h)-\varphi(a)} = \frac{f'(a+\theta_1 h)}{\varphi'(a+\theta_2 h)},$$

where $0 < \theta_1 < 1$ and $0 < \theta_2 < 1$. In general θ_1 and θ_2 are different. Is there a θ between 0 and 1 such that

$$\frac{f(a+h)-f(a)}{\varphi(a+h)-\varphi(a)} = \frac{f'(a+\theta h)}{\varphi'(a+\theta h)}?$$

$\Bigl($Apply Rolle's theorem to the function

$$\psi(x) \equiv f(x) - f(a) - [\varphi(x) - \varphi(a)]\frac{f(a+h)-f(a)}{\varphi(a+h)-\varphi(a)}\Bigr).$$

27. The trigonometric polynomial

$$a_0 + a_1 \cos x + \cdots + a_n \cos nx,$$

where the coefficients are all real and $|a_0| + |a_1| + \cdots + |a_{n-1}| < a_n$, has at least $2n$ zeros in the interval $(0, 2\pi)$.

CHAPTER III

FUNCTIONS OF MORE THAN ONE VARIABLE

30. We shall discuss here functions of two variables in particular, and leave to the reader the statement and proof of similar properties of functions of three or more variables. He will see that many of the properties of functions of one variable carry over with obvious changes to functions of more than one variable.

31. Continuity. The function $f(x, y)$ is said to be continuous at the point (x_0, y_0) if there is associated to every positive number ϵ a positive number η such that

$$|f(x, y) - f(x_0, y_0)| < \epsilon,$$

when $|x - x_0| < \eta$ and $|y - y_0| < \eta$.

This implies that in order to be continuous at the point (x_0, y_0) a function must be defined throughout a neighborhood of this point. The function is said to be continuous in a region if it is continuous at every point of the region.

If we include its boundary points in a region we say that the region is *closed*. If not all the boundary points are included we say that the region is *open*, at least in part. The region of continuity of a function may be open or closed.

The diameter of a circle is the longest straight line that can be drawn between two points on its circumference. We accordingly call the longest straight line that can be drawn between two points on the boundary of any closed region the *diameter* of the region.

THEOREM 1. *If the function $f(x, y)$ is continuous in the closed region A, there is associated to every positive number ϵ a positive number η such that if A is divided into partial regions whose diameters are all less than η, the absolute value of the difference between the values of $f(x, y)$ at any two points of the same partial region is less than ϵ.*

§ 31] FUNCTIONS OF VARIABLES

If the theorem were not true for A it would not be true for all the partial regions into which A might be divided. Suppose that we divide A into partial regions by two systems of equidistant lines parallel to the coordinate axes respectively. If there is one of these partial regions, as A_1, with respect to which the theorem is not true, we divide A_1 into partial regions in the same way. The theorem would not be true for at least one of these. Let A_2 be such a one. We could continue this process indefinitely and thus form an infinite sequence $A_1, A_2, \cdots, A_n, \cdots$ of squares or partial squares for each of which the theorem is not true. If the region A_n is the square bounded by the lines $x = a_n, x = b_n, y = c_n, y = d_n$, or a part of this square, then $b_n - a_n \to 0$ and $d_n - c_n \to 0$ as $n \to \infty$. Hence the diameter of this square approaches zero as $n \to \infty$, and a_n and b_n approach a common limit, λ, while c_n and d_n approach a common limit μ. Now the point (λ, μ) lies in the interior or upon the boundary of the region A, and therefore $f(x, y)$ is continuous at this point. There is then an $\eta > 0$ such that

$$|f(x, y) - f(\lambda, \mu)| < \frac{\epsilon}{2} \tag{1}$$

when x and y are the coordinates of a point of A for which

$$|x - \lambda| < \eta \quad \text{and} \quad |y - \mu| < \eta. \tag{2}$$

On the other hand, for a sufficiently large n the region A_n lies in the square whose center is at (λ, μ) and whose sides are equal to 2η. The coordinates of all points of A_n satisfy (2) and therefore (1). If (x_1, y_1) and (x_2, y_2) are any two such points, we have

$$|f(x_1, y_1) - f(\lambda, \mu)| < \frac{\epsilon}{2}$$

and

$$|f(x_2, y_2) - f(\lambda, \mu)| < \frac{\epsilon}{2}.$$

Therefore

$$|f(x_1, y_1) - f(x_2, y_2)| < \epsilon.$$

But this contradicts the supposition we made in regard to A_n. The theorem is therefore proved.

COROLLARY. *To every positive number ϵ there is associated a positive number η such that*

$$|f(x_1, y_1) - f(x_2, y_2)| < \epsilon$$

when the points (x_1, y_1) and (x_2, y_2) are in A and

$$|x_1 - x_2| < \eta \quad and \quad |y_1 - y_2| < \eta.$$

The proof is similar to that of §18 for functions of one variable. A function that has this property for a region A is said to be *uniformly continuous in A*. The corollary says that if a function of two variables is continuous in a closed region it is uniformly continuous there. If a function is continuous in an open region it may, or may not, be uniformly continuous there. For example, the function

$$f(x, y) = \frac{xy}{x^2 + y^2}$$

is continuous within the square bounded by the axes and the lines $x = 1$ and $y = 1$. If it were uniformly continuous in this open region, there would be an η associated with every positive ϵ in the way described. In order to see if there is such a number we consider the circle with center at the origin and radius $\frac{\eta}{2}$, where η is any positive number whatever. If ϵ is sufficiently small, there are two positive numbers, m_1 and m_2, such that

$$\left|\frac{m_1}{1 + m_1^2} - \frac{m_2}{1 + m_2^2}\right| > \epsilon.$$

Then if (x_1, y_1) is any point of the region that lies within the circle and on the line $y = m_1 x$, and (x_2, y_2) is a similar point on the line $y = m_2 x$, we have

$$|x_1 - x_2| < \eta, \quad |y_1 - y_2| < \eta$$

and

$$|f(x_1, y_1) - f(x_2, y_2)| > \epsilon,$$

since $f(x_1, y_1) = \dfrac{m_1}{1 + m_1^2}$ and $f(x_2, y_2) = \dfrac{m_2}{1 + m_2^2}$. Therefore the function is not uniformly continuous within the square. Any function that is continuous in a closed region furnishes an example of a function that is uniformly continuous in an open region.

We state without proof a few additional properties of functions of two variables that are continuous in a closed region.

(**a**) *A function of two variables that is continuous in a closed region A is bounded there—that is, it has an upper bound M and a lower bound m. The difference between these is called the oscillation of the function in A.*

(**b**) *There is a point in A at which the value of the function is equal to its upper limit in A, and a point at which the value of the function is equal to its lower limit in A.*

(**c**) *There is an unlimited number of points in A at which the value of the function is equal to any number between M and m.*

The reader should notice and explain the difference between statement (**c**) and the corresponding one concerning functions of one variable (§18).

32. Partial derivatives. If we give a fixed value to y and let x change, the function $f(x, y)$, which we suppose to be continuous and one-valued, becomes a function of x alone. In case this function of x has a derivative—that is, in case

$$\frac{f(x + \Delta x, y) - f(x, y)}{\Delta x}$$

approaches a finite limit as $\Delta x \to 0$—we call this limit the *partial derivative* of $f(x, y)$ with respect to x, and represent it by one of the symbols $f_x(x, y)$, $\dfrac{\partial f(x, y)}{\partial x}$, or $D_x f(x, y)$. The partial derivative of $f(x, y)$ with respect to y is

$$\lim_{\Delta y \to 0} \frac{f(x, y + \Delta y) - f(x, y)}{\Delta y}$$

when this limit exists. There is a notation for this corresponding to the notation for the partial derivative with respect to x.

The products

$$d_x f = \frac{\partial f}{\partial x}\Delta x \quad \text{and} \quad d_y f = \frac{\partial f}{\partial y}\Delta y$$

are called the *partial differentials* of $f(x, y)$ with respect to x and y respectively.

33. Total differentials. If $u = f(x, y)$ we have

$$\begin{aligned}\Delta u &= f(x + \Delta x, y + \Delta y) - f(x, y) \\ &= [f(x + \Delta x, y + \Delta y) - f(x, y + \Delta y)] \\ &\quad + [f(x, y + \Delta y) - f(x, y)].\end{aligned}$$

By applying the mean value theorem for functions of one variable to the differences in the square brackets we find that if f_x and f_y exist in the neighborhood of the point (x, y),

$$\Delta u = f_x(x + \theta\Delta x, y + \Delta y)\Delta x + f_y(x, y + \theta'\Delta y)\Delta y,$$

where $0 < \theta < 1$ and $0 < \theta' < 1$. If f_x and f_y are continuous at this point,[1]

$$\Delta u = [f_x(x, y) + \epsilon_1]\Delta x + [f_y(x, y) + \epsilon_2]\Delta y, \qquad (3)$$

where $\epsilon_1 \to 0$ and $\epsilon_2 \to 0$ as $\Delta x \to 0$ and $\Delta y \to 0$.

DEFINITION. The function $f(x, y)$ is said to be *differentiable* at the point (x, y) if it is uniquely defined in the neighborhood of this point and if

$$\Delta u = A\Delta x + B\Delta y + \epsilon_1\Delta x + \epsilon_2\Delta y, \qquad (4)$$

where A and B are independent of Δx and Δy and $\epsilon_1 \to 0$ and $\epsilon_2 \to 0$ with Δx and Δy.

If $f(x, y)$ is differentiable, the part $A\Delta x + B\Delta y$ is called the *total differential* of $f(x, y)$ and is represented by the symbol du:

$$du = A\Delta x + B\Delta y.$$

---

[1] Since in the mean value theorem the continuity of the derivative was not assumed, it is not necessary here to assume the continuity of both of these partial derivatives.

In this formula it is usual to represent Δx by dx and Δy by dy. Then

$$du = A\,dx + B\,dy. \tag{5}$$

It is clear that if a function is differentiable at a point, its first partial derivatives have definite finite values at this point. For if in (4) we put $\Delta y = 0$ we have

$$\frac{\Delta u}{\Delta x} = A + \epsilon_1$$

and $\lim\limits_{\Delta x \to 0} \frac{\Delta u}{\Delta x} = A = \frac{\partial u}{\partial x}$. Similarly $\lim\limits_{\Delta y \to 0} \frac{\Delta u}{\Delta y} = B = \frac{\partial u}{\partial y}$, when $\Delta x = 0$. Then we can write

$$du = \frac{\partial u}{\partial x} dx + \frac{\partial u}{\partial y} dy.$$

It does not follow that the function is differentiable at a point if its first partial derivatives have definite values there. This would follow however if these partial derivatives existed in the neighborhood of the point and one of them were continuous at the point.

THEOREM 1. *If $f(x, y)$ is differentiable at the point (x, y) it is continuous at this point.*

This is an immediate consequence of (4).

Since x and y are independent variables we can, if we wish, hold dx and dy constant. We shall find it convenient to do this. Then du is a function of x and y only and has a total differential which we call the *second differential* of u and represent by the symbol d^2u.

$$d^2u = \frac{\partial^2 u}{\partial x^2} \overline{dx^2} + 2 \frac{\partial^2 u}{\partial x \partial y} dx\,dy + \frac{\partial^2 u}{\partial y^2} \overline{dy^2}.$$

In a similar way we can form higher differentials of u, if we make suitable assumptions in regard to the partial derivatives of u of lower order. We shall see in the next article that the corresponding formulae are more complicated if x and y are not the independent variables.

34. If x and y are differentiable functions of a single variable t, u is also a function of t and we have from (3)

$$\frac{\Delta u}{\Delta t} = (f_x + \epsilon_1)\frac{\Delta x}{\Delta t} + (f_y + \epsilon_2)\frac{\Delta y}{\Delta t},$$

and therefore

$$\frac{du}{dt} = f_x\frac{dx}{dt} + f_y\frac{dy}{dt}, \qquad (6)$$

if the derivatives f_x, f_y exist in the neighborhood of (x, y), and one of them is continuous at this point.

If we differentiate each side of (6) with respect to t, we obtain the formula

$$\frac{d^2u}{dt^2} = \frac{d}{dt}(f_x)\frac{dx}{dt} + f_x\frac{d^2x}{dt^2} + \frac{d}{dt}(f_y)\frac{dy}{dt} + f_y\frac{d^2y}{dt^2}.$$

But

$$\frac{d}{dt}(f_x) = f_{x,x}\frac{dx}{dt} + f_{x,y}\frac{dy}{dt}$$

and

$$\frac{d}{dt}(f_y) = f_{y,x}\frac{dx}{dt} + f_{y,y}\frac{dy}{dt},$$

where we have represented the derivatives $\frac{\partial}{\partial x}(f_x), \frac{\partial}{\partial y}(f_x),$ $\frac{\partial}{\partial x}(f_y),$ and $\frac{\partial}{\partial y}(f_y)$ respectively by the symbols $f_{x,x}, f_{x,y}, f_{y,x},$ and $f_{y,y}.$ Hence

$$\frac{d^2u}{dt^2} = \left(f_{x,x}\frac{dx}{dt} + f_{x,y}\frac{dy}{dt}\right)\frac{dx}{dt} + f_x\frac{d^2x}{dt^2}$$

$$+ \left(f_{y,x}\frac{dx}{dt} + f_{y,y}\frac{dy}{dt}\right)\frac{dy}{dt} + f_y\frac{d^2y}{dt^2}$$

$$= f_{x,x}\left(\frac{dx}{dt}\right)^2 + 2f_{x,y}\frac{dx}{dt}\frac{dy}{dt} + f_{y,y}\left(\frac{dy}{dt}\right)^2$$

$$+ f_x\frac{d^2x}{dt^2} + f_y\frac{d^2y}{dt^2}. \qquad (7)$$

We have here assumed that $f_{x,y} = f_{y,x}$. This will be proved in §42.

If x and y are functions of several variables of which t is one, (4) would hold with $\dfrac{du}{dt}$, $\dfrac{dx}{dt}$, and $\dfrac{dy}{dt}$ replaced by $\dfrac{\partial u}{\partial t}$, $\dfrac{\partial x}{\partial t}$, and $\dfrac{\partial y}{\partial t}$ respectively:

$$\frac{\partial u}{\partial t} = f_x \frac{\partial x}{\partial t} + f_y \frac{\partial y}{\partial t}. \tag{8}$$

If in particular x and y depend upon the two variables r and s, we have in a similar way from (5)

$$\left.\begin{aligned}\frac{\partial^2 u}{\partial r^2} &= f_{x,x}\left(\frac{\partial x}{\partial r}\right)^2 + 2 f_{x,y}\frac{\partial x}{\partial r}\frac{\partial y}{\partial r} + f_{y,y}\left(\frac{\partial y}{\partial r}\right)^2 \\ &\quad + f_x \frac{\partial^2 x}{\partial r^2} + f_y \frac{\partial^2 y}{\partial r^2} \\ \text{and} & \\ \frac{\partial^2 u}{\partial s^2} &= f_{x,x}\left(\frac{\partial x}{\partial s}\right)^2 + 2 f_{x,y}\frac{\partial x}{\partial s}\frac{\partial y}{\partial s} + f_{y,y}\left(\frac{\partial y}{\partial s}\right)^2 \\ &\quad + f_x \frac{\partial^2 x}{\partial s^2} + f_y \frac{\partial^2 y}{\partial s^2}.\end{aligned}\right\} \tag{9}$$

The reader can verify that

$$\frac{\partial^2 u}{\partial r \partial s} = f_{x,x} \frac{\partial x}{\partial r}\frac{\partial x}{\partial s} + f_{x,y}\left(\frac{\partial x}{\partial r}\frac{\partial y}{\partial s} + \frac{\partial x}{\partial s}\frac{\partial y}{\partial r}\right)$$
$$+ f_{y,y}\frac{\partial y}{\partial r}\frac{\partial y}{\partial s} + f_x \frac{\partial^2 x}{\partial r \partial s} + f_y \frac{\partial^2 y}{\partial r \partial s}. \tag{10}$$

Formulae of this kind are necessary in order to introduce new independent variables in an equation involving x, y, and u and partial derivatives of u with respect to x and y; as, for example, when we wish to change from rectangular to polar coordinates in a partial differential equation. In this case we have $x = r \cos \theta$ and $y = r \sin \theta$.

EXERCISE. Show that

$$f_{x,x} + f_{y,y} = f_{r,r} + \frac{1}{r^2} f_{\theta,\theta} + \frac{1}{r} f_r.$$

35. An application in geometry. Let

$$x = f(t), \quad y = \varphi(t), \quad z = \psi(t)$$

be the parametric equations of a curve through the point $P = (x_1, y_1, z_1)$ which corresponds to $t = t_1$. If $f'(t_1)$, $\varphi'(t_1)$, and $\psi'(t_1)$ are not all zero,

$$\frac{x - f(t_1)}{f'(t_1)} = \frac{y - \varphi(t_1)}{\varphi'(t_1)} = \frac{z - \psi(t_1)}{\psi'(t_1)} \tag{11}$$

are the equations of the tangent to the curve at the point P. We assume that the curve lies on the surface

$$z = F(x, y).$$

Then we have the identity

$$\psi(t) = F[f(t), \varphi(t)]$$

and

$$\psi'(t_1) = F_x[f(t_1), \varphi(t_1)]f'(t_1) + F_y[f(t_1), \varphi(t_1)]\varphi'(t_1),$$

provided that the partial derivatives F_x and F_y exist in the neighborhood of P and one of them is continuous at P. If we eliminate $f'(t_1)$, $\varphi'(t_1)$, and $\psi'(t_1)$ from this last equation by means of (9) we obtain the equation

$$z - \psi(t_1) = F_x[f(t_1), \varphi(t_1)][x - f(t_1)] \\ + F_y[f(t_1), \varphi(t_1)][y - \varphi(t_1)],$$

or

$$z - z_1 = F_x(x_1, y_1)(x - x_1) + F_y(x_1 y_1)(y - y_1). \tag{12}$$

This is the equation of a plane through the point P and it is satisfied by every set of values of x, y, and z that satisfy (11). But the particular tangent line in question is determined by the curve on the surface to which it is tangent, and this in turn is determined by the functions $f(t)$, $\varphi(t)$, and $\psi(t)$. These functions, however, do not appear in the equation of the plane. Hence the tangent lines at P to all the curves on the surface that pass through P lie in the plane (12). This plane is therefore called the *tangent plane* to the surface at P.

EXERCISES

Find the equations of the tangent plane and the normal line of each of the following surfaces at the point indicated:

1. $x^2 + y^2 = 3z$ $(2, -1, 5/3)$.   2. $y = x \tan \dfrac{z}{2}$ $\left(-3, -3, \dfrac{\pi}{2}\right)$.
3. $xyz = -1$ $(1, 1, -1)$.
4. Find the distance from the origin to the plane $ax + by + cz + d = 0$.
5. Find the equations of the normal to the surface $z = f(x, y)$ at the point (x_1, y_1, z_1).
6. Show that the helix $x = 3\cos t$, $y = 3\sin t$, $z = t$ and the surface $xyz = 1$ have in common a point corresponding to a value of t between 0 and $\dfrac{\pi}{4}$.
7. Does the arc of the helix $x = \cos t$, $y = \sin t$, $z = t$ between the point for which $t = 0$ and the point for which $t = \dfrac{\pi}{4}$ have a point in common with this surface? Do the curve and the surface have any points in common?
8. Find $\dfrac{\partial z}{\partial x}$ and $\dfrac{\partial z}{\partial y}$ when $z = \log \sqrt{1 + x^2 + y^2}$.
9. Show that $\dfrac{\partial^2 f}{\partial x^2} + \dfrac{\partial^2 f}{\partial y^2} = 0$ if $f = \log(x^2 + y^2)$.
10. Show that $\dfrac{\partial^2 f}{\partial x^2} + \dfrac{\partial^2 f}{\partial y^2} = 0$ if $f = \arctan \dfrac{y}{x}$.
11. Show that $\dfrac{\partial^2 f}{\partial x^2} = a^2 \dfrac{\partial^2 f}{\partial y^2}$ if $f = \tan(y + ax) + \tan(y - ax)$.
12. Find $x \dfrac{\partial z}{\partial x} + y \dfrac{\partial z}{\partial y}$ when $z = f\left(\dfrac{y}{x}\right)$.
13. Find $x \dfrac{\partial z}{\partial x} - y \dfrac{\partial z}{\partial y}$ when $z = f(x\,y)$.
14. Show that $\dfrac{\partial^2 z}{\partial x^2} + \dfrac{\partial^2 z}{\partial y^2} = 0$ when $z = \varphi(x + iy) + \psi(x - iy)$.
15. Given $f(x, y, z) = \dfrac{1}{\sqrt{x^2 + y^2 + z^2}}$. Find $f_{xx} + f_{yy} + f_{zz}$.

36. Invariance of the total differential. If $\varphi(x, y)$ and $\psi(x, y)$ are differentiable functions and r and s are connected with x and y by the equations

$$r = \varphi(x, y), \qquad s = \psi(x, y),$$

which we assume can be solved for x and y in terms of r

and s, we have from (8)

$$\frac{\partial u}{\partial x} = \frac{\partial u}{\partial r}\frac{\partial r}{\partial x} + \frac{\partial u}{\partial s}\frac{\partial s}{\partial x},$$

$$\frac{\partial u}{\partial y} = \frac{\partial u}{\partial r}\frac{\partial r}{\partial y} + \frac{\partial u}{\partial s}\frac{\partial s}{\partial y}.$$

Then

$$\begin{aligned}du &= \frac{\partial u}{\partial x}dx + \frac{\partial u}{\partial y}dy \\ &= \left(\frac{\partial u}{\partial r}\frac{\partial r}{\partial x} + \frac{\partial u}{\partial s}\frac{\partial s}{\partial x}\right)dx + \left(\frac{\partial u}{\partial r}\frac{\partial r}{\partial y} + \frac{\partial u}{\partial s}\frac{\partial s}{\partial y}\right)dy \\ &= \frac{\partial u}{\partial r}\left(\frac{\partial r}{\partial x}dx + \frac{\partial r}{\partial y}dy\right) + \frac{\partial u}{\partial s}\left(\frac{\partial s}{\partial x}dx + \frac{\partial s}{\partial y}dy\right) \\ &= \frac{\partial u}{\partial r}dr + \frac{\partial u}{\partial s}ds.\end{aligned}$$

That is, the form of du is the same whether the variables with respect to which we differentiate are looked upon as independent or dependent variables. This is not true of the higher differentials of u.

37. Differentiation of equations. If we have the equation

$$u = f(x, y, z) = \text{constant}, \tag{13}$$

what is the value of the total differential du? Obviously $\Delta u = 0$. But du and Δu are not in general equal. If we assume that $f(x, y, z)$ is a differentiable function, we have $du = \frac{\partial f}{\partial x}dx + \frac{\partial f}{\partial y}dy + \frac{\partial f}{\partial z}dz$, and the value of du at any point (x, y, z) depends upon the values we assign to dx, dy, and dz. But if (13) is to be satisfied, these values are not independent. Under certain conditions which we assume to be satisfied, equation (13) determines z as a function of x and y (§169). In the expression given above for du we understand by dz the total differential of this function.

$$dz = -\frac{\frac{\partial f}{\partial x}}{\frac{\partial f}{\partial z}}dx - \frac{\frac{\partial f}{\partial y}}{\frac{\partial f}{\partial z}}dy,$$

and therefore $du = 0$. This establishes the following important theorem:

THEOREM 2. *If each member of an equation connecting the variables x, y, and z is a differentiable function of these variables, the total differentials of the two members are equal.*

The theorem holds irrespective of any additional differentiable relations connecting the variables. It also holds for any number of variables. For example, if we have two variables, x and y, connected by the equation $f(x, y) = 0$, then

$$\frac{\partial f}{\partial x} dx + \frac{\partial f}{\partial y} dy = 0, \quad \text{or} \quad \frac{dy}{dx} = -\frac{\dfrac{\partial f}{\partial x}}{\dfrac{\partial f}{\partial y}}.$$

Hence the equation of the tangent to the curve $f(x, y) = 0$ at the point (x_0, y_0) is

$$\left(\frac{\partial f}{\partial x}\right)_0 (x - x_0) + \left(\frac{\partial f}{\partial y}\right)_0 (y - y_0) = 0,$$

unless $\left(\dfrac{\partial f}{\partial x}\right)_0 = \left(\dfrac{\partial f}{\partial y}\right)_0 = 0$. A point on the curve at which $\dfrac{\partial f}{\partial x}$ and $\dfrac{\partial f}{\partial y}$ both equal zero is called a *singular point* of the curve.

38. Directional derivatives. Let $P \equiv (x, y)$ be any point on a given curve C and let s be the distance along C measured from an arbitrary point on it to P. Then x and y are functions of the single variable s and we can apply (6) to the function $u = f(x, y)$.

$$\frac{du}{ds} = \frac{\partial f}{\partial x}\frac{dx}{ds} + \frac{\partial f}{\partial y}\frac{dy}{ds}.$$

But $\dfrac{dx}{ds} = \lim\limits_{\Delta s \to 0} \dfrac{\Delta x}{\Delta s} = \lim \dfrac{\Delta x}{c} \cdot \dfrac{c}{\Delta s}$, where c is the chord connecting the two points corresponding to s and $s + \Delta s$. But $\lim \dfrac{\Delta x}{c} = \cos \alpha$, where α is the inclination of the tangent to

C at P, and $\lim \dfrac{c}{\Delta s} = 1$ (§187). Hence $\dfrac{dx}{ds} = \cos \alpha$. Similarly we find that $\dfrac{dy}{ds} = \sin \alpha$. Then

$$\frac{du}{ds} = \frac{\partial f}{\partial x} \cos \alpha + \frac{\partial f}{\partial y} \sin \alpha. \qquad (14)$$

We call $\dfrac{\partial u}{\partial s}$ the *directional derivative of u along C*. But at a given point it depends only on the direction of the tangent to C at P. We therefore call it also the *derivative of u in the direction* α.

Since $\dfrac{\partial u}{\partial s}$ is a function of α we may represent it by $\varphi(\alpha)$. It is clear that $\varphi(0) = \dfrac{\partial f}{\partial x}$ and $\varphi\left(\dfrac{\pi}{2}\right) = \dfrac{\partial f}{\partial y}$. Moreover $\varphi(\alpha) = 0$ when

$$\frac{\partial f}{\partial x} \cos \alpha + \frac{\partial f}{\partial y} \sin \alpha = 0,$$

or

$$\tan \alpha = - \frac{\dfrac{\partial f}{\partial x}}{\dfrac{\partial f}{\partial y}}.$$

But the right member of this last equation is the slope at the point P of that curve of the family $f(x, y) = c$ that passes through this point. In other words, the directional derivative of $f(x, y)$ along the curve $f(x, y) = c$ is zero. It has its maximal absolute value when

$$\varphi'(\alpha) = - \frac{\partial f}{\partial x} \sin \alpha + \frac{\partial f}{\partial y} \cos \alpha = 0,$$

or

$$\tan \alpha = \frac{\dfrac{\partial f}{\partial y}}{\dfrac{\partial f}{\partial x}}, \qquad (15)$$

since $\varphi(\alpha) = 0$ when $\alpha = \alpha_1$ and $\alpha = \alpha_1 + \pi$, α_1 being the inclination of the tangent of the curve $f(x, y) = c$. The direction α_2 for the maximum absolute value of $\varphi(\alpha)$ is therefore perpendicular to the tangent to this curve.

The curves $f(x, y) = c$ are called the *contour lines* of the surface $z = f(x, y)$. They are the orthogonal projections on the (x, y)-plane of the points on the surface that are at the same distance from the (x, y)-plane, and are of great value in showing on a map the different elevations of land in the region mapped.

It follows from (14) and (15) that when $\alpha = \alpha_2$ we have

$$\frac{\partial u}{\partial s} = \varphi(\alpha_2) = \sqrt{\left(\frac{\partial f}{\partial x}\right)^2 + \left(\frac{\partial f}{\partial y}\right)^2}.$$

Since this derivative is taken in a direction normal to the curve $f(x, y) = c$, we denote it by the symbol $\frac{\partial f}{\partial n}$. It is called the *gradient* of $f(x, y)$.

Similar results can be obtained in connection with directional derivatives of functions of three variables. We leave the details to the reader.

EXERCISES

Find the directional derivative of each of the following functions at the point and in the direction indicated:

1. $x^2 + y^2$ (2, 3). $\alpha = 60°$.
2. $x^2 - y^2$ (4, 1). $\alpha = 45°$.
3. $\cos xy$ (2, -3). $\alpha = 30°$.
4. Show that the magnitude of the gradient of $x^2 + y^2$ is constant along a circle with center at the origin.
5. Show that the magnitude of the gradient of $\frac{1}{x^2 + y^2}$ is constant along the same circles.
6. Find the direction and magnitude of the gradient of $x - y$ at the point (5, 2).
7. Find the direction and magnitude of the gradient of
$$V = \log \sqrt{x^2 + y^2}.$$
8. Find the direction and magnitude of the gradient of
$$V = \log \sqrt{x^2 + y^2 + z^2}.$$

Draw some of the contour lines of each of the following functions:

9. $x^2 + y^2 = 4z$. 10. $x^2 + y^2 = 3(z + 4)$. 11. $\dfrac{x^2}{16} + \dfrac{y^2}{9} + \dfrac{z^2}{4} = 1$.

What kind of curves are the contour lines of:

12. $x^2 + y^2 + z^2 = 1$. 13. $\dfrac{x^2}{4} + \dfrac{y^2}{9} = z$. 14. $\dfrac{x^2}{4} + \dfrac{y^2}{9} = z - 1$.

15. $\dfrac{x^2}{4} - \dfrac{y^2}{9} = 2z$. 16. $xyz = 1$. 17. $z = \dfrac{x - y}{x + y}$.

18. $z = \dfrac{x^2 + y^2 - 2}{2x}$. 19. $z = \dfrac{3x^2 - y^2 + 1}{2x^2 + y^2 - 1}$.

20. What is the direction of the gradient of $\dfrac{x^2}{4} + \dfrac{y^2}{16} - 1$ at the point $(2, -3)$?

21. Of $\dfrac{x^2}{4} - \dfrac{y^2}{8} - 1$ at the point $(2, 3)$?

22. Find the direction and magnitude of the gradient of $f = e^{-y} \sin 2x + e^{-2y} \sin x$ at the point $\left(\dfrac{\pi}{2}, 0\right)$.

39. If u is a function of x, y, and z, we define the differential of u to be

$$du = \frac{\partial u}{\partial x} dx + \frac{\partial u}{\partial y} dy + \frac{\partial u}{\partial z} dz.$$

This is analogous to the definition of the differential of a function of two variables.

Theorem 3. *If u is a function of x, y, and z, and*

$$du = X dx + Y dy + Z dz,$$

then $X = \dfrac{\partial u}{\partial x}$, $Y = \dfrac{\partial u}{\partial y}$, $Z = \dfrac{\partial u}{\partial z}$.

Since x, y, and z are independent we can hold any two of them, as y and z, fixed. Then $dy = dz = 0$ and $du = X dx$, or $\dfrac{\partial u}{\partial x} = X$.

40. Exact differentials. If

$$du = M dx + N dy,$$

where M and N are differentiable functions of x and y, then (see §39)

$$\frac{\partial u}{\partial x} = M \quad \text{and} \quad \frac{\partial u}{\partial y} = N.$$

From this it follows that

$$\frac{\partial^2 u}{\partial x \partial y} = \frac{\partial M}{\partial y} \quad \text{and} \quad \frac{\partial^2 u}{\partial y \partial x} = \frac{\partial N}{\partial x},$$

and therefore in view of §42 that

$$\frac{\partial M}{\partial y} = \frac{\partial N}{\partial x}. \tag{16}$$

This is a necessary condition that

$$M dx + N dy$$

be the total differential of some function u of x and y.

Moreover (16) is a sufficient condition for the existence of a function $u = f(x, y)$ such that

$$du = M dx + N dy.$$

For the function

$$u = \int M dx + \varphi(y),$$

where $\varphi(y)$ is an arbitrary function of y, is such that $\frac{\partial u}{\partial x} = M$, and if we can determine $\varphi(y)$ in such a way that $\frac{\partial u}{\partial y} = N$, we shall have $du = M dx + N dy$. Now

$$\frac{\partial u}{\partial y} = \frac{\partial}{\partial y} \int M dx + \varphi'(y)$$

$$= \int \frac{\partial M}{\partial y} dx + \varphi'(y).$$

Hence in order to meet the condition that $\frac{\partial u}{\partial y}$ shall be equal to N we must have

$$\varphi'(y) = N - \frac{\partial}{\partial y} \int M dx. \tag{17}$$

We could not satisfy this condition (17) if the right side contained x, since the left side is a function of y alone and x

and y are independent variables. But the right side does not contain x, as is shown by the fact that its derivative with respect to x is zero.

$$\frac{\partial}{\partial x}\left(N - \frac{\partial}{\partial y}\int M\,dx\right)$$
$$= \frac{\partial N}{\partial x} - \frac{\partial}{\partial x}\int \frac{\partial M}{\partial y}\,dx = \frac{\partial N}{\partial x} - \frac{\partial M}{\partial y} = 0.$$

From (17) we have

$$\varphi(y) = \int\left(N - \frac{\partial}{\partial y}\int M\,dx\right)dy.$$

Then

$$u = \int M\,dx + \int\left(N - \frac{\partial}{\partial y}\int M\,dx\right)dy$$

is a function of x and y such that

$$du = M\,dx + N\,dy,$$

since

$$\frac{\partial u}{\partial x} = M$$

and

$$\frac{\partial u}{\partial y} = \frac{\partial}{\partial y}\int M\,dx + N - \frac{\partial}{\partial y}\int M\,dx = N.$$

THEOREM 4. *A necessary and sufficient condition that*

$$M\,dx + N\,dy$$

be an exact differential is that

$$\frac{\partial M}{\partial y} = \frac{\partial N}{\partial x}.$$

In this proof we have assumed that $\frac{\partial}{\partial x}\int \frac{\partial M}{\partial y}\,dx = \frac{\partial M}{\partial y}$ and that $\frac{\partial}{\partial y}\int M\,dx = \int \frac{\partial M}{\partial y}\,dx$. For a justification of these assumptions see §§67 and 68. We add the corresponding theorem for functions of three variables:

§40] FUNCTIONS OF VARIABLES

Theorem 5. *A necessary and sufficient condition that*

$$Xdx + Ydy + Zdz$$

be an exact differential is that

$$\frac{\partial X}{\partial y} = \frac{\partial Y}{\partial x}, \quad \frac{\partial Y}{\partial z} = \frac{\partial Z}{\partial y}, \quad \frac{\partial Z}{\partial x} = \frac{\partial X}{\partial z}.$$

In the first place, this condition is necessary, for if there is a function $u = f(x, y, z)$ such that

$$du = Xdx + Ydy + Zdz,$$

then by Theorem 3

$$X = \frac{\partial u}{\partial x}, \quad Y = \frac{\partial u}{\partial y}, \quad Z = \frac{\partial u}{\partial z},$$

and from this it follows that

$$\frac{\partial X}{\partial y} = \frac{\partial Y}{\partial x}, \quad \frac{\partial Y}{\partial z} = \frac{\partial Z}{\partial y}, \quad \frac{\partial Z}{\partial x} = \frac{\partial X}{\partial z},$$

since

$$\frac{\partial X}{\partial y} = \frac{\partial^2 u}{\partial x \partial y}, \quad \frac{\partial Y}{\partial x} = \frac{\partial^2 u}{\partial y \partial x}, \quad \frac{\partial Y}{\partial y} = \frac{\partial^2 u}{\partial y \partial z},$$

$$\frac{\partial Z}{\partial y} = \frac{\partial^2 u}{\partial z \partial y}, \quad \frac{\partial Z}{\partial x} = \frac{\partial^2 u}{\partial z \partial x}, \quad \frac{\partial X}{\partial z} = \frac{\partial^2 u}{\partial x \partial z}.$$

(See §42.) Moreover the condition is sufficient. In order to see that this is so we regard z for the moment as a constant. We are entitled to do this since it is one of the independent variables. Then we know from Theorem 4 and the given conditions that there is a function $\varphi(x, y, z)$ such that

$$d\varphi = Xdx + Ydy.$$

We now put

$$u = f(x, y, z) = \varphi(x, y, z) + \psi(z), \quad (18)$$

where $\psi(z)$ is an arbitrary function, and seek to determine $\psi(z)$ in such a way as to make

$$du = Xdx + Ydy + Zdz.$$

It follows from (18) that

$$du = Xdx + Ydx + \left[\frac{\partial \varphi}{\partial z} + \psi'(z)\right]dz.$$

Hence it will be sufficient so to determine $\psi(z)$ as to have

$$\frac{\partial \varphi}{\partial z} + \psi'(z) = Z,$$

or

$$\psi'(z) = Z - \frac{\partial \varphi}{\partial z}. \qquad (19)$$

It will be impossible to satisfy (19) unless the right member is free from x and y. Now

$$\frac{\partial}{\partial x}\left(Z - \frac{\partial \varphi}{\partial z}\right) = \frac{\partial Z}{\partial x} - \frac{\partial^2 \varphi}{\partial z \partial x} = \frac{\partial Z}{\partial x} - \frac{\partial}{\partial z}\frac{\partial \varphi}{\partial x}.$$

But $\dfrac{\partial \varphi}{\partial x} = X$. Hence

$$\frac{\partial}{\partial x}\left(Z - \frac{\partial \varphi}{\partial z}\right) = \frac{\partial Z}{\partial x} - \frac{\partial X}{\partial z} = 0$$

by virtue of the conditions of the theorem that are assumed. In a similar way we can see that

$$\frac{\partial}{\partial y}\left(Z - \frac{\partial \varphi}{\partial z}\right) = \frac{\partial Z}{\partial y} - \frac{\partial Y}{\partial z} = 0.$$

Then $Z - \dfrac{\partial \varphi}{\partial z}$ is free from x and y and (19) will be satisfied by the function

$$\psi(z) = \int\left(Z - \frac{\partial \varphi}{\partial z}\right)dz.$$

For this choice of the arbitrary function $\psi(z)$

$$u = \varphi(x, y, z) + \psi(z)$$

is a function whose total differential is

$$Xdx + Ydy + Zdz.$$

This completes the proof of Theorem 5.

EXERCISES

Point out the exact differentials from the following and find the functions of which they are the differentials:

1. $\dfrac{2x}{x^2+y^2}dx + \dfrac{2y}{x^2+y^2}dy.$
2. $\dfrac{2x}{x^2-y^2}dx + \dfrac{2y}{x^2-y^2}dy.$
3. $\dfrac{xdx - ydy}{x^2+y^2}.$
4. $-e^{-x}\sin y\,dx + e^{-x}\cos y\,dy.$
5. $\log yz\,dx + \dfrac{zx}{y}dy + \dfrac{xy}{z}dz.$
6. $\dfrac{y^2dx}{x(x-y)^2} - \dfrac{ydy}{(x-y)^2}.$

41. Euler's theorem on homogeneous functions. Suppose that a function $f(x_1, x_2, \cdots, x_n)$ of the n variables $x_1, x_2, \cdots, x_n$ is such that

$$f(tx_1, tx_2, \cdots, tx_n) = t^m f(x_1, x_2, \cdots, x_n), \qquad (20)$$

where t is indeterminate. We then say that f is *homogeneous of degree m* in the variables $x_1, x_2, \cdots, x_n$. For example, the functions $x^2 + y^2$, $\sqrt{x^2 + y^2}$, $\dfrac{1}{\sqrt{x^2+y^2}}$, and $\dfrac{x^{1/2}}{x^2+y^2}$ are homogeneous of degrees 2, 1, -1, and $-\dfrac{3}{2}$, respectively; while $\dfrac{\sin x}{x}$ is not homogeneous.

Equation (20) can be written in the form

$$f(u_1, u_2, \cdots, u_n) = t^m f(x_1, x_2, \cdots, x_n),$$

where $u_i = tx_i$ ($i = 1, 2, \cdots, n$). Differentiation with respect to t gives us

$$\frac{\partial f(u_1, u_2, \cdots, u_n)}{\partial u_1}\frac{\partial u_1}{\partial t} + \frac{\partial f}{\partial u_2}\frac{\partial u_2}{\partial t} + \cdots + \frac{\partial f}{\partial u_n}\frac{\partial u_n}{\partial t}$$
$$= m t^{m-1} f(x_1, x_2, \cdots, x_n).$$

For $t = 1$ we have

$$x_1 \frac{\partial f}{\partial x_1} + x_2 \frac{\partial f}{\partial x_2} + \cdots + x_n \frac{\partial f}{\partial x_n} = mf,$$

where now the arguments on each side are $x_1, x_2, \cdots, x_n$, since $\dfrac{\partial u_i}{\partial t} = x_i$. This proves what is known as *Euler's Theorem:*

THEOREM 6. *If $f(x_1, x_2, \cdots, x_n)$ is homogeneous of degree m, then*

$$x_1 \frac{\partial f}{\partial x_1} + x_2 \frac{\partial f}{\partial x_2} + \cdots + x_n \frac{\partial f}{\partial x_n} = mf.$$

42. The order of differentiation. If $f(x, y)$ is defined in a given region and if f_x, f_y, and $f_{x,y}$ all exist at a point (x, y) of this region, the second derivative $f_{y,x}$ may, or may not, exist at this point; and if it exists it may, or may not, be equal to $f_{x,y}$. The following theorem, which is due to Schwarz, gives a set of sufficient conditions for the existence of $f_{y,x}$ and of its equality with $f_{x,y}$.

THEOREM 7. *If f_x, f_y, and $f_{x,y}$ all exist in the neighborhood of the point (x, y) and if $f_{x,y}$ is continuous at this point, then $f_{y,x}$ exists at the point and $f_{x,y} = f_{y,x}$.*

The function

$$\varphi(x) = f(x, y + k) - f(x, y)$$

for a fixed y and a fixed k satisfies the conditions of the mean value theorem for functions of one variable. Hence by a repeated application of the theorem

$$\varphi(x + h) - \varphi(x) = h[f_x(x + \theta h, y + k) - f_x(x + \theta h, y)]$$
$$= hk f_{x,y}(x + \theta h, y + \theta' k),$$

where $0 < \theta < 1$ and $0 < \theta' < 1$. Then, since by hypothesis $f_{x,y}$ is continuous at the point (x, y),

$$\varphi(x + h) - \varphi(x) = hk[f_{x,y}(x, y) + \epsilon],$$

where $\epsilon \to 0$ as h and $k \to 0$. In view of the definition of $\varphi(x)$ this is equivalent to the equation

$$f(x + h, y + k) - f(x + h, y) - f(x, y + k) + f(x, y)$$
$$= hk[f_{x,y}(x, y) + \epsilon].$$

Then

$$\frac{f(x + h, y + k) - f(x + h, y)}{k} - \frac{f(x, y + k) - f(x, y)}{k}$$
$$= h[f_{x,y}(x, y) + \epsilon].$$

In the limit as $k \to 0$ we have
$$f_y(x + h, y) - f_y(x, y) = h[f_{x,y}(x, y) + \epsilon_h],$$
where ϵ_h is the limit, for a fixed h, of ϵ as $k \to 0$. We know that, for a fixed h sufficiently small in absolute value, ϵ approaches a limit as $k \to 0$, since, by hypothesis, f_y exists in the neighborhood of the point (x, y). Then
$$\frac{f_y(x + h, y) - f_y(x, y)}{h} = f_{x,y}(x, y) + \epsilon_h.$$
In the limit as $h \to 0$ we have
$$f_{y,x}(x, y) = f_{x,y}(x, y).$$

Under the conditions of the theorem then it is **immaterial** whether we differentiate first with respect to x and then with respect to y, or first with respect to y and then with respect to x. We can apply the theorem repeatedly to higher derivatives and to functions of more than two variables. The conclusion is that under certain conditions
$$\frac{\partial^{(n)} f}{\partial x^p \partial y^q \partial z^r \cdots} \quad (p + q + r + \cdots = n)$$
is the same regardless of the order of the differentiations, provided that there are p differentiations with respect to x, q with respect to y, and so on.

We have assumed the existence of only one of the second derivatives, but we also assumed the continuity of this derivative. There is a theorem of Young's to the effect that we can omit this assumption of continuity if we add the assumption that all the second derivatives exist.[1]

EXERCISES

Show that $\dfrac{\partial^2 f}{\partial x \partial y} = \dfrac{\partial^2 f}{\partial y \partial x}$ for each of the following functions:

1. $x^2 - y^2$. 2. $\dfrac{x + y}{x - y}$. 3. $\log \sqrt{x^2 + 2y^2}$. 4. $e^{x+y} \sin(x - y)$.

5. $\arccos \dfrac{x}{y}$. 6. $\arcsin \dfrac{y}{x}$. 7. $\log \dfrac{x^2 - y^2}{x^2 + y^2}$.

[1] See, for example, de la Vallée-Poussin, *Cours d'analyse infinitésimale*, Vol. 1, 5th ed., pp. 116–117.

MISCELLANEOUS EXERCISES

1. Show that the equation of the tangent plane to the surface $ax^2 + by^2 + cz^2 = 1$ at the point (x_0, y_0, z_0) can be written in the form $ax_0x + by_0y + cz_0z = 1$.

2. Prove that $Mdx + Ndy$ is exact if $Pdx + Qdy$ and $(M + P)dx + (N + Q)dy$ are.

3. Show that if $u = f(x, y, z)$ is a differentiable function of x, y, and z, and z is a differentiable function $\varphi(x, y)$ of x and y, then $f[x, y, \varphi(x, y)]$ is a differentiable function of x and y.

4. Find the directional derivative of the function $F(x, y, z)$ along the curve $x = f(t)$, $y = \varphi(t)$, $z = \psi(t)$.

5. Show that the gradient of the function $f(x, y)$ in terms of polar coordinates is $\sqrt{\varphi_r^2 + \frac{1}{r^2}\varphi_\theta^2}$, where $f(x, y) = \varphi(r, \theta)$.

6. Show that in polar coordinates the directional derivative along a radius vector is $\dfrac{\partial f}{\partial r}$.

7. Show that in polar coordinates the directional derivative along the normal to a radius vector is $\dfrac{1}{r}\dfrac{\partial f}{\partial \theta}$.

8. Find the directional derivative in a direction that makes an angle α with the radius vector. (Use polar coordinates.)

9. Show that the square of the gradient is equal to the sum of the squares of the directional derivatives in two mutually perpendicular directions.

10. Prove that if $f(x, y) = xy\dfrac{x^2 - y^2}{x^2 + y^2}$ when x and y are not both equal to zero and $f(0, 0) = 0$, then $f(x, y)$ is continuous at the origin.

11. Find the first partial derivatives of $f(x, y)$ at the origin.

12. Prove that if $z = \arctan\dfrac{y}{x}$, then

$$(1 + q^2)r - 2pqs + (1 + p^2)t = 0,$$

where

$$p = \frac{\partial z}{\partial x}, \quad q = \frac{\partial z}{\partial y}, \quad r = \frac{\partial^2 z}{\partial x^2}, \quad s = \frac{\partial^2 z}{\partial x \partial y}, \quad t = \frac{\partial^2 z}{\partial y^2}.$$

13. Prove that

$$\frac{1}{y^2}\left(\frac{\partial^2 z}{\partial x^2} + 2xy^2\frac{\partial z}{\partial x} + 2(y - y^3)\frac{\partial z}{\partial y} + x^2y^2z\right) = \frac{\partial^2 z}{\partial u^2} + 2uv^2\frac{\partial z}{\partial u} + 2(v - v^3)\frac{\partial z}{\partial v} + u^2v^2z,$$

in case $x = uv$ and $y = \dfrac{1}{v}$.

14. The function $f(x, y)$ is defined as follows:

$$f(x, y) = \frac{x}{y} \text{ when } |x| \leq |y|,$$

$$f(x, y) = \frac{y}{x} \text{ when } |y| \leq |x|, \text{ except at the origin};$$

$$f(0, 0) = 0.$$

Show that the function is continuous everywhere, except at the origin; and that at the origin it is a continuous function of x alone, and of y alone.

CHAPTER IV

TAYLOR'S EXPANSION WITH THE REMAINDER. APPLICATIONS

43. In studying the behavior of a function $f(x)$ in the neighborhood of a point $x = a$ it is frequently desirable to have an approximate expression for the function in terms of a polynomial in powers of $x - a$. This is due to the fact that polynomials are comparatively simple functions. Much depends upon the degree of approximation. It is therefore important to be able to estimate the difference between the given function and the polynomial used.

Now it is well known that any polynomial $f(x)$ of degree n can be written in the form

$$f(x) = f(a) + f'(a)(x - a) + \cdots + f^{(n)}(a)\frac{(x-a)^n}{n!},$$

or

$$f(a + h) = f(a) + f'(a)h + \cdots + f^{(n)}(a)\frac{h^n}{n!}, \quad (1)$$

where $h = x - a$. If $f(x)$ is not a polynomial, the two members of (1) are not identically equal, and we represent the difference by $R_n(a + h)$. The polynomial

$$f(a) + f'(a)(x - a) + \cdots + f^{(n)}(a)\frac{(x-a)^n}{n!}$$

and its first n derivatives are equal to the function $f(x)$ and its first n derivatives for $x = a$. We wish to form an estimate of the difference between $f(x)$ and this polynomial—that is, an estimate of the value of $R_n(x)$. For this purpose we consider the function

$$F(x) = f(a + h) - f(x) - \frac{a + h - x}{1}f'(x) - $$
$$\cdots - \frac{(a + h - x)^n}{n!}f^{(n)}(x) - \frac{(a + h - x)^p}{h^p}R_n(a + h), \quad (2)$$

§43] EXPANSION WITH THE REMAINDER

where p is an arbitrary positive integer. We assume that $f(x)$ and its first $n + 1$ derivatives exist in the closed interval $(a, a + h)$. Then $F(x)$ has a derivative for every x in this interval, and it vanishes for $x = a$ by virtue of (2). It also obviously vanishes for $x = a + h$. We have therefore by Rolle's theorem

$$F'(a + \theta h) = 0 \qquad (0 < \theta < 1).$$

It is easy to verify that

$$F'(x) = -\frac{(a + h - x)^n}{n!} f^{(n+1)}(x)$$
$$+ \frac{p(a + h - x)^{p-1}}{h^p} R_n(a + h).$$

Then

$$F'(a + \theta h) = -\frac{h^n(1 - \theta)^n}{n!} f^{(n+1)}(a + \theta h)$$
$$+ \frac{ph^{p-1}(1 - \theta)^{p-1}}{h^p} R_n(a + h) = 0,$$

and therefore

$$R_n(a + h) = \frac{h^{n+1}(1 - \theta)^{n-p+1}}{n! \cdot p} f^{(n+1)}(a + \theta h). \qquad (3)$$

The right member of (3) is then the remainder after $n + 1$ terms in the expansion of $f(a + h)$ in powers of h or the difference between the function and the polynomial. It has many forms according to the value we assign to p. For $p = n + 1$ we have

$$R_n = \frac{h^{n+1}}{(n + 1)!} f^{(n+1)}(a + \theta h).$$

This form is due to Lagrange. For $p = 1$ we get a form which was given by Cauchy:

$$R_n = \frac{h^{n+1}(1 - \theta)^n}{n!} f^{(n+1)}(a + \theta h).$$

The right member of the equation

$$f(a + h) = f(a) + f'(a)h + \cdots + f^{(n)}(a)\frac{h^n}{n!} + R_n \qquad (4)$$

is called *Taylor's series with a remainder* for the function $f(x)$.

44. Only one expansion. In this expansion with either the Lagrange or the Cauchy form of the remainder the coefficients of the different powers of h, except the $(n + 1)$th power, are constants, and the coefficient of the $(n + 1)$th power is bounded in the interval $(a, a + h)$ if $f^{(n+1)}(x)$ is. Under these circumstances the expansion is unique. For, if

$$c_0 + c_1 h + \cdots + c_n h^n + c_{n+1} h^{n+1}$$
$$= C_0 + C_1 h + \cdots + C_n h^n + C_{n+1} h^{n+1},$$

where the coefficients are all constants, except c_{n+1} and C_{n+1}, which are bounded functions of h, we can put $h = 0$. The equation then becomes $c_0 = C_0$. Then

$$c_1 + c_2 h + \cdots + c_{n+1} h^n = C_1 + C_2 h + \cdots + C_{n+1} h^n$$

for $h \neq 0$. As $h \to 0$ the two members of this equation approach c_1 and C_1 respectively. Hence $c_1 = C_1$. In a similar way we see that $c_i = C_i$ $(i = 1, 2, \cdots, n)$. Then finally
$$c_{n+1} h = C_{n+1} h$$
and
$$c_{n+1} = C_{n+1}.$$

45. Infinite series. If the function has derivatives of all orders in the interval $(a, a + h)$, the expansion (4) can be continued indefinitely. This leads to an infinite series which converges in case R_n approaches a finite limit as $n \to \infty$. If this limit is zero for all values of x in the interval, the series converges to $f(x)$ in this interval, and we can write the expansion in either of the forms:

$$f(a + h) = f(a) + f'(a)h + \cdots + f^{(n)}(a)\frac{h^n}{n!} + \cdots,$$
$$f(x) = f(a) + f'(a)(x - a) + \cdots$$
$$+ f^{(n)}(a)\frac{(x - a)^n}{n!} + \cdots. \quad (5)$$

The second of these forms is the same as the first with $x - a$

written in place of h. The infinite series in the right member of either one is known as *Taylor's series*. If we take $a = 0$ we have

$$f(x) = f(0) + f'(0)x + \cdots + f^{(n)}(0)\frac{x^n}{n!} + \cdots. \qquad (6)$$

This is known as *MacLaurin's series*. It gives the expansion of $f(x)$ in the powers of x.

The argument of § 44 can be used to show that there is only one expansion of $f(x)$ in a power series in $x - a$.

It is in general difficult to determine the convergence behavior of R. But in many of the simpler functions that one usually meets, this can be done with comparative ease. We shall see some illustrations of this in the following articles.

46. Expansion of log $(1 + x)$. We cannot expand $\log x$ in powers of x inasmuch as it is discontinuous at $x = 0$. But we can expand $\log (a + x)$ in powers of x if $a \neq 0$. We take $a = 1$.

$$f(x) = \log(1 + x),$$
$$f'(x) = \frac{1}{1 + x},$$
$$\cdots\cdots\cdots\cdots\cdots$$
$$\cdots\cdots\cdots\cdots\cdots$$
$$f^{(n)}(x) = (-1)^{n-1}\frac{(n-1)!}{(1+x)^n},$$
$$\cdots\cdots\cdots\cdots\cdots\cdots$$
$$\cdots\cdots\cdots\cdots\cdots\cdots$$

$$\log(1 + x) = x - \frac{x^2}{2} + \cdots + (-1)^{n-1}\frac{x^n}{n}$$
$$+ (-1)^n \frac{x^{n+1}}{(n+1)(1 + \theta x)^{n+1}}, \qquad (7)$$

where $0 < \theta < 1$. Since $\log(x + 1)$ has derivatives of all orders this expansion can be continued indefinitely. The resulting infinite series converges for $-1 < x \leqq 1$, and diverges for $x = -1$ and for $|x| > |$ (§§119, 121, and 122). We conclude from this that R_n approaches a finite limit as

$n \to \infty$ for $|x| < 1$ and for $x = 1$. As a matter of fact, this limit is zero, as shown by the following argument: For $x = 1$ we have, using the Lagrange form of the remainder,

$$|R_n| = \frac{1}{n+1} \cdot \frac{1}{(1+\theta)^{n+1}}.$$

This approaches zero as $n \to \infty$. For $-1 < x < 1$ we use Cauchy's form of the remainder

$$R_n = \frac{x^{n+1}(1-\theta)^n}{(1+\theta x)^{n+1}} = x^{n+1}\left(\frac{1-\theta}{1+\theta x}\right)^n \cdot \frac{1}{1+\theta x}.$$

Since θ is positive and less than 1, $\frac{1-\theta}{1+\theta x}$ is also. Hence $R_n \to 0$ as $n \to \infty$.

We have then, for $-1 < x \leqq 1$,

$$\log(1+x) = x - \frac{x^2}{2} + \cdots + (-1)^{n-1}\frac{x^n}{n} + \cdots. \quad (8)$$

Since this series is valid for $x = 1$ we have

$$\log 2 = 1 - \frac{1}{2} + \cdots + (-1)^{n-1}\frac{1}{n} + \cdots.$$

If in (8) we change the sign of x we obtain the series

$$\log(1-x) = -x - \frac{x^2}{2} - \cdots - \frac{x^n}{n} - \cdots.$$

Hence

$$\log\frac{1+x}{1-x} = 2\left(\frac{x}{1} + \frac{x^3}{3} + \cdots + \frac{x^{2n+1}}{2n+1} + \cdots\right) \quad (9)$$

and this series converges for $-1 < x < 1$. But $\frac{1+x}{1-x} = N$ if $x = \frac{N-1}{N+1}$. When N is greater than 1 this value of x is positive and less than 1. Hence (9) can be used to compute the logarithm of any number greater than 1. It is subject to the practical disadvantage, however, that it converges slowly. For purposes of computation therefore it would be desirable to transform it into a more rapidly convergent series. This can readily be done. We can, for example, put $x = \dfrac{1}{2N+1}$, or $\dfrac{1+x}{1-x} = \dfrac{N+1}{N}$. Then we have

$$\log(N+1) = \log N + 2\left[\frac{1}{2N+1} + \frac{1}{3(2N+1)^3} + \cdots\right].$$

This is a rapidly converging series which gives $\log(N+1)$ when we know $\log N$.

47. Euler's constant. As $n \to \infty$ the sum $S_n = 1 + \tfrac{1}{2} + \cdots + \dfrac{1}{n}$ and $\log n$ both increase without limit. But the difference approaches a limit C, as we shall show.

$$S_n - \log n = \left(1 - \log \tfrac{2}{1}\right) + \left(\tfrac{1}{2} - \log \tfrac{3}{2}\right) + \cdots + \left(\tfrac{1}{n} - \log \tfrac{n+1}{n}\right) + \log \tfrac{n+1}{n}.$$

Now for $n = 1$ we have from (7)

$$\log(1+x) = x - \frac{x^2}{2(1+\theta x)^2}.$$

If in this formula we put $x = \dfrac{1}{p}$, where p is a positive integer, we obtain the formula

$$\frac{1}{p} - \log \frac{p+1}{p} = \frac{1}{2p^2\left(1 + \dfrac{\theta}{p}\right)^2} < \frac{1}{p^2}.$$

Hence the series

$$\Sigma\left(\frac{1}{n} - \log \frac{n+1}{n}\right)$$

converges to a limit C (Chap. IX, Th. 3). Moreover $\log \dfrac{n+1}{n} \to 0$ as $n \to \infty$. Hence

$$S_n - \log n \to C.$$

This limit C is called *Euler's* or *Mascheroni's Constant*. Its value is approximately 0.57721566. It is of great importance in some branches of analysis.

48. Other examples.

(a) $\quad \sin x = x - \dfrac{x^3}{3!} + \cdots + (-1)^n \dfrac{x^{2n+1}}{(2n+1)!} + R_{2n+1}$

where $|R_{2n+1}| = \left| \dfrac{x^{2n+2}}{(2n+2)!} \sin \theta x \right|$ or $\left| \dfrac{x^{2n+2}}{(2n+2)!} \cos \theta x \right|$,

according to the value of n. For any value of x both of these expressions approach zero as $n \to \infty$. Hence

$$\sin x = x - \dfrac{x^3}{3!} + \cdots + (-1)^n \dfrac{x^{2n+1}}{(2n+1)!} + \cdots$$

for all values of x.

(b) We can see in a similar way that

$$\cos x = 1 - \dfrac{x^2}{2!} + \cdots + (-1)^n \dfrac{x^{2n}}{(2n)!} + \cdots$$

for all values of x.

(c) $\quad e^x = 1 + x + \cdots + \dfrac{x^n}{n!} + \dfrac{e^{\theta x}}{(n+1)!} x^{n+1}.$

The remainder here approaches zero for all values of x as $n \to \infty$.

(d) In the case of the function $\tan x$ the nth derivative is very complicated for large values of n, and it is not practicable to determine the expansion in this direct way. It is better to resort to indirect methods (see § 149).

(e) The expansion of $(1 + x)^m$ when $|x| < |$ is of great importance. We assume that m is not a positive integer.

$\quad f(x) = (1 + x)^m,$
$\quad f'(x) = m(1 + x)^{m-1},$
$\quad \cdots \cdots \cdots \cdots \cdots$
$\quad \cdots \cdots \cdots \cdots \cdots$
$\quad f^{(n)}(x) = m(m-1) \cdots (m-n+1)(1+x)^{m-n},$
$\quad \cdots \cdots \cdots \cdots \cdots \cdots \cdots \cdots \cdots$

§48] EXPANSION WITH THE REMAINDER

Hence

$$(1+x)^m = 1 + mx + \frac{m(m-1)}{2!}x^2 + \cdots$$
$$+ \frac{m(m-1)\cdots(m-n+1)}{n!}x^n + R_n,$$

where the Cauchy form of R_n is

$$R_n = \frac{m(m-1)\cdots(m-n)}{n!}x^{n+1}\left(\frac{1-\theta}{1+\theta x}\right)^n(1+\theta x)^{m-1}.$$

If $|x| < 1$ then $0 < \frac{1-\theta}{1+\theta x} < 1$. The ratio of the factor

$$\frac{m(m-1)\cdots(m-n)}{n!}x^{n+1}$$

to

$$\frac{m(m-1)\cdots(m-n+1)}{(n-1)!}x^n$$

is $\frac{m-n}{n}x$, and this approaches $-x$ which is less than 1 in absolute value. Hence

$$\frac{m(m-1)\cdots(m-n)}{n!}x^{n+1} \to 0$$

as $n \to \infty$. It follows that $R_n \to 0$ under the same circumstances and that therefore for $|x| < 1$

$$(1+x)^m = 1 + mx + \cdots$$
$$+ \frac{m(m-1)\cdots(m-n+1)}{n!}x^n + \cdots.$$

EXERCISES

Find the first four terms in the expansion in powers of x of each of the following functions:

1. $\log(1+e^x)$. 2. $e^x \cos x$. 3. $e^{\sin x}$. 4. $\arcsin x$.
5. $\frac{x}{e^x+1}$. 6. $\log(1+\sin x)$. 7. $\log(1-x+x^2)$.

8. Find an upper bound for the difference between $x - \frac{x^3}{6}$ and $\sin x$.

9. Show that the coefficient of x^2 in the expansion of $e^{ax}\cos ax$ in powers of x is zero.

49. Indeterminate forms. If $f(x)$ and $\varphi(x)$ both equal zero when $x = a$ the symbol $\dfrac{f(a)}{\varphi(a)}$ has no meaning. We say that it is *indeterminate*. The fraction $\dfrac{f(x)}{\varphi(x)}$ may however approach a limit as $x \to a$. If $f(x)$ and $\varphi(x)$ both have derivatives in the neighborhood of $x = a$, and the pth derivative is the first derivative of $f(x)$ that does not vanish for $x = a$, while the qth one is the first one of $\varphi(x)$ that does not vanish here, we can write

$$\frac{f(x)}{\varphi(x)} = \frac{f^{(p)}(a + \theta\overline{x-a})\dfrac{(x-a)^p}{p!}}{\varphi^{(q)}(a + \theta_1\overline{x-a})\dfrac{(x-a)^q}{q!}}.$$

If $p = q$, $\dfrac{f(x)}{\varphi(x)} \to \dfrac{f^{(p)}(a)}{\varphi^{(p)}(a)}$; if $p > q$, $\dfrac{f(x)}{\varphi(x)} \to 0$; if $p < q$,

$\left|\dfrac{f(x)}{\varphi(x)}\right| \to \infty$ as $x \to a$. We assume that $f^{(p)}(x)$ and $\varphi^{(q)}(x)$ are continuous at $x = a$.

If $f(x) \to 0$ and $\varphi(x) \to 0$ when $x \to \infty$, we can put $x = \dfrac{1}{t}$ and consider the limit of $\dfrac{f\left(\dfrac{1}{t}\right)}{\varphi\left(\dfrac{1}{t}\right)}$ as $t \to 0$. This gives us

$$\lim_{t \to 0} \frac{f\left(\dfrac{1}{t}\right)}{\varphi\left(\dfrac{1}{t}\right)} = \lim \frac{\dfrac{1}{t^2}f'\left(\dfrac{1}{t}\right)}{\dfrac{1}{t^2}\varphi'\left(\dfrac{1}{t}\right)} = \lim_{x \to \infty} \frac{f'(x)}{\varphi'(x)}.$$

By way of illustration we evaluate the following indeterminate forms:

(a) $\dfrac{\sin x}{x}$ for $x = 0$. Here $f(x) = \sin x$ and $\varphi(x) = x$.

§ 50]   EXPANSION WITH THE REMAINDER   73

Hence $\dfrac{f'(x)}{\varphi'(x)} \to 1$ as $x \to 0$, and therefore $\dfrac{\sin x}{x} \to 1$.

(b) $\dfrac{1 - \cos x}{x}$ for $x = 0$. $\dfrac{f'(x)}{\varphi'(x)} = \dfrac{\sin x}{1} \to 0$.

(c) $\dfrac{1 - \cos x}{x^2}$ for $x = 0$. $f'(0) = \varphi'(0) = 0$; $f''(0) = 1$,

$\varphi''(0) = 2$. Hence $\dfrac{1 - \cos x}{x^2} \to \dfrac{1}{2}$.

We cannot say that $\dfrac{f(x)}{\varphi(x)}$ does not approach a limit A if $\dfrac{f'(x)}{\varphi'(x)}$ does not. Suppose, for example, that $f(x) = x^2 \sin \dfrac{1}{x}$ when $x \neq 0$ and $f(0) = 0$, while $\varphi(x) = \sin x$. Then $\dfrac{f(x)}{\varphi(x)}$ is indeterminate at $x = 0$ and both numerator and denominator have derivatives in the neighborhood of this point.

$$f'(x) = 2x \sin \dfrac{1}{x} - \cos \dfrac{1}{x} \quad (x \neq 0),$$
$$f'(x) = 0 \quad (x = 0),$$
$$\varphi'(x) = \cos x \quad \text{for all values of } x.$$

Hence the fraction $\dfrac{f'(x)}{\varphi'(x)}$ does not approach a limit as $x \to 0$, and yet $\dfrac{f(x)}{\varphi(x)} = \dfrac{x}{\sin x} \cdot x \sin \dfrac{1}{x} \to 0$.

50. The indeterminate form $\dfrac{\infty}{\infty}$. If $f(x)$ and $\varphi(x)$ both increase indefinitely in absolute value as $x \to a$ and neither has a finite value for $x = a$, the symbol $\dfrac{f(a)}{\varphi(a)}$ is meaningless. But it may be that $\dfrac{f(x)}{\varphi(x)}$ approaches a finite limit as $x \to a$. We assume that $f(x)$ and $\varphi(x)$ both have derivatives in the neighborhood of $x = a$, except at a, which do not both vanish for any value of x in this neighborhood.

Let x and x_0 be two values of x in this neighborhood such that $f'(x)$ and $\varphi'(x)$ exist and do not simultaneously vanish for either of these values or any value between them. This requires that x and x_0 be on the same side of the point a. By the law of the mean we have (§ 23)

$$\frac{f(x) - f(x_0)}{\varphi(x) - \varphi(x_0)} = \frac{f'(\xi)}{\varphi'(\xi)},$$

where ξ lies between x and x_0. Now

$$\frac{f(x) - f(x_0)}{\varphi(x) - \varphi(x_0)} = \frac{f(x)}{\varphi(x)} \cdot \frac{1 - \dfrac{f(x_0)}{f(x)}}{1 - \dfrac{\varphi(x_0)}{\varphi(x)}}.$$

Hence

$$\frac{f(x)}{\varphi(x)} = \frac{f'(\xi)}{\varphi'(\xi)} \cdot \frac{1 - \dfrac{\varphi(x_0)}{\varphi(x)}}{1 - \dfrac{f(x_0)}{f(x)}}.$$

If $\dfrac{f'(x)}{\varphi'(x)}$ approaches a limit A as $x \to a$, then $\dfrac{f(x)}{\varphi(x)}$ approaches the same limit. To see this consider that x and x_0, and therefore ξ, can be taken as close to a as we wish. If we hold x_0 fixed and let $x \to a$, the fraction

$$\frac{1 - \dfrac{\varphi(x_0)}{\varphi(x)}}{1 - \dfrac{f(x_0)}{f(x)}} \to 1.$$

Hence[1] $\dfrac{f(x)}{\varphi(x)} \to A$ as $x \to a$. It follows also that if $\dfrac{f'(x)}{\varphi'(x)}$ increases in absolute value without limit, $\dfrac{f(x)}{\varphi(x)}$ does also.

Nothing essential in this article needs to be changed if $a = \infty$.

EXERCISES

Find the limit as $x \to 0$ of each of the following fractions:

1. $\dfrac{e^x - 1}{x}$. 2. $\dfrac{e^x - 1}{x^2}$. 3. $\dfrac{\sin x - x}{\tan x - x}$. 4. $\dfrac{\sin x - x}{x^3}$.

[1] See de la Vallée-Poussin, *Cours d'analyse infinitésimale*, Vol. 1, 5th ed., pp. 94 and 95.

Determine the following limits:

5. $\lim\limits_{x \to \pi/2} \dfrac{x \sin x - \dfrac{\pi}{2}}{\cos x}$. 6. $\lim\limits_{x \to -1} \dfrac{\sin 2\pi x}{1+x}$. 7. $\lim\limits_{x \to \pi/4} \dfrac{\cot x - 1}{x - \dfrac{\pi}{4}}$.

8. $\lim\limits_{x \to 0} \dfrac{1 - \cos x}{\tan^2 x}$.

9. Find the limit of $\dfrac{\log x}{x}$ as $x \to \infty$.

Determine the following limits:

10. $\lim\limits_{x \to \infty} \dfrac{\log x}{1 + x^2}$. 11. $\lim\limits_{x \to \infty} x e^{-x^2}$. 12. $\lim\limits_{x \to \infty} x^n e^{-x^2}$ $(n > 0)$.

13. $\lim\limits_{x \to \infty} \dfrac{\log x}{x^{1/2}}$. 14. $\lim\limits_{x \to 0} x^m \log x$ $(m > 0)$.

15. Show that $f'(0) = f''(0) = 0$ in case $f(x) = e^{-1/x^2}$ when $x \neq 0$ and $f(0) = 0$.

51. Applications to geometry. Some of the properties of the curve $y = f(x)$ in the neighborhood of the point whose abscissa is a can be determined by a comparison with the curve

$$Y = f(a) + f'(a)(x - a) + \cdots + f^{(n)}(a) \dfrac{(x - a)^n}{n!} = P_n(x).$$

As we have seen,

$$y = f(a) + f'(a)(x - a) + \cdots + f^{(n)}(a) \dfrac{(x - a)^n}{n!}$$
$$+ \dfrac{f^{(n+1)}(a + \theta h)(x - a)^{n+1}}{(n+1)!},$$

where $0 < \theta < 1$. For $n = 1$ the first of these curves is the tangent line of the other one at the point $(a, f(a))$, and we have

$$y - Y = f''(a + \theta h) \dfrac{(x - a)^2}{2!}.$$

This shows us that $y = f(x)$ lies above its tangent on both sides of the point of contact if $f''(a) > 0$, and below the tangent on both sides if $f''(a) < 0$. If $f''(a) = 0$, let $f^{(n)}(x)$ be the first derivative of $f(x)$ after the first one that does not vanish for $x = a$. Then

$$y - Y = f^{(n)}(a + \theta h)\frac{(x-a)^n}{n!},$$

and the curve and its tangent cross if n is odd, and do not cross if n is even.

DEFINITION. A point on the curve at which the tangent crosses the curve is called a *point of inflection*. So that if n is odd the point on the curve whose abscissa is a is a point of inflection.

The next approximation is the curve $Y = P_2(x)$. This is a parabola with a vertical axis that is tangent to the curve $y = f(x)$ at the point $(a, f(a))$. The difference between the ordinates to the given curve and the parabola corresponding to the abscissa $a + h$ is an infinitesimal of order three or higher in comparison with h, while the difference between the ordinates to the curve and its tangent is in general an infinitesimal of order two. In this sense the parabola is closer to the curve in the neighborhood of the point whose abscissa is a than the tangent is. And in general if $n > m$ the curve $Y = P_n(x)$ lies closer to the given curve than the curve $Y = P_m(x)$.

52. Maxima and minima of functions of one variable. Suppose that the function $f(x)$ has a maximum or a minimum at $x = a$. Then $f(a + h) - f(a)$ will not change sign as h varies through sufficiently small absolute values. If the function has a derivative at $x = a$ the fractions

$$\frac{f(a+h) - f(a)}{h} \quad \text{and} \quad \frac{f(a-h) - f(a)}{-h}$$

approach $f'(a)$ as a common limit as h approaches zero through positive values. But the two fractions cannot have the same sign because of the hypothesis concerning the numerators. Hence $f'(a) = 0$. That is, in order that $f(x)$ have a maximum or a minimum at $x = a$ it is necessary that $f'(a) = 0$. If $f(a + h) - f(a) > 0$, the function has a minimum at $x = a$; if this difference is negative the function has a maximum. If there is a maximum or a minimum at a point we say that the function has an *extremum* there. But the function does not necessarily have an extremum at $x = a$ when $f'(a) = 0$. In order to examine this point more closely we make use of Taylor's series with a remainder.

We assume that $f(x)$ has a second derivative in the neighborhood of the point $x = a$ which is continuous at $x = a$ and does not vanish there, while $f'(a) = 0$. Then we have

$$f(x) - f(a) = f''(a + \theta h)\frac{h^2}{2}.$$

If $f''(a) < 0$ we have a maximum, and a minimum if $f''(a) > 0$. If $f''(a) = 0$, let $f^{(n)}(x)$ be the first derivative that does not vanish at $x = a$, and assume that it is continuous at this point. Then

$$f(x) - f(a) = f^{(n)}(a + \theta h)\frac{h^n}{n!}.$$

If n is odd, $f(x) - f(a)$ changes sign with h and the function has a point of inflection at $x = a$. But if n is even, $f(x) - f(a)$ has a fixed sign in the neighborhood of this point and there is an extremum—a maximum in case $f^{(n)}(a) < 0$ and a minimum in case $f^{(n)}(a) > 0$.

EXERCISES

Find the maximum and the minimum points, and the points of inflection, of each of the following curves. Draw the curves:
1. $y = x^2 - 3x + 10$. 2. $y = x^4$. 3. $y = x^3 + 6x$. 4. $y = \sin x$.
5. $y = \sin x + \cos x$. 6. $y = (x + 1)^2(x - 2)$.
7. $y^2 = (x - 1)^2(x - 2)$. 8. $y = \dfrac{x^3}{x^2 + 9}$. 9. $y = \dfrac{4 - x^2}{4 + x^2}$.

53. Generalized Taylor series. We can readily derive Taylor Series for functions of any limited number of variables. We give the details of this derivation for functions of two variables.

For this purpose we assume that the function $f(x, y)$ is

continuous in the neighborhood of the point (a, b), that all of its partial derivatives of order not exceeding n exist and are continuous in this neighborhood, and that the partial derivatives of order $n + 1$ exist here. For any x and y in the neighborhood there is an h and a k such that

$$\frac{x - a}{h} = \frac{y - b}{k}.$$

If we denote this common value by t we have $x = a + ht$, $y = b + kt$. Then

$$f(x, y) = f(a + ht, b + kt)$$

is a function of a single variable t, say, $\varphi(t)$. As t varies the point (x, y) moves along the straight line $\frac{x - a}{h} = \frac{y - b}{k}$. MacLaurin's Series for $\varphi(t)$ is

$$\varphi(t) = \varphi(0) + \varphi'(0)t + \cdots + \varphi^{(n)}(0)\frac{t^n}{n!}$$
$$+ \varphi^{(n+1)}(\theta t)\frac{t^{n+1}}{(n+1)!}. \quad (10)$$

Now

$$\varphi'(t) = \frac{\partial f}{\partial x}h + \frac{\partial f}{\partial y}k,$$
$$\varphi''(t) = \frac{\partial^2 f}{\partial x^2}h^2 + 2\frac{\partial^2 f}{\partial x \partial y}hk + \frac{\partial^2 f}{\partial y^2}k^2,$$

and in general

$$\varphi^{(n)}(t) = \left(\frac{\partial f}{\partial x}h + \frac{\partial f}{\partial y}k\right)^{(n)},$$

where the parentheses around the exponent indicate that the binomial $\frac{\partial f}{\partial x}h + \frac{\partial f}{\partial y}k$ is to be raised to the nth power and in the result $\left(\frac{\partial f}{\partial x}\right)^i\left(\frac{\partial f}{\partial y}\right)^{n-i}$ is to be replaced by $\frac{\partial^n f}{\partial x^i \partial y^{n-i}}$ for all values of i from 0 to n inclusive. This general result can be established by induction. The variables in the partial derivatives that occur in these formulae are $a + ht$

§ 54] EXPANSION WITH THE REMAINDER

and $b + kt$. For $t = 0$ we have

$$\varphi^{(n)}(0) = \left[\frac{\partial f(a, b)}{\partial x}h + \frac{\partial f(a, b)}{\partial y}k\right]^{(n)}.$$

Moreover

$$\varphi^{(n+1)}(\theta t) = \left(\frac{\partial f}{\partial x}h + \frac{\partial f}{\partial y}k\right)^{(n+1)},$$

where in the derivatives x and y are to be replaced by $a + \theta ht$ and $b + \theta kt$, respectively. If in (10) we put $t = 1$ we obtain the formula

$$f(x, y) = f(a + h, b + k) = f(a, b) + \left(\frac{\partial f}{\partial x}h + \frac{\partial f}{\partial y}k\right)$$

$$+ \cdots + \frac{\left(\frac{\partial f}{\partial x}h + \frac{\partial f}{\partial y}k\right)^{(n)}}{n!} + \frac{\left(\frac{\partial f}{\partial x}h + \frac{\partial f}{\partial y}k\right)^{(n+1)}}{(n+1)!}. \quad (11)$$

In the expression for the remainder x and y are to be replaced by $a + \theta h$ and $b + \theta k$ respectively, while in the other parts of the right member of (11) the variables are to be replaced by a and b respectively. We can write (11) in the alternative form

$$f(x, y) = f(a, b) + \left[\frac{\partial f}{\partial x}(x - a) + \frac{\partial f}{\partial y}(y - b)\right] + \cdots$$

$$+ \frac{\left[\frac{\partial f}{\partial x}(x - a) + \frac{\partial f}{\partial y}(y - b)\right]^{(n)}}{n!}$$

$$+ \frac{\left[\frac{\partial f}{\partial x}(x - a) + \frac{\partial f}{\partial y}(y - b)\right]^{(n+1)}}{(n+1)!}, \quad (12)$$

with the same understanding as before as to what is to be put for x and y in the various partial derivatives in the right member.

54. Maxima and minima of functions of two variables. If the point (x_0, y_0) is a maximum or a minimum point of the function $f(x, y)$ the difference $f(x_0 + h, y_0 + k) - f(x, y)$ will not change sign as h and k vary provided they remain sufficiently small in absolute value. If we put $k = 0$ we conclude as in § 52 that $\dfrac{\partial f(x_0, y_0)}{\partial x} = 0$; if we put $h = 0$ we see that $\dfrac{\partial f(x_0, y_0)}{\partial y} = 0$. Hence in order that $f(x, y)$ have a maximum or a minimum at (x_0, y_0) it is necessary that

$$\frac{\partial f(x_0, y_0)}{\partial x} = 0 \quad \text{and} \quad \frac{\partial f(x_0, y_0)}{\partial y} = 0. \qquad (13)$$

We assume the existence of the partial derivatives of $f(x, y)$ of the first three orders in the neighborhood of the point (x_0, y_0) and also that the third derivatives are bounded in this neighborhood. Now if the function has a maximum or a minimum at (x_0, y_0)

$$\Delta = f(x_0 + h, y_0 + k) - f(x_0, y_0)$$

$$= \frac{1}{2!}\left(\frac{\partial^2 f}{\partial x^2}h^2 + 2\frac{\partial^2 f}{\partial x \partial y}hk + \frac{\partial^2 f}{\partial y^2}k^2\right)_{\substack{x=x_0 \\ y=y_0}}$$

$$+ \frac{1}{3!}\left(\frac{\partial f}{\partial x}h + \frac{\partial f}{\partial y}k\right)^{(3)}_{\substack{x=x_0+\theta h \\ y=y_0+\theta k}}, \quad (14)$$

and we can draw a circle with center at (x_0, y_0) and a sufficiently small radius r to assure us that Δ shall not have different signs within the circle. The point $(x_0 + h, y_0 + k)$ will be inside this circle if $h = \rho \cos \theta$, and $k = \rho \sin \theta$, where $0 < \rho < r$ and $0 \leq \theta < 2\pi$. With these values substituted for h and k respectively (14) becomes

$$\Delta = \frac{\rho^2}{2}\left[A \cos^2 \theta + 2B \cos \theta \sin \theta + C \sin^2 \theta + \frac{\rho}{3}L\right], \quad (15)$$

where we have written A, B, and C for $\dfrac{\partial^2 f}{\partial x_0^2}$, $\dfrac{\partial^2 f}{\partial x_0 \partial y_0}$, and $\dfrac{\partial^2 f}{\partial y_0^2}$ respectively and L is a function of the coordinates that is bounded in the neighborhood of (x_0, y_0). This expression for Δ shows that for sufficiently small values of ρ the sign of Δ is the same as the sign of

$$A \cos^2 \theta + 2B \sin \theta \cos \theta + C \sin^2 \theta, \qquad (16)$$

provided that A, B, and C are not all zero. But this expression can be written in the form

$$\frac{(C \sin \theta + B \cos \theta)^2 + (AC - B^2) \cos^2 \theta}{C}. \qquad (17)$$

There are three cases to be considered:

(a) $AC - B^2 > 0$. In this case the expression (17) has the same sign as C, and therefore Δ has this sign. We have then a maximum if $C < 0$ and a minimum if $C > 0$. Obviously $C \neq 0$ and A has the same sign as C.

(b) $AC - B^2 < 0$. In this case (17) has the same sign as C when $\theta = \frac{\pi}{2}$ and the opposite sign when $C \sin \theta + B \cos \theta = 0$, or $\tan \theta = -\frac{B}{C}$, provided that $C \neq 0$. In this case there is neither a maximum nor a minimum. If $C = 0$ the same conclusion follows. For if $A \neq 0$ we can write (16) in the form

$$\frac{(A \cos \theta + B \sin \theta)^2 - B^2 \cdot \sin^2 \theta}{A}.$$

This has one sign or the other according to the value of θ. If $A = 0$, $C = 0$, (16) has the form $2 B \sin \theta \cos \theta$, and this value can be either positive or negative.

(c) $AC - B^2 = 0$. In this case (16) vanishes for certain values of θ but does not have different signs as θ varies. The situation is much more complicated than in the other cases, and will not be discussed here. We also omit a discussion of the case $A = B = C = 0$, because of the complications that appear.

55. Lagrange multipliers. Consider the problem of determining the rectangle of maximum area that can be inscribed in the ellipse $\frac{x^2}{a^2} + \frac{y^2}{b^2} = 1$. In the first place, the sides of the rectangle will be parallel to the axes of the ellipse. Why? Then if (x, y) is the upper right-hand corner of the rectangle the area is $A = 4xy$. But x and y

are connected by the equation $\frac{x^2}{a^2} + \frac{y^2}{b^2} = 1$. Hence

$$A = 4\frac{b}{a}x\sqrt{a^2 - x^2}.$$

The problem is in this way reduced to a problem of finding the maximum value of a function of one variable. In general, if we wish to find an extreme value of $f(x, y)$ subject to the condition that $\varphi(x, y) = 0$ we first solve the latter equation for one of the variables in terms of the other one; say, y in terms of x. If this gives us $y = g(x)$ we have $f(x, y) = f[x, g(x)]$, and the condition $f'(x) = 0$ of § 52 is the same as

$$f_x + f_y g'(x) = 0.$$

Moreover $g(x)$ satisfies the equation $\varphi_x + \varphi_y g'(x) = 0$. Hence

$$f_x + \lambda\varphi_x + (f_y + \lambda\varphi_y)g'(x) = 0,$$

where λ is an undetermined multiplier. We can take advantage of the fact that λ is undetermined and select it in such a way as to satisfy the equation $f_y + \lambda\varphi_y = 0$. Then $f_x + \lambda\varphi_x = 0$. Moreover $\varphi(x, y) = 0$. Hence if (ξ, η) is a point at which $f(x, y)$ is a maximum or a minimum subject to the given conditions, we can determine it by solving the equations

$$\begin{aligned} f_x(x, y) + \lambda\varphi_x(x, y) &= 0, \\ f_y(x, y) + \lambda\varphi_y(x, y) &= 0, \\ \varphi(x, y) &= 0, \end{aligned} \qquad (18)$$

for x, y, and λ.

The first two of these equations give necessary conditions for an extreme value of $f(x, y) + \lambda\varphi(x, y)$.

In the example cited at the beginning of this article we have

$$4y + \frac{2\lambda x}{a^2} = 0, \qquad 4x + \frac{2\lambda y}{b^2} = 0, \qquad \frac{x^2}{a^2} + \frac{y^2}{b^2} - 1 = 0.$$

If we eliminate λ from the first two of these equations we get $\dfrac{x^2}{a^2} = \dfrac{y^2}{b^2}$. Hence $x = \dfrac{a}{\sqrt{2}}$, $y = \dfrac{b}{\sqrt{2}}$, and $A = 2ab$. There is nothing in this discussion up to this point to justify us in concluding that $2ab$ is the maximum area we are looking for. But we observe further that $A = 0$ when $x = 0$ and when $x = a$, and we know that there is a value of x between these two values for which A is a maximum, since it is not negative. The discussion shows that $x = \dfrac{a}{\sqrt{2}}$ is the only value of this kind that satisfies the conditions. We conclude therefore that $2ab$ is in fact the maximum area.

The undetermined multiplier λ is called a *Lagrange multiplier*.

There is nothing in equations (18) to indicate that we eliminated y and considered x as the independent variable. If we had eliminated x and considered y as the independent variable these equations would have been the same. Herein lies the advantage in the use of this multiplier.

In order to be sure that the solution of $\varphi(x, y) = 0$ for either x or y in terms of the other will have the properties we rely upon in our discussion we assume that φ_x and φ_y do not both vanish for the same set of values of x and y (see §169).

56. A generalization. Suppose that we wish to find an extreme value of the function $f(x, y, z)$ subject to the condition that $\varphi(x, y, z) = 0$. If $\varphi_z \neq 0$ we know from the reference just made that this equation determines z as a function of x and y, $z = g(x, y)$. Then
$$f(x, y, z) = f[x, y, g(x, y)],$$
and the conditions for an extreme value are
$$\frac{\partial f}{\partial x} = f_x + f_z \frac{\partial z}{\partial x} = 0$$
and
$$\frac{\partial f}{\partial y} = f_y + f_z \frac{\partial z}{\partial y} = 0.$$

Moreover
$$\varphi_x + \varphi_z \frac{\partial z}{\partial x} = 0,$$
$$\varphi_y + \varphi_z \frac{\partial z}{\partial y} = 0.$$

Then
$$f_x + \lambda \varphi_x + (f_z + \lambda \varphi_z) \frac{\partial z}{\partial x} = 0,$$
$$f_y + \lambda \varphi_y + (f_z + \lambda \varphi_z) \frac{\partial z}{\partial y} = 0.$$

If we determine λ from the equation $f_z + \lambda \varphi_z = 0$, the coordinates of any extreme point will satisfy the equations,

$$\begin{aligned} f_x + \lambda \varphi_x &= 0, \\ f_y + \lambda \varphi_y &= 0, \\ \varphi(x, y, z) &= 0. \end{aligned} \qquad (19)$$

Then we have
$$f_x - f_z \frac{\varphi_x}{\varphi_z} = 0,$$
$$f_y - f_z \frac{\varphi_y}{\varphi_z} = 0;$$

and these are equivalent to the equations

$$\frac{\partial f}{\partial x} = 0 \quad \text{and} \quad \frac{\partial f}{\partial y} = 0,$$

where z is considered as the function of x and y that is determined from the auxiliary condition.

This method of Lagrange multipliers can be extended to an arbitrary number of variables subject to any number of conditions less than the number of variables.

EXERCISES

1. Find the minimum value of the function $\frac{x^2}{9} + \frac{y^2}{4} - 5$.

2. Does the surface $z = xy$ have a minimum or a maximum at the origin?

3. What is the maximum value of z in case $(x - 2)^2 + (y + 5)^2 + (z - 3)^2 = 7$?

4. Determine the rectangular parallelopiped of given volume and minimum surface.

5. What is the triangle for which the product of the sines of its angles is a maximum?

6. What is the triangle for which the sum of the sines of its angles is a maximum? Is there a minimum for this sum? A lower limit?

7. Is there a triangle for which the sum of the cosines of its angles is a maximum or a minimum?

8. Show that if $0 < \alpha + \beta + \gamma = a < 3\pi$, then $\sin \alpha + \sin \beta + \sin \gamma$ is a maximum when $\alpha = \beta = \gamma = \dfrac{a}{3}$.

9. Show that under the same conditions $\cos \alpha + \cos \beta + \cos \gamma$ is a maximum or a minimum when $\alpha = \beta = \gamma = \dfrac{a}{3} \neq (2n+1)\dfrac{\pi}{2}$, according to the value of a.

10. In general, if $x + y + z = a$, then $f(x) + f(y) + f(z)$ is a maximum or a minimum when $x = y = z = \dfrac{a}{3}$ according to the sign of $f''\left(\dfrac{a}{3}\right)$, which is assumed to be different from zero.

11. How close does the line $\dfrac{x+1}{2} = \dfrac{y-1}{3} = z$ come to the origin?

12. Which of the coordinate axes comes closest to this line?

13. How close does the curve $x^2 - 4xy + 5y^2 - 2x + y + 1 = 0$ come to the y-axis?

14. What is its greatest distance from the y-axis?

15. What is the rectangle of minimum perimeter that can be inscribed in a circle?

16. Find the largest parallelopiped that can be inscribed in the ellipsoid $\dfrac{x^2}{a^2} + \dfrac{y^2}{b^2} + \dfrac{z^2}{c^2} = 1$.

MISCELLANEOUS EXERCISES

1. Show that of all rectangles with a given area the square has the least perimeter.

2. Which of these rectangles has the least diagonal?

Which of the following surfaces has a maximum or a minimum at the origin?

3. $z = x^2 + y^2$. **4.** $z = x^2 - y^2$. **5.** $z^2 = x^2 - y^2$.

6. $\dfrac{x^2}{4} + \dfrac{y^2}{9} + z = 0$.

7. Find the shortest distance between the lines $x = y + 1$, $z = 2y - 3$ and $x = y - 2$, $z = y + 1$.

8. Show that if $P(x)$ is a polynomial every multiple root of the equation $P(x) = 0$ is a root of the equation $P'(x) = 0$.

9. Show that $\tan x \geq x$ in the interval $\left(0, \dfrac{\pi}{2}\right)$, and that $\sin x \leq x$.

10. Show that $\dfrac{\sin x}{x}$ steadily decreases as x increases from 0 to $\dfrac{\pi}{2}$.

Establish the following inequalities for $0 < x < 1$:

11. $0 \leq x - \log(1 + x) < \dfrac{x^2}{2}$. **12.** $x - \log(1 + x) > \dfrac{x^2}{3 + x}$.

13. Find the path of a ray of light from A to B in case these points are on opposite sides of the x-axis, the velocity of light from A to the axis being v_1 and v_2 from the axis to B. Assume that the path is such as to make the time from A to B a minimum.

14. Find a point on the x-axis such that the sum of its distances from two fixed points in the (x, y)-plane and on the same side of the axis is a minimum. Show that when this distance is a minimum the lines from the point on the axis to the given points make equal angles with the axis.

15. Prove that the curve $y = ax^3 + bx^2 + cx + d$ has not more than one point of inflection.

16. If the illumination at a point from a light varies inversely as the square of the distance from the light to the point, determine the position of the point P on the line from A to B in order that the illumination at P due to lights at A and B shall be a minimum, the light at A being eight times as intense as that at B.

17. When x battery cells each of voltage E and internal resistance r are arranged in series with $\dfrac{n}{x}$ rows in parallel, the current sent through a resistance R amounts to $\dfrac{nxE}{x^2 r + nR}$ amperes. How many of the cells must be in series in order that the current shall be a maximum?

18. If a current of C amperes is passing through a conductor whose resistance is r ohms per mile, the waste per mile is $W = C^2 r + \dfrac{A}{r}$, where A is a constant. For a given C what should the resistance be in order that the waste be a minimum?

19. What is the shortest distance from the origin to the curve $y = 3x^2 + 5$?

20. Find the points on the surface $xyz = 1$ that are nearest to the origin.

21. Show that the shortest distance from a fixed point to the surface $f(x, y, z) = 0$ is along a normal to the surface. Assume that the fixed point is at the origin.

EXPANSION WITH THE REMAINDER

22. Find the points on the circle $x^2 + y^2 = 1$ that are at the greatest and the least distances from the line $x + y = 5$.

23. Find the points on the sphere $x^2 + y^2 + z^2 = 1$ that are nearest to the plane $x + y + z = 5$, and those that are farthest from it.

24. What is the shortest distance from the ellipse $\dfrac{x^2}{9} + \dfrac{y^2}{4} = 1$ to the line $x + y = 5$?

25. Find the maximum rectangular parallelopiped in the first octant that has three faces in the coordinate planes and one vertex in the plane $x + 2y + 3z = 1$.

26. Prove that the sum of the squares of the distances of a point within a triangle to the vertices is least when the point is at the intersection of the medians.

CHAPTER V

THE DEFINITE INTEGRAL

57. An appeal to geometry. Consider the curve $y = f(x)$, where $f(x)$ is a positive continuous function in the interval (a, b), and divide the interval into a set of n partial intervals by the points $x_0 = a, x_1, x_2, \cdots, x_n = b$. Upon each partial interval (x_{i-1}, x_i) as base erect the rectangles R_i and r_i whose altitudes are respectively the maximum and the minimum values M_i and m_i of $f(x)$ in the partial interval. Unless we do violence to our intuition we must define the area of the portion of the plane bounded by the curve, the x-axis, and the ordinates $x = a$ and $x = b$ as a number between $\sum R_i$ and $\sum r_i$, for every possible method of subdivision of the interval (a, b). We shall see that there is only one such number—that is, only one number A such that $\sum r_i \leq A \leq \sum R_i$ for all possible methods of subdivision of (a, b). We therefore take this number A as the definition of the area in question.

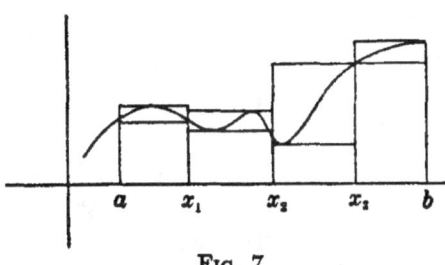

FIG. 7

We have made this appeal to geometry because it brings to the reader's attention in a simple and natural way the sums $\sum r_i = \sum m_i(x_i - x_{i-1})$ and $\sum R_i = M_i(x_i - x_{i-1})$, which are of great importance in the analytical theory we shall now discuss.

58. Let $f(x)$ be any function of x that is bounded in (a, b). We form the partial intervals $(a, x_1), (x_1, x_2), \cdots, (x_{n-1}, b)$, as before; and then the sums

$$s = \sum m_i(x_i - x_{i-1}) \quad \text{and} \quad S = \sum M_i(x_i - x_{i-1}),$$

where m_i is the lower limit of $f(x)$ in the interval (x_{i-1}, x_i) and M_i is the upper limit. We shall call S an *upper sum* and s a *lower sum*. We cannot say that m_i and M_i are the minimum and maximum values respectively of $f(x)$ in (x_{i-1}, x_i), since the function may not have maximum and minimum values in these partial intervals.

In the first place, $s \leq S$. Moreover for every set of partial intervals the corresponding upper sum is equal to, or greater than, $m(b - a)$, where m is the lower limit of $f(x)$ in (a, b). Similarly every lower sum is equal to, or less than, $M(b - a)$, where M is the upper limit of $f(x)$ in (a, b). There is therefore a number I such that no upper sum is less than I, and for every $\epsilon > 0$ there is an upper sum less than $I + \epsilon$; and a number I' such that no lower sum is greater than I', and there is a lower sum greater than $I' - \epsilon$ (see § 3).

59. Consecutive methods of subdivision. If we subdivide the partial intervals (x_{i-1}, x_i) in any way we shall get a new subdivision of the whole interval which will be said to be *consecutive* to the original one. In forming a subdivision consecutive to a given one it is not necessary that every one of the original partial intervals be subdivided.

THEOREM 1. *If α' represents a subdivision consecutive to the subdivision α of the interval (a, b) with $\sum$ and S the respective upper sums for the function $f(x)$ and σ and s the respective lower sums, then $\sum \leq S$ and $\sigma \geq s$.*

Consider first the effect on S of inserting a single point y in the partial interval (x_{i-1}, x_i). If M_i' and M_i'' are the upper limits of $f(x)$ in the respective intervals (x_{i-1}, y) and (y, x_i), it is clear that neither M_i' nor M_i'' exceeds M_i, and therefore $M_i(x_i - x_{i-1}) \geq M_i'(y - x_{i-1}) + M_i''(x_i - y)$. Since this is true for every new point inserted, it follows that $S \geq \sum$. It can be shown in a similar way that $s \leq \sigma$.

THEOREM 2. *If S and s and S' and s' are the upper and lower sums corresponding to any two methods of subdivision of (a, b) then $S \geq s'$ and $s \leq S'$.*

If we superpose the two subdivisions, we get a subdivision

that is consecutive to either of the original ones; and if Σ and σ are the sums corresponding to this new subdivision, then

$$\Sigma \leqq S, \qquad \sigma \geqq s, \qquad \Sigma \leqq S', \qquad \sigma \geqq s'.$$

Since $\Sigma \geqq \sigma$, we have

$$s \leqq \sigma \leqq \Sigma \leqq S' \qquad \text{and} \qquad s' \leqq \sigma \leqq \Sigma \leqq S.$$

If we had $I' > I$, there would be an $s' > I$ and therefore no S less than $I + \epsilon$, in case $\epsilon < s' - I$. Hence $I' \leqq I$.

60. Darboux's theorem. *If we increase the number of partial intervals indefinitely in such a way as to make their maximum length approach zero, S will approach I and s will approach I'.*

We make the provisional supposition that $f(x) > 0$ in (a, b). If ϵ is an arbitrary positive number there is a method of subdivision $a, x_1, x_2, \cdots, x_{n-1}, b$ for which $S < I + \epsilon$. Let S' be the corresponding sum with respect to a subdivision every one of whose partial intervals is less than a certain number η. The contribution to S' from those intervals that do not include any of the points $x_1, x_2, \cdots, x_{n-1}$ within their interiors is less than $I + \epsilon$. The number of remaining intervals does not exceed $n - 1$, and their contribution to S' does not exceed $(n - 1)M\eta$. Hence

$$S' < I + \epsilon + (n - 1)M\eta.$$

If now $(n - 1)M\eta < \epsilon$, that is, if $\eta < \dfrac{\epsilon}{M(n - 1)}$, then

$$S' < I + 2\epsilon.$$

This means that by taking the partial intervals sufficiently small the corresponding S' can be made to differ from I by an arbitrarily small amount.

We have assumed that $f(x)$ is positive in (a, b). If it is not, there is a constant C such that $f(x) + C$ is positive throughout the interval. If Σ is the sum for $f(x) + C$ corresponding to S for $f(x)$, then $\Sigma = S + C(b - a)$. The lower limit of all possible Σ's is $I + C(b - a)$. Then

there is a number η such that
$$\Sigma < I + C(b - a) + \epsilon$$
when all the partial intervals are less than η. For all such subdivisions $S < I + \epsilon$. A similar discussion with respect to s and I' would complete the proof of the theorem.

61. Integrable functions. If $I = I'$ for a function $f(x)$ and an interval (a, b) we say that $f(x)$ is *integrable in* (a, b). The common value of I and I' in the case of an integrable function is called *the definite integral of $f(x)$ from a to b* and is represented by the symbol $\int_a^b f(x)dx$. We have assumed that $a \neq b$. In order to provide for this excluded case we supplement this definition by the agreement that $\int_a^a f(x)dx$ shall be zero.

Not every bounded function is integrable. Consider for example the function defined as follows for the interval $(0, 1)$:
$$f(x) = 0 \quad \text{when } x \text{ is rational,}$$
$$f(x) = 1 \quad \text{when } x \text{ is irrational.}$$

Since there is a rational number and an irrational number in every interval (see §2), we have $S = 1$ and $s = 0$ for any subdivision of $(0, 1)$. Hence $I = 1$ and $I' = 0$, and the function is not integrable in the given interval. This raises the questions as to whether there are any integrable functions, and if there are, how they can be recognized. The following discussion will bring to light large classes of functions that are integrable.

Theorem 3. *A necessary and sufficient condition that $f(x)$ be integrable in the interval (a, b) is that for every positive number ϵ there shall be a number η such that $S - s$ is less than ϵ when each of the partial intervals is less than η.*

For, if $f(x)$ is integrable, S and s have a common limit I, and there is a number η so small that $S - I$ and $I - s$ are both less than $\frac{\epsilon}{2}$ when each of the partial intervals is less than η. Then $S - s$ is less than ϵ. This proves that the condition is necessary.

On the other hand, if $S - s < \epsilon$ for every sufficiently small η, then

$$S - s = (S - I) + (I - I') + (I' - s) < \epsilon.$$

Now none of these differences is negative. Hence $I - I' < \epsilon$. But I and I' are fixed numbers and ϵ is arbitrary. Hence $I = I'$ and $f(x)$ is integrable.

From this result we can derive an important property of continuous functions. It was shown in §18 that if $f(x)$ is continuous in (a, b) there is an η corresponding to every positive ϵ such that the difference of the values of $f(x)$ at any two points of this interval whose distance apart does not exceed η is less than ϵ in absolute value. This means that if each of the partial intervals is less than η in length then, for every i, $M_i - m_i < \epsilon$, and

$$S - s = \sum_{i=1}^{n} (M_i - m_i)(x_i - x_{i-1}) < \epsilon(b - a).$$

Hence every function of x that is continuous in (a, b) is integrable in (a, b).

Suppose now that $f(x)$ is a bounded monotonically increasing function in (a, b). Then

$$S = f(x_1)(x_1 - a) + f(x_2)(x_2 - x_1) + \cdots + f(b)(b - x_{n-1}),$$
$$s = f(a)(x_1 - a) + f(x_1)(x_2 - x_1) + \cdots + f(x_{n-1})(b - x_{n-1}).$$

Hence

$$S - s = (x_1 - a)[f(x_1) - f(a)] + (x_2 - x_1)[f(x_2) - f(x_1)] + \cdots + (b - x_{n-1})[f(b) - f(x_{n-1})].$$

If all the partial intervals are less than η in length, we have, since none of the differences in the right member of this equation is negative,

$$S - s < \eta\{[f(x_1) - f(a)] + [f(x_2) - f(x_1)] + \cdots + [f(b) - f(x_{n-1})]\} = \eta[f(b) - f(a)].$$

If we take $\eta < \dfrac{\epsilon}{f(b) - f(a)}$ we have $S - s < \epsilon$. We can argue in a similar way in the case of a monotonically decreasing function. Hence *every function that is bounded and monotonic in (a, b) is integrable in (a, b).*

62. Properties of the definite integral.

(a) *If $f(x)$ is integrable in (a, b) the limit of the sum*

$$S_n = \sum_{i=1}^{n} f(\xi_i)(x_i - x_{i-1}), \qquad (x_0 = a, \ x_n = b)$$

where $x_{i-1} \leq \xi_i \leq x_i$ is equal to $\int_a^b f(x)dx$. For $S \geq S_n \geq s$ and S and s have the definite integral as a common limit (see §10, Ex. 1).

(b) *If $f(x)$ is integrable in (a, b), $|f(x)|$ is also integrable there, and* $\left|\int_a^b f(x)dx\right| \leq \int_a^b |f(x)|dx.$

The first part of this statement follows immediately from Theorem 3; the second part from the inequality

$$\left|\int_a^b f(x)dx\right| = |\lim \sum f(\xi_i)(x_i - x_{i-1})|$$
$$\leq \lim \sum |f(\xi_i)| \, |x_i - x_{i-1}|.$$

DEFINITION. *If a number δ_i is associated to every partial interval (x_{i-1}, x_i) of (a, b), and if for every positive number ϵ there is a positive number η such that $|\delta_i| < \epsilon$ for every i when every partial interval is less than η, we shall say that the δ_i approach zero uniformly.*

(c) *If the δ_i approach zero uniformly the sum*

$$S' = \sum_{i=1}^{n} [f(\xi_i) + \delta_i](x_i - x_{i-1})$$

approaches the limit $\int_a^b f(x)dx$. This can be seen as follows: There is a positive number η such that

$$\left|\sum_{i=1}^{n} f(\xi_i)(x_i - x_{i-1}) - \int_a^b f(x)dx\right| < \epsilon$$

and
$$|\delta_i| < \epsilon,$$
when $x_i - x_{i-1} < \eta$ for every i; and we have
$$S' - \int_a^b f(x)dx = \left[\sum_{i=1}^n f(\xi_i)(x_i - x_{i-1}) - \int_a^b f(x)dx\right] + \sum_{i=1}^n \delta_i(x_i - x_{i-1}).$$

If each of the partial intervals is less than η
$$\left|S' - \int_a^b f(x)dx\right| < \epsilon + \epsilon(b-a).$$

In other words,
$$\lim \sum [f(\xi_i) + \delta_i](x_i - x_{i-1})$$
$$= \lim \sum f(\xi_i)(x_i - x_{i-1}) = \int_a^b f(x)dx.$$

This formula enables us to find the limits of certain sums that appear in many problems in geometry and physics.

(d) *If $f(x)$ is integrable in (a, b) it is integrable in (α, β), where $a \leq \alpha < \beta \leq b$.* Let I_1, I_2, I_3 be the limits of the upper sums relative to the respective intervals (a, α), (α, β), and (β, b), and I_1', I_2', I_3' the limits of the lower sums. Now $I = I_1 + I_2 + I_3$ and $I' = I_1' + I_2' + I_3'$. Moreover $I = I'$ since $f(x)$ is integrable in (a, b). Then $I_1 + I_2 + I_3 = I_1' + I_2' + I_3'$, or
$$(I_1 - I_1') + (I_2 - I_2') + (I_3 - I_3') = 0.$$
None of the differences in the left member of this equation is negative. They are therefore all zero, and $f(x)$ is integrable in (α, β).

(e) *If $f(x)$ is integrable in (a, b) and C is a constant, $Cf(x)$ is also integrable with* $\int_a^b Cf(x)dx = C\int_a^b f(x)dx$. For
$$\int_a^b f(x)dx = \lim f\sum (\xi_i)(x_i - x_{i-1}),$$

and therefore

$$C\int_a^b f(x)dx = C \lim \sum f(\xi_i)(x_i - x_{i-1})$$
$$= \lim \sum Cf(\xi_i)(x_i - x_{i-1}) = \int_a^b Cf(x)dx.$$

(f) $\int_a^b f(x)dx = -\int_b^a dx.$ For,

$$\int_a^b f(x)dx = \lim \sum f(\xi_i)(x_i - x_{i-1})$$
$$= -\lim \sum f(\xi_i)(x_{i-1} - x_i) = -\int_b^a f(x)dx.$$

(g) $\int_a^b f(x)dx = \int_a^c f(x)dx + \int_c^b f(x)dx.$ This is immediately obvious if $a < c < b$, and we have only to apply (f) in order to see its validity in case c is not in the interval (a, b). The formula is therefore general.

63. **Theorem 4.** *A function $f(x)$ is integrable in (a, b) if it is bounded and has only a finite number of discontinuities in the interval.*

Suppose that $|f(x)| < M$ in (a, b). If there is only one discontinuity, $x = c$, in the interval, choose a positive number $\delta < \dfrac{\epsilon}{8M}$, where ϵ is an arbitrary positive number. The contribution from the interval $(c - \delta, c + \delta)$ to the difference $S - s$ is less than $2M \cdot 2\delta$, and therefore less than $\dfrac{\epsilon}{2}$. The integrand is continuous in the intervals $(a, c - \delta)$ and $(c + \delta, b)$. There is therefore an $\eta > 0$ such that when these last intervals are divided into partial intervals each of length less than η the value of $S - s$ for these two intervals corresponding to this subdivision is less than $\dfrac{\epsilon}{2}$. For the whole interval we have therefore $S - s < \epsilon$, and $f(x)$ is integrable in (a, b).

If there are n points of discontinuity, $c_1, c_2, \cdots, c_n$, we enclose them in the partial intervals $(c_i - \delta_i, c_i + \delta_i)$

($i = 1, 2, 3, \cdots, n$), selected in such a way that

$$2 \sum_{i=1}^{n} \delta_i < \frac{\epsilon}{4M}.$$

Then the part of $S - s$ that is due to these partial intervals is less than $2M \cdot \frac{\epsilon}{4M} = \frac{\epsilon}{2}$. The integrand is continuous in the intervals $(a, c_1 - \delta_1), (c_1 + \delta_1, c_2 - \delta_2), \cdots, (c_n + \delta_n, b)$. There is therefore an $\eta > 0$ such that when these last intervals are divided into partial intervals each of length less than η, the value of $S - s$ for these intervals is less than $\frac{\epsilon}{2}$. It will then be less than ϵ for (a, b), and $f(x)$ is integrable in (a, b).

EXERCISE. Modify the details of this argument to fit the case in which $c_1 = a$ or $c_n = b$.

64. Volume of a solid of revolution. It follows from the definition of integral in § 61 and the definition of area in § 57 that if $f(x)$ is continuous in (a, b) the area A bounded by the curve $y = f(x)$, the x-axis, and the ordinates $x = a$, $x = b$ is given by the formula

$$A = \int_a^b f(x)dx. \qquad (1)$$

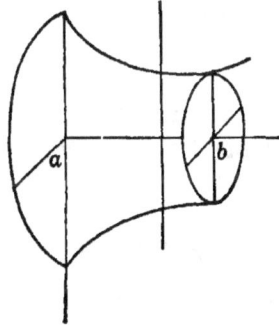

Fig. 8

We arrive at another simple geometric application of the definite integral in the following way: Consider the solid generated by revolving about the x-axis the area bounded by the curve $y = f(x)$, the x-axis, and the ordinates $x = a, y = b$; and divide it into n partial solids by the planes $x = x_1, x = x_2, \cdots, x = x_{n-1}$, where $a < x_1 < x_2 < \cdots < x_{n-1} < b$. Now we wish to define volume in such a way that the total volume V shall be equal to $\sum \Delta_i V$, where $\Delta_i V$ is the volume of the ith partial solid, and the $\Delta_i V$ for all methods of subdivisions of the interval (a, b) shall satisfy the inequalities

$$\pi m_i^2(x_i - x_{i-1}) \leqq \Delta_i V \leqq \pi M_i^2(x_i - x_{i-1}),$$

where M_i is the maximum absolute value of any ordinate of the generating curve in the interval (x_{i-1}, x_i) and m_i is the minimum absolute value of any such ordinate. This requires that

$$\pi \sum m_i^2(x_i - x_{i-1}) \leqq V \leqq \pi \sum M_i^2(x_i - x_{i-1}).$$

But the only number between these extremes for all methods of subdivision of the interval (a, b) is

$$\pi \int_a^b y^2 dx.$$

Hence by definition

$$V = \pi \int_a^b y^2 dx. \tag{2}$$

65. First mean value theorem for integrals.

THEOREM 5. *If $f(x)$ and $\varphi(x)$ are continuous in the interval (a, b) and $\varphi(x)$ does not change sign in the interval, then*

$$\int_a^b f(x)\varphi(x)dx = f(\xi) \int_a^b \varphi(x)dx,$$

where $a < \xi < b$.

If M and m are the maximum and minimum values respectively of $f(x)$ in (a, b), and $\varphi(x)$ is positive or zero in this interval, then

$$\int_a^b [M - f(x)]\varphi(x)dx \geqq 0, \qquad \int_a^b [f(x) - m]\varphi(x)dx \geqq 0;$$

and therefore

$$M \int_a^b \varphi(x)dx \geqq \int_a^b f(x)\varphi(x)dx \geqq m \int_a^b \varphi(x)dx.$$

Hence

$$\int_a^b f(x)\varphi(x)dx = \mu \int_a^b \varphi(x)dx,$$

where $M \geqq \mu \geqq m$. By Theorem 6, Ch. II, there is a value ξ between a and b such that $f(\xi) = \mu$. Hence

$$\int_a^b f(x)\varphi(x)dx = f(\xi)\int_a^b \varphi(x)dx.$$

If $\varphi(x)$ were negative or zero throughout the interval the inequalities in this discussion would all be reversed. This would not affect the conclusion.

If we put $\varphi(x) = 1$, we get the important formula

$$\int_a^b f(x)dx = f(\xi)(b - a).$$

This is obvious geometrically.

66. The second mean value theorem for integrals. The following theorem concerning integrals whose integrands are products of two integrable functions is known as *the second mean value theorem*, and is of importance in certain cases.

THEOREM 6. *If $f(x)$ is continuous and monotonic in (a, b) while $\varphi(x)$ is continuous and does not change its sign more than a finite number of times in this interval, then*

$$\int_a^b f(x)\varphi(x)dx = f(a)\int_a^\xi \varphi(x)dx + f(b)\int_\xi^b \varphi(x)dx$$

where $a \leqq \xi \leqq b$.

We assume first that $f(x)$ is monotonically decreasing and non-negative in (a, b). We can divide the whole interval into a finite number of partial intervals $(x_i - x_{i-1})$ $(i = 1, 2, \cdots, n)$, where $x_0 = a$ and $x_n = b$, such that $\varphi(x)$ does not change sign in any one of them. Then

$$\int_a^b f(x)\varphi(x)dx = \sum_{i=1}^n \int_{x_{i-1}}^{x_i} f(x)\varphi(x)dx.$$

By the first mean value theorem

$$\int_{x_{i-1}}^{x_i} f(x)\varphi(x)dx = \mu_i \int_{x_{i-1}}^{x_i} \varphi(x)dx,$$

where $f(x_{i-1}) \geq \mu_i \geq f(x_i)$. If we denote the function $\int_a^x \varphi(x)dx$ by $F(x)$ we have

$$\int_{x_{i-1}}^{x_i} f(x)\varphi(x)dx = \mu_i[F(x_i) - F(x_{i-1})],$$

and therefore

$$\int_a^b f(x)\varphi(x)dx = \sum_{i=1}^n \mu_i[F(x_i) - F(x_{i-1})]$$
$$= [f(a) - \mu_1]F(a) + \sum_{i=2}^n (\mu_{i-1} - \mu_i)F(x_{i-1}) + \mu_n F(b).$$

We have added the term $f(a)F(a)$ to the right side. This is permissible since $F(a) = 0$. By hypothesis no one of the differences $f(a) - \mu_1, \mu_1 - \mu_2, \cdots, \mu_{n-1} - \mu_n$, nor μ_n is negative. Hence

$$\int_a^b f(x)\varphi(x)dx$$
$$= M[f(a) - \mu_1 + (\mu_1 - \mu_2) + \cdots + (\mu_{n-1} - \mu_n) + \mu_n]$$
$$= Mf(a),$$

where M is between the greatest and the least of $F(a)$, $F(x_1), \cdots, F(b)$. But $F(x)$ is continuous in (a, b) (see § 67) and therefore there is in this interval a ξ for which $M = F(\xi)$. This gives us the formula

$$\int_a^b f(x)\varphi(x)dx = f(a)\int_a^\xi \varphi(x)dx. \tag{A}$$

If $f(x)$ increases monotonically in (a, b) and is never negative there, we can show in a similar way that

$$\int_a^b f(x)\varphi(x)dx = f(b)\int_\xi^b \varphi(x)dx, \tag{B}$$

where $a \leq \xi \leq b$.

If now $f(x)$ decreases monotonically in (a, b) and is not restricted as to sign, we put $g(x) = f(x) - f(b)$. This makes $g(x)$ a non-negative monotonically decreasing function in (a, b). Then by A we have

$$\int_a^b g(x)\varphi(x)dx = g(a)\int_a^\xi \varphi(x)dx,$$

where $a \leq \xi \leq b$. This last equation can be written as follows:

$$\int_a^b f(x)\varphi(x)dx = [f(a) - f(b)]\int_a^\xi \varphi(x)dx + f(b)\int_a^b \varphi(x)dx,$$

or

$$\int_a^b f(x)\varphi(x)dx = f(a)\int_a^\xi \varphi(x)dx + f(b)\int_\xi^b \varphi(x)dx.$$

If $f(x)$ increases monotonically in (a, b) we put $g(x) = f(x) - f(a)$ and apply B.

67. The definite integral as a function of one of its limits.
If $f(x)$ is integrable in (a, b) it is integrable in any interval contained in (a, b) (see § 62, d). If then $a \leq x \leq b$, the integral $\int_a^x f(x)dx$ is a function of x which we shall desigate by the symbol $F(x)$. Now if $a \leq x + h \leq b$ then

$$F(x + h) = \int_a^{x+h} f(x)dx.$$

Hence

$$F(x + h) - F(x) = \int_a^{x+h} f(x)dx - \int_a^x f(x)dx$$

$$= \int_x^{x+h} f(x)dx = \mu h,$$

where μ lies between the upper and the lower bounds of $f(x)$ in (a, b). It follows that

$$F(x + h) - F(x) \to 0$$

as $h \to 0$. This means that $F(x)$ is a continuous function of x in the interval (a, b) whether $f(x)$ is continuous there or not. But if $f(x)$ is continuous there we have, by the mean value theorem for integrals,

$$F(x + h) - F(x) = f(\xi)h,$$

where ξ lies between x and $x + h$. Hence
$$\lim_{h \to 0} \frac{F(x+h) - F(x)}{h} = \lim_{\xi \to x} f(\xi) = f(x).$$

The integral is therefore a function whose derivative is the integrand.

68. Differentiation under the integral sign. If $f(x, \alpha)$ is a continuous function of x and α when $a \leqq x \leqq b$ and $\alpha_0 \leqq \alpha \leqq \alpha_1$, the integral $\int_a^b f(x, \alpha) dx$ is a function of α. We represent it by the symbol $F(\alpha)$. In order to show that it is a continuous function of α we assume first that a and b are independent of α. Now

$$F(\alpha + \Delta\alpha) - F(\alpha) = \int_a^b [f(x, \alpha + \Delta\alpha) - f(x, \alpha)] dx,$$

where α and $\alpha + \Delta\alpha$ are in the interval (α_0, α_1). By hypothesis $f(x, \alpha)$ is continuous in the closed region described above. It is therefore uniformly continuous there (§ 31). Then to any $\epsilon > 0$ there is associated an $\eta > 0$ such that when $|\Delta\alpha| < \eta$ we have

$$|f(x, \alpha + \Delta\alpha) - f(x, \alpha)| < \epsilon$$

for every set of values of x and α under consideration. It follows that

$$|F(\alpha + \Delta\alpha) - F(\alpha)| < \int_a^b \epsilon \, dx = \epsilon(b - a)$$

and that therefore $F(\alpha)$ is continuous in the given interval.

In order to discuss the existence of a derivative of $F(\alpha)$ we make the further assumption that $f_\alpha(x, \alpha)$ is also a continuous function of x and α when $a \leqq x \leqq b$ and $\alpha_0 \leqq \alpha \leqq \alpha_1$. Then by virtue of the mean value theorem for derivatives

$$\begin{aligned}f(x, \alpha + \Delta\alpha) - f(x, \alpha) &= \Delta\alpha f_\alpha(x, \alpha + \theta\Delta\alpha)\\ &= \Delta\alpha[f_\alpha(x, \alpha) + \epsilon_1], \quad 0 < \theta < 1,\end{aligned}$$

where ϵ_1 approaches zero along with $\Delta\alpha$. Now ϵ_1 can by virtue of the uniform continuity of $f_\alpha(x, \alpha)$ be made uniformly less than an arbitrary positive number by making the absolute value of $\Delta\alpha$ sufficiently small. Hence from the equation

$$\frac{F(\alpha + \Delta\alpha) - F(\alpha)}{\Delta\alpha} = \int_a^b f_\alpha(x, \alpha)dx + \int_a^b \epsilon_1 dx$$

we can conclude that

$$\frac{dF}{d\alpha} = \int_a^b f_\alpha(x, \alpha)dx. \tag{3}$$

The result is somewhat different if the limits of integration are also differentiable functions of α. In this case we have

$$F(\alpha + \Delta\alpha) - F(\alpha) = \int_{a+\Delta a}^{b+\Delta b} f(x, \alpha + \Delta\alpha)dx - \int_a^b f(x, \alpha)dx$$

$$= \int_a^b [f(x, \alpha + \Delta\alpha) - f(x, \alpha)]dx$$

$$+ \int_b^{b+\Delta b} f(x, \alpha + \Delta\alpha)dx - \int_a^{a+\Delta a} f(x, \alpha + \Delta\alpha)dx,$$

where Δa and Δb are the increments caused in a and b respectively by the increment $\Delta\alpha$ in α. If we apply the law of the mean for integrals to the last two of these integrals, we get

$$\frac{F(\alpha + \Delta\alpha) - F(\alpha)}{\Delta\alpha} = \int_a^b \frac{f(x, \alpha + \Delta\alpha) - f(x, \alpha)}{\Delta\alpha}dx$$

$$+ \frac{\Delta b}{\Delta\alpha}f(b + \theta\Delta b, \alpha + \Delta\alpha) - \frac{\Delta a}{\Delta\alpha}f(a + \theta'\Delta a, \alpha + \Delta\alpha),$$

where $0 < \theta < 1$ and $0 < \theta' < 1$. If we take the limit of each side of this equation as $\Delta\alpha \to 0$ we get the formula

$$\frac{dF}{d\alpha} = \int_a^b f_\alpha(x, \alpha)dx + \frac{db}{d\alpha}f(b, \alpha) - \frac{da}{d\alpha}f(a, \alpha). \tag{4}$$

This is the general formula for differentiation under the integral sign. It includes (3) as a special case.

We can conclude that $F(\alpha)$ is a continuous function of α even when we assume merely that a and b are continuous functions of α and make no assumption at all as to the existence of a partial derivative of $f(x, \alpha)$ with respect to α.

The question of the integration under the integral sign will be considered in another place (§§ 97 and 98).

69. The indefinite integral. Let G be any function whose derivative is $f(x)$. Then $F(x)$ and $G(x)$ have the same derivative and therefore differ by a constant. That is, $F(x) = G(x) + C$. But $F(a) = 0$. Hence $C = -G(a)$ and

$$\int_a^x f(x)dx = F(x) = G(x) - G(a).$$

Any function such as $G(x)$ whose derivative is $f(x)$ is called the *indefinite integral* of $f(x)$ and is represented by the symbol

$$\int f(x)dx.$$

This is the same as the symbol for the definite integral with the exception that here no limits of integration are indicated. The appropriateness of the term " indefinite " is due to the fact that such an integral is determined only up to an additive constant. The preceding formula shows that the knowledge of any indefinite integral enables us to determine the value of the corresponding definite integral.

EXERCISES

1. Find the area of the ellipse $\dfrac{x^2}{a^2} + \dfrac{y^2}{b^2} = 1$.
2. Find the area of the curve $y^2 = (x - 1)(5 - x)$.
3. Find the area cut off from the parabola $y^2 = 10x$ by the line $y = 2x$.

Find the areas of the curves bounded as follows:

4. By the x-axis and one arch of the curve $y = \sin x$.
5. By the curve $x^{2/3} + y^{2/3} = a^{2/3}$ (the astroid). Draw the curve.
6. By one arch of the cycloid $x = a(\theta - \sin \theta)$, $y = a(1 - \cos \theta)$ and the x-axis.

7. The hyperbola $x^2 - y^2 = 2a^2$ divides the ellipse $\dfrac{x^2}{4} + \dfrac{y^2}{2} = a^2$ into three parts. Find the area of each part.

8. Find the volume generated by revolving the ellipse $\dfrac{x^2}{9} + \dfrac{y^2}{16} = 1$ about its major axis.

9. Find the volume generated by revolving the same ellipse about its minor axis.

10. What is the volume of a spherical cap of height h on a sphere of radius r?

11. What is the volume formed by revolving one arch of the curve $y = \sin x$ about the x-axis?

12. Determine the volume formed by revolving one arch of the cycloid of Exercise 6 about the x-axis.

13. Revolve this arch about its maximum ordinate and find the volume generated.

14. Rotate the area bounded by the axes and the curve $x^{1/2} + y^{1/2} = a^{1/2}$ about the y-axis and find the volume generated.

15. Find the volume of the anchor ring generated by revolving the circle $(x - a)^2 + (y - b)^2 = r^2$ ($b > r$) about the x-axis.

The pressure on a vertical plane area immersed in a liquid is given by the formula $P = w \displaystyle\int_a^b xy\,dy$, where x is the width of the area at the depth y, a and b are the depths of the top and bottom, respectively, and w is the weight of a cubic unit of the liquid (see Osgood, *Differential and Integral Calculus*, p. 161).

16. Find the pressure on a rectangle immersed vertically in a liquid, one side of the rectangle lying in the surface.

17. Find the pressure on a vertical triangle with its base in the surface.

18. Find the pressure on the triangle when one vertex is at the surface.

19. Find the pressure on a vertical semi-circle with its bounding diameter in the surface.

20. A vertical circle of radius one foot has its center three feet below the surface. What is the pressure on it?

21. A dam is in the form of an isosceles trapezoid 80 feet across the top, 50 feet across the bottom, and 25 feet high. Find the pressure on it in tons when the water is flush with the top. A cubic foot of water weighs $62\tfrac{1}{2}$ pounds.

22. A reservoir has a circular bulkhead two feet in diameter in its side wall. What is the pressure on this bulkhead when the level of the water is 20 feet above its center?

70. Approximate integration.

Normally we evaluate the definite integral $\int_a^b f(x)dx$ by first finding the indefinite integral $\int f(x)dx$. But when we cannot do this we must resort to other methods. We can for example take advantage of the fact that the definite integral is equal to the area bounded by the curve, the x-axis, and the ordinates $x = a$ and $x = b$. There are various ways of getting at least the approximate value of this area, and therefore of the definite integral. One way of doing this is to replace the integrand $f(x)$ by a function $\varphi(x)$ that differs but little from $f(x)$ and whose indefinite integral can be found.

Suppose for example that the interval is of length $2h$ and that c is its center point. Then $a = c - h$ and $b = c + h$. It is possible to determine the constants A, B, and C in such a way that the curve $y = Ax^2 + Bx + C$ shall pass through the points $[c - h, f(c - h)]$, $[c, f(c)]$, and $[c + h, f(c + h)]$. The area under this curve between the ordinates $x = c - h$ and $x = c + h$ is

$$S = \frac{A[(c+h)^3 - (c-h)^3]}{3} + \frac{B[(c+h)^2 - (c-h)^2]}{2} + C[(c+h) - (c-h)]$$

$$= \frac{h}{3}(6Ac^2 + 2Ah^2 + 6Bc + 6C).$$

But since the parabola passes through the given points we have

$$A(c-h)^2 + B(c-h) + C = f(c-h),$$
$$Ac^2 + Bc + C = f(c),$$
$$A(c+h)^2 + B(c+h) + C = f(c+h).$$

And therefore

$$6Ac^2 + 2Ah^2 + 6Bc + 6C = f(c-h) + 4f(c) + f(c+h).$$

Hence

$$S = \frac{h}{3}[f(c-h) + 4f(c) + f(c+h)].$$

This gives us a rough approximation to the value of the definite integral $\int_a^b f(x)dx$. We can get a better approximation by dividing interval (a, b) into $2n$ ($n > 1$) equal partial intervals (x_{i-1}, x_i) with $x_0 = a$ and $x_{2n} = b$, and then dealing with each successive pair of intervals in the way we have dealt with the whole interval. This leads to the following rule for computing the approximate value of $\int_a^b f(x)dx$:

$$S = (y_0 + 4y_1 + 2y_2 + 4y_3 + \cdots$$
$$+ 2y_{2n-2} + 4y_{2n-1} + y_{2n})\frac{h}{3},$$

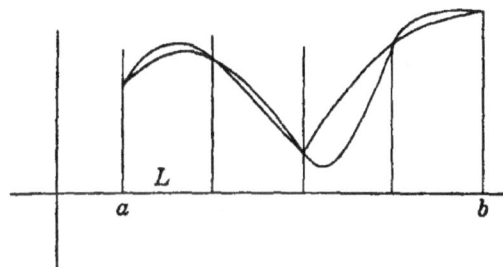

Fig. 9

where h is the length of each of the partial intervals and $y_i = f(x_i)$. This is known as *Simpson's rule* for the approximate evaluation of a definite integral.

71. Determination of the error. We wish now to form an estimate of the error due to the use of this rule. That is, we want to determine as nearly as possible the value of the difference

$$\int_a^b f(x)dx - S.$$

In the cases we would naturally be considering in this connection we could not hope to get its exact value.

The method we shall use assumes that the first four derivatives of $f(x)$ exist in (a, b) and that $f^{\mathrm{IV}}(x)$ is continuous here. Consider the value of S for $n = 1$. Then, if we

§ 71] THE DEFINITE INTEGRAL 107

denote the difference by $\varphi(h)$,

$$\varphi(h) = \int_{c-h}^{c+h} f(x)dx - \frac{h}{3}[f(c-h) + 4f(c) + f(c+h)].$$

If we denote the indefinite integral by $F(x)$, then

$$\varphi(h) = F(c+h) - F(c-h)$$
$$- \frac{h}{3}[f(c-h) + 4f(c) + f(c+h)],$$

$$\varphi'(h) = f(c+h) + f(c-h)$$
$$- \frac{1}{3}[f(c-h) + 4f(c) + f(c+h)]$$
$$- \frac{h}{3}[-f'(c-h) + f'(c+h)]$$
$$= \frac{2}{3}f(c+h) + \frac{2}{3}f(c-h) - \frac{4}{3}f(c)$$
$$- \frac{h}{3}[-f'(c-h) + f'(c+h)],$$

$$\varphi''(h) = \frac{1}{3}f'(c+h) - \frac{1}{3}f'(c-h)$$
$$- \frac{h}{3}[f''(c-h) + f''(c+h)],$$

$$\varphi'''(h) = -\frac{h}{3}[f'''(c+h) - f'''(c-h)].$$

If we apply the mean value theorem to the difference $f'''(c+h) - f'''(c-h)$ we find that

$$\varphi'''(h) = -\frac{2h^2}{3}f^{\text{IV}}(\xi),$$

where $c - h < \xi < c + h$. From this expression for $\varphi'''(h)$ we work back to a new expression for $\varphi(h)$ by taking advantage of the fact that $\varphi(h)$, $\varphi'(h)$, and $\varphi''(h)$ are all equal to zero when $h = 0$. We have

$$\varphi''(h) - \varphi''(0) = \varphi''(h) = \int_0^h \varphi'''(h)dh$$
$$= -\int_0^h \frac{2h^2}{3} f^{\text{IV}}(\xi)dh.$$

We cannot evaluate this last integral exactly inasmuch as ξ and therefore also $f^{IV}(\xi)$ is an undetermined function of h. But we know that the integrand is integrable since it is equal to $\varphi'''(h)$ and that we can apply the first mean value theorem for integrals to it since h^2 is not negative. We have then

$$\varphi''(h) = -\frac{2}{9}h^3 f^{IV}(\xi).$$

But the ξ that occurs here is not the same as the ξ that occurs in the expression for $\varphi'''(h)$. We use the same symbol for them since all we know about either one is that it lies between $c - h$ and $c + h$. Then, in the same way,

$$\varphi'(h) - \varphi'(0) = \varphi'(h) = \int_0^h \varphi''(h)dh = -\frac{1}{18}h^4 f^{IV}(\xi)$$

and

$$\varphi(h) - \varphi(0) = \varphi(h) = \int_0^h \varphi'(h)dh = -\frac{1}{90}h^5 f^{IV}(\xi).$$

This gives us an expression for the error due to the use of Simpson's rule with two partial intervals. If we use $2n$ partial intervals, the error is of the form

$$-\frac{1}{90}h^5[f^{IV}(\xi_1) + f^{IV}(\xi_2) + \cdots + f^{IV}(\xi_n)]$$

where ξ_i lies in the ith pair of partial intervals. But

$$\frac{f^{IV}(\xi_1) + f^{IV}(\xi_2) + \cdots + f^{IV}(\xi_n)}{n}$$

lies between the greatest and the least value of $f^{IV}(x)$ in the interval. Then since $f^{IV}(x)$ is continuous in this interval there is a value ξ in the interval such that

$$f^{IV}(\xi_1) + \cdots + f^{IV}(\xi_n) = nf^{IV}(\xi).$$

The expression for the total error E is therefore of the form

$$E = -\frac{1}{90}h^5 n f^{IV}(\xi).$$

But $2nh = b - a$. Hence

$$E = -\frac{1}{180} h^4 (b-a) f^{IV}(\xi).$$

Since this value is the product of a bounded factor by the fourth power of h, it approaches zero rapidly as h approaches zero. We can therefore get a good approximation by the use of Simpson's rule.

EXERCISES

Use Simpson's Rule to get an approximate value of each of the following integrals:

1. $\int_0^{\pi/2} \sin x\, dx$. Use six intervals. Test the accuracy of your result by the formula given in the text and by direct evaluation of the integral.

2. $\int_0^2 e^{-x^2} dx$. Use six intervals and then ten.

3. $\int_0^1 \frac{\log(1+x)}{1+x^2} dx$. Use eight intervals.

4. $\int_0^{\pi/2} \frac{d\theta}{\sqrt{1 - k^2 \sin^2 \theta}}$, $k = 0.3$. Use six intervals.

5. $\int_0^2 e^{x^2} dx$. Use eight intervals.

6. $\int_0^{\pi/2} \frac{\sin x}{x} dx$. Use six intervals.

7. Find an approximate value of log 3 by applying Simpson's Rule to the integral $\int_1^3 \frac{dx}{x}$.

8. $\int_0^1 \frac{dx}{1+x^2} = \frac{\pi}{4}$. Get an approximate value of π by applying Simpson's Rule to this integral. Check the accuracy of your result against the known value of π.

9. Find an approximate value of π by applying Simpson's Rule to the integral $\int_0^{1/2} \frac{dx}{\sqrt{1-x^2}}$.

72. Applications of the definite integral. (a) *Center of mass.* If a particle with the mass m is at the distance d

from a straight line, the product dm is called the *moment of the particle* with respect to the line. The moment of a system of particles with respect to a line is the sum of their several moments with respect to this line. This notion of moment of a system of particles with respect to a line, or axis, can be readily extended to a curve, an area, or a volume. Consider the arc of the curve $y = \varphi(x)$ between the two points $A \equiv (a, c)$ and $B \equiv (b, d)$ and suppose it divided up into partial arcs, the ith one of which is of length $\Delta_i s$. If $\rho(s)$ is the density of the arc at the point whose distance from a fixed point on the arc is s, the limit of the sum $\Sigma y(s_i)\rho(s_i)\Delta_i s$ as the maximum value of the $\Delta_i s$ approaches zero is called the *moment I_x* of the arc AB with respect to the x-axis. We assume that $\rho(s)$ is a continuous function of s along this arc. This limit is equal to the integral $\int y(s)\rho(s)ds$. Hence

$$I_x = \int y(s)\rho(s)ds.$$

Similarly

$$I_y = \int x(s)\rho(s)ds.$$

The point $(\bar{x}, \bar{y})$, where

$$\bar{x} = \frac{\int x(s)\rho(s)ds}{\int \rho(s)ds}, \qquad \bar{y} = \frac{\int y(s)\rho(s)ds}{\int \rho(s)ds},$$

is called the *center of mass* of the arc. It is the point at which the whole mass of the arc could be concentrated without affecting the moment of the mass with respect to either axis.

If we have a bounded region S of area A in the xy-plane with a distribution of matter of density ρ, its moments with respect to the axes are

$$I_x = \iint y\rho\, dx dy \qquad \text{and} \qquad I_y = \iint x\rho\, dx dy.$$

§ 72] THE DEFINITE INTEGRAL 111

These moments would be unchanged if the whole mass were concentrated at the point $(\bar{x}, \bar{y})$, where

$$\bar{x} = \frac{\iint x\rho\,dx\,dy}{\iint \rho\,dx\,dy} \quad \text{and} \quad \bar{y} = \frac{\iint y\rho\,dx\,dy}{\iint \rho\,dx\,dy}.$$

This point is called the *center of mass* of the area. Similarly the center of mass of a volume distribution is $(\bar{x}, \bar{y}, \bar{z})$, where

$$\bar{x} = \frac{\iiint x\rho\,dx\,dy\,dz}{\iiint \rho\,dx\,dy\,dz}, \quad \bar{y} = \frac{\iiint y\rho\,dx\,dy\,dz}{\iiint \rho\,dx\,dy\,dz},$$

$$\bar{z} = \frac{\iiint z\rho\,dx\,dy\,dz}{\iiint \rho\,dx\,dy\,dz}.$$

(b) *Moment of inertia.* The kinetic energy of a mass m moving with the linear velocity v is by definition $\tfrac{1}{2}mv^2$. Hence the kinetic energy of such a particle rotating with the angular velocity ω around an axis at a distance d is $\tfrac{1}{2}md^2\omega^2$, since in this case $v = d\omega$. The coefficient of $\tfrac{1}{2}\omega^2$ in this expression is by definition the *moment of inertia* I of the particle with respect to the axis. That is, $I = md^2$. The moment of inertia of a system of particles with respect to an axis is by definition the sum of the moments of the several particles with respect to this axis. We can extend this notion to a continuous distribution of matter along a curve in a way that is now familiar to the reader. Thus, if the moment is taken with respect to the x-axis,

$$I = \int \rho y^2\,ds = \int \rho y^2 \sqrt{1 + \left(\frac{dy}{dx}\right)^2}\,dx, \qquad (5)$$

where ρ is the density of the distribution at the point on the curve whose abscissa is x.

If we have a homogeneous plane area, its moment of inertia with respect to the x-axis is

$$\rho \int_c^d Xy^2 dy,$$

where X is the width of the area parallel to the x-axis at a distance y from this axis.

The moment of inertia of a plane area about an axis perpendicular to its plane is sometimes spoken of as the moment of inertia of the area about the point in which the axis cuts the plane.

(c) *Mean value.* If the arc AB is homogeneous—that is, if ρ is a constant—we have

$$\bar{x} = \frac{\int x(s) ds}{s}, \qquad \bar{y} = \frac{\int y(s) ds}{s},$$

where s is the length of the arc. We call these integrals the *mean values* of the abscissae and ordinates respectively of the curve with respect to the length of arc. In general, if y is an integrable function of x, $\dfrac{\int_{x_0}^{x_1} y dx}{x_1 - x_0}$ is called the mean value of y with respect to x as x varies from x_0 to x_1. It is important to note that this mean value is dependent upon the variable with respect to which it is taken. Consider, for example, the upper half of the circle $x^2 + y^2 = 1$. The mean value of y with respect to x is $\dfrac{1}{2}\int_{-1}^{1} \sqrt{1 - x^2}\, dx = \dfrac{\pi}{4}$, while its mean value with respect to the length of arc is

$$\frac{1}{\pi}\int_0^\pi y ds = \frac{1}{\pi}\int_{-1}^{1} y\sqrt{1 + \frac{x^2}{y^2}}\, dx = \frac{2}{\pi}.$$

EXERCISES

1. Find the mean value with respect to x of $\cos x$ in the interval $\left(0, \dfrac{\pi}{2}\right)$.

2. Of $\sin^2 x$ in the interval $(0, \pi)$.

THE DEFINITE INTEGRAL

3. Given $s = a \cos nt$ (simple harmonic motion). Find the relation between the mean kinetic energy with respect to the time over a period and the maximum kinetic energy.

4. The relation between pressure and volume in the case of a certain expanding gas is $pv^{1.3} = 400$. Find the mean pressure with respect to the volume as the volume increases from 2 to 10.

5. Find the center of mass of a homogeneous wire in the position of the upper half of the ellipse $\dfrac{x^2}{9} + \dfrac{y^2}{16} = 1$. Determine $\bar{x}$ without any integration.

Find the center of mass of each of the following arcs:

6. The upper half of the circle $x^2 + y^2 = 9$.
7. The right half of the curve $x^{2/3} + y^{2/3} = 1$.
8. The complete arch of the cycloid.
9. The part of the catenary $y = \dfrac{a}{2}(e^{x/a} + e^{-x/a})$ between the points whose abscissae are c and $-c$.

10. Find the mean value of the ordinate in the upper half of the ellipse $\dfrac{x^2}{a^2} + \dfrac{y^2}{b^2} = 1$ with respect to x.

11. Is the mean value of the square of the ordinate the square of the mean value of the ordinate?

12. Find the mean ordinate in the arch of the cycloid with respect to the abscissa.

13. Find the center of mass of the upper half of the homogeneous circle $x^2 + y^2 = 4$.

14. Find the center of mass of the segment of the homogeneous parabola $y^2 = 6x$ cut off by the latus rectum.

15. Find the center of mass of a homogeneous triangle.

16. Find the center of mass of the left half of the homogeneous sphere $x^2 + y^2 + z^2 = 9$.

17. Of the homogeneous segment of $y^2 + z^2 = 4x$ cut off by the plane $x = 5$.

Find the moment of inertia of each of the following homogeneous areas:

18. A circle about a diameter.
19. The ellipse $x^2 + \dfrac{y^2}{9} = 1$ about its major axis.
20. The same ellipse about its minor axis.
21. A rectangle about one side.
22. A rectangle about a diagonal.
23. Find the moment of inertia of an arc of the catenary $y = \dfrac{a}{2}(e^{x/a} + e^{-x/a})$ about the x-axis.

24. A vessel in the shape of a right circular cylinder is closed at the top and weighs half as much as the water it can contain. How much water should be put into it to bring the center of mass as low as possible?

MISCELLANEOUS EXERCISES

1. Give a suitable definition of the area A bounded by the curve $\rho = f(\theta)$ and the two radii vectors $\theta = \alpha$, $\theta = \beta$. Then show that
$$A = \tfrac{1}{2}\int_\alpha^\beta r^2 d\theta.$$

2. Show that $f(x)\sin x$ and $f(x)\cos x$ are integrable in the interval (a, b) if $f(x)$ is integrable here.

3. What function is defined by the equation $f(x) = \int_0^x f(t)dt + 1$? By $f(x) = \int_0^x f(t)dt$? Assume that $f(x)$ is continuous.

4. If $f(x)$ and $\varphi(x)$ are integrable in (a, b), $f(x) \pm \varphi(x)$ are also integrable in this interval.

5. If $f(x)$ and $\varphi(x)$ are integrable in (a, b), $f(x)\varphi(x)$ is also integrable in this interval.

6. In the proof of Theorem 5 it was assumed that
$$\int_a^b [M - f(x)]\varphi(x)dx = \int_a^b M\varphi(x)dx - \int_a^b f(x)\varphi(x)dx.$$
Prove this.

7. The function $f(x, y, y')$ is defined by the equation
$$f(x, y, y') = \int_0^{y'} (y' - t)\Phi(t, y - xt)dt,$$
where Φ is an arbitrary differentiable function of its two arguments. Show that $\dfrac{\partial^2 f}{\partial y'^2} = \Phi(y', y - xy')$.

8. Show that if for the interval (a, b) the function $f(x)$ is such that for all possible subdivisions of (a, b) $\sum(M_i - m_i) < \alpha$, where α is a constant, then $f(x)$ is integrable in (a, b).

A function with this property is said to have *limited total variation* in (a, b).

9. It is obvious on physical grounds that the position of the center of mass of a curve is independent of the position of the coordinate axes. Show analytically that this is the case.

10. Show that the area between the catenary $y = \dfrac{a}{2}(e^{x/a} + e^{-x/a})$, the axes, and the ordinate $x = b$ is double the area of the triangle formed by this ordinate, the tangent to the catenary at the point whose abscissa is b, and the perpendicular upon this tangent from the foot of this ordinate.

11. Prove that of all the curves of the one-parameter family $y = f(x) + \alpha$ $(a \leq x \leq b)$ that one has the least moment of inertia about the x-axis whose center of mass is on this axis.

12. Show in two ways that in the case of curves of the form $y = ax^3 + bx^2 + cx + d$ Simpson's rule gives the exact value of the definite integral.

13. Show from the expression for E that the two definite integrals $\int_a^b f_1(x)dx$ and $\int_a^b f_2(x)dx$ are equal in case $f_1(x)$ and $f_2(x)$ are polynomials of the fourth degree in x with the same coefficient for x^4 and with $f_1(a) = f_2(a)$, $f_1\left(\frac{a+b}{2}\right) = f_2\left(\frac{a+b}{2}\right)$, and $f_1(b) = f_2(b)$. Verify your conclusion by direct computation.

14. Prove that if the bounded function $f(x)$ is positive and continuous in (a, b) then $\int_a^b f(x)dx \cdot \int_a^b \frac{dx}{f(x)} \geq (b-a)^2$ and that the equality sign holds when $f(x)$ is a constant.

CHAPTER VI

INDEFINITE INTEGRALS

73. The definite integral has been defined as the limit of a sum. But we are in general unable to express this sum in a form whose limit we can determine and it is therefore impossible for us to evaluate the definite integral in a direct way. We can however attack the problem in an indirect way by *taking advantage of the properties of the indefinite integral* as shown in § 69.

It was shown there that if we know any function $F(x)$ whose derivative is $f(x)$ we can compute the value of the definite integral $\int_a^b f(x)dx$ by means of the formula $\int_a^b f(x)dx = F(b) - F(a)$. The problem of determining the indefinite integral of a given function is therefore of importance, and we shall devote the present chapter to a consideration of certain details connected with it.

In some cases the problem is an extremely simple one when we have in mind the rules for finding derivatives. For example, we recognize immediately that $\cos x$ is the derivative of $\sin x$, and that therefore

$$\int \cos x\, dx = \sin x + C.$$

We have here a formula for integration. Similar formulae will come to the mind of the reader.

Simple rules for integration will become evident from a consideration of the rules for differentiation. For example, the derivative of the sum of two functions is the sum of their derivatives, and therefore the integral of the sum of two functions is the sum of their integrals. Also the derivative of the product of a constant and a function is the product of the constant and the derivative of the function, and therefore the integral of the product of a constant and a function is the product of the constant and the integral of the function. This is expressed in symbols thus

$$\int Cf(x)dx = C \int f(x)dx \qquad \text{(see § 62, e)}.$$

It appears from this that a constant factor can be written on whichever side of the integral sign we find the more convenient. But this is not true of factors that depend upon the variable of integration. Thus,

$$4 \int \sin x\, dx = \int 4 \sin x\, dx,$$

but

$$\int x \sin x\, dx \neq x \int \sin x\, dx.$$

74. Integration of rational functions. There are many functions whose integrals cannot be found by a direct application of these simple rules. This is true in fact of most functions. The integral of a polynomial can be written down immediately. It is another polynomial. But the integral of a rational function $\frac{P(x)}{Q(x)}$, where P and Q are relatively prime polynomials and Q is not a constant, is not in general a rational function.

If the degree of $P(x)$ is not less than the degree of $Q(x)$ we have

$$\frac{P(x)}{Q(x)} = M(x) + \frac{P_1(x)}{Q(x)},$$

where $M(x)$ and $P_1(x)$ are polynomials and $P_1(x)$ is of lower degree than $Q(x)$. Since the integration of the polynomial $M(x)$ offers no difficulty, the fraction $\frac{P_1(x)}{Q(x)}$ is the only part that requires discussion. We therefore assume that in the original fraction $\frac{P(x)}{Q(x)}$ the degree of the numerator is less than the degree of the denominator.

Let D_1 be the highest common divisor of Q and Q'; D_2 the highest common divisor of D_1 and D_1'; and, in general, D_i the highest common divisor of D_{i-1} and D'_{i-1}. If k is the least number for which D_k is constant, Q contains some linear factors to the kth power and none to any higher power. Moreover D_{k-1} is the product of all the linear factors of Q that occur to the kth power. The quotient $\dfrac{Q(x)}{D^k_{k-1}(x)}$ therefore contains all the linear factors of Q to their original powers except those that occur to the kth power in Q. By considering this quotient in the same way we can determine the second highest power of the linear factors that occur in Q and the product of the linear factors that occur to this power. And this process can be continued until we have considered all the linear factors of Q. Thus we can write

$$Q = Q_1 \cdot Q_2^2 \cdot \ \cdots \ \cdot Q_k^k,$$

where $Q_1, Q_2, \cdots, Q_k$ are polynomials without multiple roots such that no two of them have a common factor. Some of these polynomials, except the last one, may be constants. Since Q_1 and $Q_2^2 \cdot Q_3^3 \cdot \ \cdots \ \cdot Q_k^k$ are relatively prime, there are two polynomials A_1 and B, such that [1]

$$BQ_1 + A_1 Q_2^2 \cdot \ \cdots \ \cdot Q_k^k = P.$$

Hence

$$\frac{P}{Q} = \frac{A_1}{Q_1} + \frac{B}{Q_2^2 \cdot \ \cdots \ \cdot Q_k^k}.$$

If A_1 and Q_1 were not relatively prime, we should have

$$\frac{A_1}{Q_1} = \frac{\overline{A}_1}{\overline{Q}_1}$$

where $\overline{Q}_1$ is of lower degree than Q_1. This would make

$$\frac{P}{Q} = \frac{\overline{P}}{\overline{Q}}$$

where $\overline{Q}$ is of lower degree than Q. But this contradicts the

[1] See, for example, Bôcher, *Introduction to Higher Algebra*, p. 193.

hypothesis that P and Q are relatively prime. Likewise B and $Q_2^2 \cdot \ldots \cdot Q_k^k$ are relatively prime, and we can therefore proceed in the same way with the fraction $\dfrac{B}{Q_2^2 \cdot \ldots \cdot Q_k^k}$. If we continue the process through k steps we obtain the result

$$\frac{P}{Q} = \frac{A_1}{Q_1} + \frac{A_2}{Q_2^2} + \cdots + \frac{A_k}{Q_k^k},$$

where A_i is a polynomial relatively prime to Q_i.

We have then to consider the problem of integrating expressions of the form $\dfrac{P(x)}{Q^n(x)}$, where $Q(x)$ is relatively prime to its derivative. We consider first the case for which $n > 1$. There are two polynomials A and B such that

$$BQ + AQ' = P.$$

Then

$$\int \frac{P dx}{Q^n} = \int \frac{BQ + AQ'}{Q^n} dx = \int \frac{B dx}{Q^{n-1}} + \int \frac{AQ'}{Q^n} dx.$$

We can integrate the last of these integrals by parts. To do this we put $u = A$ and $dv = \dfrac{Q' dx}{Q^n}$. Then $du = A' dx$ and $v = -\dfrac{1}{(n-1)Q^{n-1}}$ and

$$\int \frac{AQ' dx}{Q^n} = \frac{-A}{(n-1)Q^{n-1}} + \frac{1}{n-1} \int \frac{A' dx}{Q^{n-1}}.$$

Hence

$$\int \frac{P dx}{Q^n} = \frac{-A}{(n-1)Q^{n-1}} + \int \frac{B_1 dx}{Q^{n-1}},$$

where B_1 is a polynomial that can easily be determined. If $n > 2$ we can continue this reduction. We finally come to the result that

$$\int \frac{P dx}{Q^n} = R(x) + \int \frac{P_1 dx}{Q},$$

where $R(x)$ is a rational function and $P_1(x)$ is a polynomial whose degree is less than the degree of Q.

This brings us to the consideration of integrals of the form $\int \frac{P dx}{Q}$, where Q is relatively prime to its derivative. We assume that the coefficients of Q are real. Then if it has any imaginary factors they occur in pairs of conjugates of the form $x - a - bi$ and $x - a + bi$, where a and b are real. The product of these, namely $(x - a)^2 + b^2$, has real coefficients. We can then, by a method with which we assume the reader is familiar, express the integrand as the sum of fractions of one or the other of the forms

$$\frac{A}{x-a}, \qquad \frac{Bx+C}{(x-a)^2+b^2},$$

where A, B, and C are constants. Now

$$\int \frac{dx}{x-a} = \log(x-a),$$

and

$$\int \frac{(Bx+C)dx}{(x-a)^2+b^2} = \frac{B}{2}\log[(x-a)^2+b^2] + \frac{Ba+C}{b}\arctan\frac{x-a}{b}.$$

We thus see that the integral of a rational fraction is the sum of a rational fraction and certain terms involving logarithms and the inverse tangent. It does not contain any other kind of term, and may not contain all of these.

In some cases in which the integrand is not a rational function of the variable of integration, the introduction of a new variable will give a rational integrand. For example, if we put $\tan\frac{x}{2} = t$,

$$\int R(\sin x, \cos x)dx = 2\int R\left(\frac{2t}{1+t^2}, \frac{1-t^2}{1+t^2}\right)\frac{dt}{1+t^2}$$
$$= \int R_1(t)dt,$$

where R denotes a rational function of its two arguments and $R_1(t)$ is a rational function of t.

EXERCISES

Integrate the following:

1. $\int \dfrac{dx}{3 + 2\cos x}$. 2. $\int \dfrac{dx}{\cos \beta - \cos x}$. 3. $\int \dfrac{\sin x + \cos x}{\sin x - \cos x} dx$.

4. $\int \dfrac{dx}{1 + \sin x - \cos x}$. 5. $\int \dfrac{dx}{1 + 2\sin x + 5\cos x}$.

6. $\int y\,dx$, where $x + y^3 - 3y^2 + 3y - 2 = 0$.

7. $\int y\,dx$, where $1 + y^3 + xy - x = 0$.

8. Find the product of the repeated factors of $x^6 + 5x^5 + 4x^4 - 10x^3 - 11x^2 + 5x + 6$; and the product of the unrepeated factors.

9. Find the product of the unrepeated factors of $x^6 + 2x^5 - 8x^4 - 14x^3 + 11x^2 + 28x + 12$; of the factors that occur to exactly the second degree; of those that occur to the third degree.

75. Unicursal curves. If $f(x, y)$ is a polynomial in x and y and y is a root of the equation $f(x, y) = 0$, the integral $\int R(x, y)\,dx$, where $R(x, y)$ is a rational function of x and y, can under certain circumstances be reduced to integrals of rational functions. We refer to the case in which the coordinates of the points of the curve $f(x, y) = 0$ are rational functions of a parameter. Such a curve is called a *unicursal curve*. If

$$x = \varphi(t), \qquad y = \psi(t),$$

where $\varphi(t)$ and $\psi(t)$ are rational functions of t, we have

$$\int R(x, y)\,dx = \int R[\varphi(t), \psi(t)]\varphi'(t)\,dt.$$

The new integrand is a rational function of t.

Consider for example the lemniscate

$$(x^2 + y^2)^2 = a^2(x^2 - y^2).$$

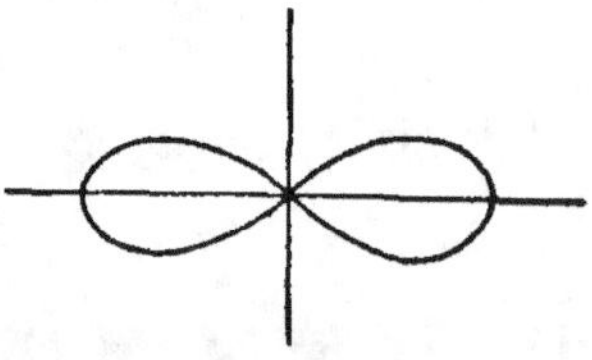

Fig. 10

The line $y = \lambda x$ cuts it in two points besides the origin.

These two points are

$$x = \frac{\pm a\sqrt{1-\lambda^2}}{1+\lambda^2}, \qquad y = \lambda x,$$

and by suitably choosing λ we can get any point of the curve from these equations. But the expression for x is not a rational function of λ. If however we put

$$\frac{1-\lambda}{1+\lambda} = \frac{a^2}{t^2},$$

we obtain

$$x = \frac{\pm a^2 t(t^2 + a^2)}{t^4 + a^4}, \qquad y = \frac{\pm a^2 t(t^2 - a^2)}{t^4 + a^4}.$$

These equations express x and y rationally in terms of t.

The curve

$$y^2 = (x-1)^2(x-2)$$

is another example of a unicursal curve. If we put $y = t(x-1)$ and eliminate y, we get $x = t^2 + 2$, and hence $y = t^3 + t$. The real points of the curve correspond to real values of t, except the conjugate point $(1, 0)$. For this point we have $t = \sqrt{-1}$.

76. Elliptic integrals. We now consider other integrals of the form $\int R(x, y) dx$, where $R(x, y)$ is a rational function of x and y, and the variables are connected by the equation

$$f(x, y) = 0, \qquad (1)$$

$f(x, y)$ being a polynomial in x and y. Such integrals are known as *Abelian integrals*. A general discussion of these integrals is beyond the scope of this book, and we shall therefore confine ourselves to certain special cases.

We suppose that (1) is of the form

$$y^2 - P(x) = 0, \qquad (2)$$

where $P(x)$ is a polynomial in x of degree p without multiple roots. If p is odd ($= 2q - 1$) we can introduce a new variable in such a way as to make the resulting integral similar to the original one with the polynomial corresponding to $P(x)$ of degree $p + 1$. Put $x = a + \dfrac{1}{t}$ where a is any number such that $P(a) \neq 0$. Then, by Taylor's theorem,

$$P(x) = P(a) + P'(a)\frac{1}{t} + \cdots + \frac{P^{(2q-1)}(a)}{(2q-1)!} \cdot \frac{1}{t^{2q-1}} = \frac{Q(t)}{t^{2q}},$$

where $Q(t)$ is a polynomial of degree $p + 1$, since $P(a) \neq 0$. From this we get

$$y = \sqrt{P(x)} = \frac{\sqrt{Q(t)}}{t^q}$$

and

$$R(x, y) = R_1[t, \sqrt{Q(t)}],$$

where $R_1[t, \sqrt{Q(t)}]$ is also a rational function of its arguments. Moreover dx is the product of dt and a rational function of t and $Q(t)$ has no multiple roots.

If p is even $(= 2q)$ and $P(a) = 0$, the same substitution gives us

$$P(x) = P'(a)\frac{1}{t} + \cdots + \frac{P^{(2q)}(a)}{(2q)!} \cdot \frac{1}{t^{2q}} = \frac{Q(t)}{t^{2q}},$$

where $Q(t)$ is of degree $p - 1$. This substitution is not necessarily real.

If $p = 2$ the curve $y^2 = P(x)$ is unicursal and we know how to rationalize the integrand. If $p = 3$ or 4 the integral

$$\int R[x, \sqrt{P(x)}]dx$$

is something new which is of considerable importance in many applications. It is called an *elliptic integral*.

EXAMPLES.

(a) Consider the problem of finding the length s of an arc of the ellipse

$$\frac{x^2}{a^2} + \frac{y^2}{b^2} = 1.$$

We obtain s from the formula

$$s = \int \sqrt{1 + \left(\frac{dy}{dx}\right)^2} \, dx = \int \frac{\sqrt{(a^2 - x^2)(a^2 - e^2 x^2)}}{a^2 - x^2} \, dx$$

(see § 186).

The integral appearing here is an elliptic integral. If we put $x = a \sin \varphi$

$$s = a \int \sqrt{1 - e^2 \sin^2 \varphi} \, d\varphi. \tag{3}$$

(b) An elliptic integral of a different kind appears in connection with the problem of the motion of a simple pendulum. Let l be the length of the pendulum and θ the angle it makes with the vertical at the time t. If s is the distance along the path of the bob measured from its lowest position to its position at the time t, and m is the mass of the bob, we have (neglecting the mass of the arm)

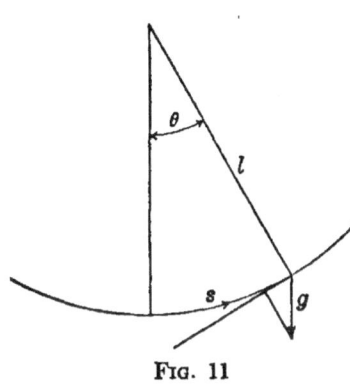

Fig. 11

$$s = l\theta,$$

$$m \frac{d^2 s}{dt^2} = ml \frac{d^2 \theta}{dt^2} = - mg \sin \theta,$$

$$2 \frac{d\theta}{dt} \frac{d^2 \theta}{dt^2} = - 2 \frac{g}{l} \sin \theta \frac{d\theta}{dt}.$$

By integration

$$\left(\frac{d\theta}{dt}\right)^2 = \frac{2g}{l} \cos \theta + C.$$

If $\theta = \alpha$ and $\frac{d\theta}{dt} = 0$ when $t = 0$, we have $C = -\frac{2g}{l} \cos \alpha$. Hence

$$t = \sqrt{\frac{l}{2g}} \int \frac{d\theta}{\sqrt{\cos \theta - \cos \alpha}} = \frac{1}{2} \sqrt{\frac{l}{g}} \int \frac{d\theta}{\sqrt{\sin^2 \frac{\alpha}{2} - \sin^2 \frac{\theta}{2}}}.$$

If we put $\sin \frac{\theta}{2} = \sin \frac{\alpha}{2} \sin \varphi$, we have

$$t = \sqrt{\frac{l}{g}} \int \frac{d\varphi}{\sqrt{1 - \sin^2 \frac{\alpha}{2} \sin^2 \varphi}} = \sqrt{\frac{l}{g}} \int \frac{d\varphi}{\sqrt{1 - k^2 \sin^2 \varphi}},$$

where $k = \sin \frac{\alpha}{2}$. This substitution gives a real value for φ for all values of θ for which $-\alpha \leqq \theta \leqq \alpha$, and these are the only values of θ we have to consider. If we take 0 and φ for the limits of integration,

$$t = \sqrt{\frac{l}{g}} \int_0^\varphi \frac{d\varphi}{\sqrt{1 - k^2 \sin^2 \varphi}} \tag{4}$$

and it is clear that $t = 0$ when $\varphi = 0$, that is, when $\theta = 0$, and the bob is lowest.

The integral in the right member of (4), namely,

$$F(k, \varphi) = \int_0^\varphi \frac{d\varphi}{\sqrt{1 - k^2 \sin^2 \varphi}} \quad (0 < k < 1), \tag{5}$$

is known as the *Elliptic Integral of the First Kind* in Legendre's form with the *modulus* k and the *amplitude* φ.

$$F\left(k, \frac{\pi}{2}\right) = K = \int_0^{\pi/2} \frac{d\varphi}{\sqrt{1 - k^2 \sin^2 \varphi}} \tag{6}$$

and is known as the *Complete Elliptic Integral of the First Kind*. The integral in the right member of (3) is the *Elliptic Integral of the Second Kind* in Legendre's form with the modulus k and the amplitude φ;

$$E(k, \varphi) = \int_0^\varphi \sqrt{1 - k^2 \sin^2 \varphi}\, d\varphi \quad (0 < k < 1). \tag{7}$$

$$E = \int_0^{\pi/2} \sqrt{1 - k^2 \sin^2 \varphi}\, d\varphi \tag{8}$$

is the *Complete Elliptic Integral of the Second Kind*.

The substitution $x = \sin \varphi$ reduces (5), (6), (7), and (8) to (5'), (6'), (7'), and (8'), respectively:

$$F(k, \varphi) = \int_0^x \frac{dx}{\sqrt{(1-x^2)(1-k^2x^2)}}, \tag{5'}$$

$$K = \int_0^1 \frac{dx}{\sqrt{(1-x^2)(1-k^2x^2)}}, \tag{6'}$$

$$E(k, \varphi) = \int_0^x \frac{\sqrt{1-k^2x^2}}{\sqrt{1-x^2}} dx, \tag{7'}$$

$$E = \int_0^1 \frac{\sqrt{1-k^2x^2}}{\sqrt{1-x^2}} dx. \tag{8'}$$

If φ is real, this restricts x to the range $-1 \leq x \leq 1$. But these integrands are real also when $|x| > \frac{1}{k}$.

77. Numerical computation. We have seen that the integral of a rational function may consist of a rational part together with logarithmic terms and terms involving the arc tangent. No other kinds of terms appear. But when we come to the elliptic integrals the situation is entirely different. We cannot express such an integral in terms of a finite number of these functions, nor in terms of these together with a finite number of the other trigonometric and inverse trigonometric functions and exponential functions. This fact makes the problem of computing the numerical value of an elliptic integral more difficult than the corresponding problem in the case of a rational function. We shall explain two ways of making this computation.

(a)

$$\frac{1}{\sqrt{1-k^2 \sin^2 \varphi}} = 1 + \frac{1}{2} k^2 \sin^2 \varphi + \cdots$$
$$+ \frac{1 \cdot 3 \cdots (2n-1)}{2 \cdot 4 \cdots 2n} k^{2n} \sin^{2n} \varphi + \cdots.$$

d'Alembert's ratio test (§ 119) shows that this series converges when $|k \sin \varphi| < 1$. But by hypothesis $0 < k < 1$.

Hence the series converges for all values of φ. Moreover the convergence is uniform within any interval for φ. We can therefore integrate term by term (see § 123), and

$$F(k, \varphi) = \int_0^\varphi \frac{d\varphi}{\sqrt{1 - k^2 \sin^2 \varphi}} = \varphi + \frac{1}{2} k^2 \int_0^\varphi \sin^2 \varphi \, d\varphi$$

$$+ \cdots + \frac{1 \cdot 3 \cdots (2n-1)}{2 \cdot 4 \cdots 2n} k^{2n} \int_0^\varphi \sin^{2n} \varphi \, d\varphi + \cdots.$$

We can proceed in a similar way for the evaluation of

$$E(k, \varphi) = \int_0^\varphi \sqrt{1 - k^2 \sin^2 \varphi} \, d\varphi.$$

(b) We introduce a new variable of integration φ_1 which is connected with the original one by the equation

$$\tan \varphi = \frac{\sin 2\varphi_1}{k + \cos 2\varphi_1}. \tag{9}$$

This is *Landen's transformation*. Starting from (9) we derive the following relations:

$$\sin^2 \varphi = \frac{\sin^2 2\varphi_1}{1 + k^2 + 2k \cos 2\varphi_1},$$

$$\sqrt{1 - k^2 \sin^2 \varphi} = \frac{1 + k \cos 2\varphi_1}{\sqrt{1 + k^2 + 2k \cos 2\varphi_1}},$$

$$\sec^2 \varphi \, d\varphi = \frac{2(1 + k \cos 2\varphi_1)}{(k + \cos 2\varphi_1)^2} d\varphi_1,$$

and

$$\sec^2 \varphi = \frac{1 + k^2 + 2k \cos 2\varphi_1}{(k + \cos 2\varphi_1)^2}.$$

From these it follows that

$$d\varphi = \frac{2(1 + k \cos 2\varphi_1)}{1 + k^2 + 2k \cos 2\varphi_1} d\varphi_1 \tag{10}$$

and

$$\frac{d\varphi}{\sqrt{1 - k^2 \sin^2 \varphi}} = \frac{2 d\varphi_1}{\sqrt{1 + 2k + k^2 - 4k \sin^2 \varphi_1}}$$

$$= \frac{2}{1 + k} \frac{d\varphi_1}{\sqrt{1 - \frac{4k}{(1 + k)^2} \sin^2 \varphi_1}}.$$

We see from (9) that we can take $\varphi_1 = 0$ when $\varphi = 0$, and therefore

$$\int_0^\varphi \frac{d\varphi}{\sqrt{1 - k^2 \sin^2 \varphi}} = \frac{2}{1 + k} \int_0^{\varphi_1} \frac{d\varphi_1}{\sqrt{1 - k_1^2 \sin^2 \varphi_1}},$$

where $k_1 = \dfrac{2\sqrt{k}}{1 + k}$. In other words,

$$F(k, \varphi) = \frac{2}{1 + k} F(k_1, \varphi_1). \tag{11}$$

Now $1 - 2\sqrt{k} + k = (1 - \sqrt{k})^2 > 0$, and therefore

$$k_1 = \frac{2\sqrt{k}}{1 + k} < 1.$$

Moreover, since $0 < k < 1$, $4 > k(1 + k)^2$. That is, $4k > k^2(1 + k)^2$, or

$$k_1 = \frac{2\sqrt{k}}{1 + k} > k.$$

It follows from (9) that $k \sin \varphi = \sin (2\varphi_1 - \varphi)$. Now $\varphi_1 = 0$ when $\varphi = 0$, and φ_1 increases when φ increases, as shown by (10). Hence as φ increases from zero to π, φ_1 increases from zero to $\dfrac{\pi}{2}$. Moreover $0 < \varphi_1 < \varphi$. Hence the transformation (9) changes an elliptic integral of the first kind, $F(k, \varphi)$, into a constant times an elliptic integral of the same kind with a greater modulus and a smaller amplitude.

By successive applications of transformation (9) we get

$$F(k, \varphi) = \frac{2}{1 + k} \cdot \frac{2}{1 + k_1} \cdot \cdots \cdot \frac{2}{1 + k_{n-1}} \cdot F(k_n, \varphi_n).$$

But

$$k_i = \frac{2\sqrt{k_{i-1}}}{1 + k_{i-1}}, \quad \text{or} \quad \frac{k_i}{\sqrt{k_{i-1}}} = \frac{2}{1 + k_{i-1}}.$$

Hence
$$F(k, \varphi) = k_n \sqrt{\frac{k_1 \cdot k_2 \cdot \cdots \cdot k_{n-1}}{k}} F(k_n, \varphi_n). \quad (12)$$

Now
$$\frac{1 - k_{n+1}}{1 - k_n} = \frac{1 - \dfrac{2\sqrt{k_n}}{1 + k_n}}{1 - k_n} = \frac{(1 - \sqrt{k_n})^2}{1 - k_n^2} = \frac{1 - \sqrt{k_n}}{1 + \sqrt{k_n}} \cdot \frac{1}{1 + k_n}.$$

This ratio is less than 1 for any n and decreases as n increases. Hence $\lim \dfrac{1 - k_{n+1}}{1 - k_n} < 1$. But $k_n < k_{n+1} < 1$. Hence k_{n+1} approaches a limit $l \leq 1$ (Ch. I, Th. 7). Then

$$\frac{1 - k_{n+1}}{1 - k_n} = \frac{1 - (l - \epsilon)}{1 - (l - \epsilon) + k_{n+1} - k_n},$$

where $k_{n+1} = l - \epsilon$. This would require that $\dfrac{1 - k_{n+1}}{1 - k_n} \to 1$ if l were less than 1, since $k_{n+1} - k_n \to 0$. Hence

$$\lim (1 - k_{n+1}) = 0,$$

or $k_{n+1} \to 1$. Then $\sqrt{1 - k_n^2 \sin^2 \varphi} \to \sqrt{1 - \sin^2 \varphi}$. Furthermore $\varphi_1, \varphi_2, \cdots, \varphi_n, \cdots$ is a monotonically decreasing sequence with $\varphi_n > 0$ for every n. Hence φ_n approaches a limit $\Phi \,(\geq 0)$ as n increases indefinitely. Now [1]

$$\left| \int_0^{\varphi_n} \frac{d\varphi}{\sqrt{1 - k_n^2 \sin^2 \varphi}} - \int_0^{\Phi} \frac{d\varphi}{\sqrt{1 - \sin^2 \varphi}} \right|$$

$$\leq \left| \int_0^{\varphi_n} \left(\frac{1}{\sqrt{1 - k_n^2 \sin^2 \varphi}} - \frac{1}{\sqrt{1 - \sin^2 \varphi}} \right) d\varphi \right|$$

$$+ \left| \int_{\varphi_n}^{\Phi} \frac{d\varphi}{\sqrt{1 - \sin^2 \varphi}} \right|. \quad (13)$$

---

[1] The reader should bear in mind that $\int_0^{\varphi_n} \dfrac{d\varphi_n}{\sqrt{1 - k_n^2 \sin^2 \varphi_n}}$ is the same as $\int_0^{\varphi_n} \dfrac{d\varphi}{\sqrt{1 - k_n^2 \sin^2 \varphi}}$.

Moreover

$$\frac{1}{\sqrt{1 - k_n^2 \sin^2 \varphi}} - \frac{1}{\sqrt{1 - \sin^2 \varphi}}$$
$$= \frac{\sqrt{1 - \sin^2 \varphi} - \sqrt{1 - k_n^2 \sin^2 \varphi}}{\sqrt{1 - k_n^2 \sin^2 \varphi} \cdot \sqrt{1 - \sin^2 \varphi}}$$
$$= \frac{(k_n^2 - 1) \sin^2 \varphi}{\sqrt{1 - k_n^2 \sin^2 \varphi} \cdot \sqrt{1 - \sin^2 \varphi} \times (\sqrt{1 - k_n^2 \sin^2 \varphi} + \sqrt{1 - \sin^2 \varphi})} \quad (14)$$

For a fixed $\varphi > \Phi$ the right member of (14), being a continuous function of k_n, which approaches 1, approaches zero as $n \to \infty$. That is, for any $\epsilon > 0$ there is an N such that when $n > N$ the absolute value of this right member is less than ϵ; and it becomes still less as φ decreases. It follows that the absolute values of the two integrals in the right member of (13) can each be made less than an arbitrary positive number by taking n sufficiently large. Hence

$$F(k_n, \varphi_n) = \int_0^{\varphi_n} \frac{d\varphi}{\sqrt{1 - k_n^2 \sin^2 \varphi}} \to \int_0^{\Phi} \frac{d\varphi}{\sqrt{1 - \sin^2 \varphi}}$$
$$= \log \tan\left(\frac{\pi}{4} + \frac{\Phi}{2}\right).$$

We conclude from this equation and (12) that

$$k_1 \cdot k_2 \cdot \ \cdots \ \cdot k_n \to L \neq 0$$

as $n \to \infty$, and that

$$F(k, \varphi) = \log \tan\left(\frac{\pi}{4} + \frac{\Phi}{2}\right) \sqrt{\frac{k_1 \cdot k_2 \cdot \ \cdots}{k}}. \quad (15)$$

Formula (15) lends itself readily to numerical computation, especially if k is near to 1, since then we need to apply transformation (9) only a few times.

78. We give the numerical details for one case.

A pendulum two feet long is pulled around until it makes an angle of 120° with the vertical and is then released.

§ 78] INDEFINITE INTEGRALS

How long will it take to rise from its lowest position to a position making an angle of 50° with the vertical?

$\alpha = 120°$
$k = \sin 60° = 0.86603$
$\theta = 50°$

$\sin \varphi = \dfrac{\sin 25°}{\sin 60°}$

$\log \sin 25°$	$= 9.62595$
$\log \sin 60°$	$= 9.93753$
$\log \sin \varphi$	$= 9.68842$

$\varphi = 29° \, 12' \, 33''$
$\sin (2\varphi_1 - \varphi) = k \sin \varphi$
$\quad\quad\quad\quad\quad = \sin 25°$
$2\varphi_1 - \varphi = 25°$
$\varphi_1 = 27° \, 6' \, 17''$

$\log 2$	$= 0.30103$
$\log \sqrt{k}$	$= 9.96877$
$\operatorname{colog} (1 + k)$	$= 9.72908$
$\log k_1$	$= 9.99888$

$\log (1 + k) = 0.27092$

$k_1 = \dfrac{2\sqrt{k}}{1 + k}$

$k_1 = 1$ (approximately)

The value of k_1 is so close to 1 we shall get a good approximation by assuming it equal to 1. On this basis we take

$\Phi = \varphi_1 = 27° \, 6' \, 17''$

$\dfrac{\pi}{4} + \dfrac{\Phi}{2} = 58° \, 33' \, 9''$

$\log \tan \left(\dfrac{\pi}{4} + \dfrac{\Phi}{2} \right)$
$\quad\quad\quad\quad = 0.21357$

$\log \log \tan \left(\dfrac{\pi}{4} + \dfrac{\Phi}{2} \right)$	$= 9.32954$
$\operatorname{colog} \mu$	$= 0.36222$
$\log \sqrt{k_1}$	$= 9.99944$
$\operatorname{colog} \sqrt{k}$	$= 0.03123$

$\dfrac{g}{l} = 16.1$

$\log 16.1 = 1.20683$
$\log \sqrt{16.1} = 0.60341$

$\log F(k, \varphi)$	$= 9.72243$
$\operatorname{colog} \sqrt{16.1}$	$= 9.39659$
$\log t$	$= 9.11902$
t	$= 0.1315$

The μ in this computation is the modulus of the common system of logarithms. Its value is 0.43429.

These two methods of computation are complementary in that (a) can be used to the better advantage when k is near to zero and (b) to the better advantage when k is near to one.

EXERCISES

1. Show that every curve of the second degree is unicursal.

2. Show that $\int \dfrac{dx}{\sqrt{\cos \alpha - \cos x}}$ is an elliptic integral.

3. What kind of an integral arises in finding the length of an arc of a hyperbola?

4. What kind of an integral arises in finding the length of an arc of the lemniscate $(x^2 + y^2)^2 - 2a^2(x^2 - y^2) = 0$?

5. Find the total length of the ellipse $\dfrac{x^2}{4} + \dfrac{y^2}{9} = 1$.

In each of the following exercises expand the integrand into a power series and then integrate the sum of the first three terms:

6. $\displaystyle\int_0^{1/2} \dfrac{dx}{\sqrt{(1-x^2)(1-\frac{3}{4}x^2)}}.$ 7. $\displaystyle\int_0^{3/4} \dfrac{dx}{\sqrt{(1-x^2)(1-\frac{1}{4}x^2)}}.$

Expand the integrand in each of the following integrals in powers of $\sin \varphi$ and then integrate the sum of the first three terms:

8. $\displaystyle\int_0^{\pi/2} \dfrac{d\varphi}{\sqrt{1 - \frac{1}{4}\sin^2 \varphi}}.$ 9. $\displaystyle\int_0^{\pi/3} \sqrt{1 - \frac{1}{2}\sin^2 \varphi}\, d\varphi.$

10. $\displaystyle\int_0^{\pi/4} \dfrac{d\varphi}{\sqrt{1 - \frac{1}{3}\sin^2 \varphi}}.$

11. A pendulum three feet long is pulled up until it makes an angle of 45° with the vertical and then released. How long will it take to rise from its lowest position to a position making an angle of 40° with the vertical?

12. Apply the transformation $x = \dfrac{1}{\gamma z + \delta}$ to the integral
$$\int \dfrac{dx}{\sqrt{(1-x^2)(1-k^2 x^2)}}.$$

13. Apply the transformation $x = \dfrac{\alpha z + \beta}{\gamma z + \delta}$ $(\alpha \delta - \beta \gamma \leqq 0)$ to the integral $\displaystyle\int \sqrt{ax^4 + bx^3 + cx^2 + dx + e}\, dx.$

CHAPTER VII

IMPROPER AND INFINITE INTEGRALS

79. Improper integrals. In our discussion of simple integrals we assumed that the integrand was bounded in the closed interval of integration, and in many cases we assumed that it was continuous. If now $f(x)$ is continuous for $a + h \leqq x \leqq b$ $(h > 0)$, but $f(x) \to \infty$ as $x \to a$, the integral $\int_{a+h}^{b} f(x)dx$ may, or may not, have a definite limit when $h \to +0$. If it has, we use the symbol $\int_{a}^{b} f(x)dx$ to denote this limit and we call the limit the *improper integral* of $f(x)$ in (a, b).

If we denote the indefinite integral $\int f(x)dx$ by $F(x)$, we have

$$\int_{a+h}^{b} f(x)dx = F(b) - F(a + h);$$

and it is necessary merely to determine whether $F(a + h)$ has a limit, or not, as h approaches zero through positive values, in order to know whether the improper integral exists or not. Consider, for example, the integral

$$\int_{a+h}^{b} \frac{dx}{\sqrt{x - a}} \quad (b > a).$$

The integrand increases without limit as x approaches a through values greater than a, and

$$\int_{a+h}^{b} \frac{dx}{\sqrt{x - a}} = 2\sqrt{x - a} \Big]_{a+h}^{b}$$
$$= 2\sqrt{b - a} - 2\sqrt{h} \to 2\sqrt{b - a} \quad \text{as} \quad h \to +0.$$

Hence the symbol $\int_{a}^{b} \frac{dx}{\sqrt{x - a}}$ represents an improper integral. On the other hand,

$$\int_{a+h}^{b} \frac{dx}{x-a} = \log(x-a)\Big]_{a+h}^{b}$$
$$= \log(b-a) - \log h \to +\infty \quad \text{as} \quad h \to +0,$$

and therefore $\int_{a}^{b} \frac{dx}{x-a}$ does not represent an improper integral.

It is customary to say that in the former case the improper integral converges and in the latter diverges. But strictly speaking it is a certain proper integral that converges or diverges, and in the case of divergence there is, in accordance with our definition, no improper integral present.

More generally,

$$\int_{a+h}^{b} \frac{dx}{(x-a)^{\mu}} = \frac{(x-a)^{1-\mu}}{1-\mu}\Big]_{a+h}^{b}$$
$$= \frac{(b-a)^{1-\mu}}{1-\mu} - \frac{h^{1-\mu}}{1-\mu} \quad (\mu \neq 1).$$

This approaches a finite limit or not as $h \to +0$ according as $1 - \mu > 0$, or $1 - \mu < 0$, that is, according as $\mu < 1$ or $\mu > 1$.

80. Tests for convergence. We have seen how we can decide whether the improper integral exists or not if we know the corresponding indefinite integral. If we do not have this information there is a perfectly general test that we can apply, but like the corresponding test for the convergence of an infinite series (§ 116), it is difficult to handle. We shall therefore confine our discussion to certain special tests that are applicable in most of the cases that present themselves in practice and are easy to use.

THEOREM 1. *If the integral* $\int_{a+h}^{b} \varphi(x)dx \ (b > a)$ *approaches a limit as h approaches zero through positive values, and if $\varphi(x)$ is positive or zero throughout (a, b), the integral*

$\int_{a+h}^{b} f(x)dx$ approaches a limit under the same circumstances, provided that $|f(x)| \leq \varphi(x)$ throughout the interval. If $\int_{a+h}^{b} \varphi(x)dx$ diverges and $f(x) \geq \varphi(x)$ throughout the interval, then $\int_{a+h}^{b} f(x)dx$ diverges.

If $f(x) \geq 0$ for $a < x \leq b$, we have

$$\int_{a+h}^{b} f(x)dx \leq \int_{a+h}^{b} \varphi(x)dx.$$

Now $\int_{a+h}^{b} f(x)dx$ does not decrease as $h \to 0$ and is always less than the limit of $\int_{a+h}^{b} \varphi(x)dx$. It therefore approaches a limit. The situation is essentially the same if $f(x) \leq 0$ for $a < x \leq b$. If $f(x)$ changes sign only a finite number of times in (a, b), or only a finite number of times in the neighborhood of a, we can choose c so near to a in this interval that $f(x) \leq 0$ or $f(x) \geq 0$ for $a < x \leq c$. Then the preceding argument applies. But if $f(x)$ changes sign an infinite number of times in the neighborhood of a, we must proceed differently. Now

$$\int_{a+h}^{b} f(x)dx = \int_{a+h}^{b} [f(x) + |f(x)|]dx - \int_{a+h}^{b} |f(x)|dx.$$

The first integrand on the right is positive or zero in (a, b) and does not exceed $2\varphi(x)$, while the second integrand is positive or zero and does not exceed $\varphi(x)$. Hence both of these integrals converge, and therefore the integral in the left member converges. The last statement of the theorem is obvious.

It follows immediately from the theorem that $\int_{a+h}^{b} f(x)dx$ converges if $\int_{a+h}^{b} |f(x)|dx$ does. In this case we say that $\int_{a}^{b} f(x)dx$ converges *absolutely*.

81. Practical rule.
We can now formulate a practical rule for determining whether $\int_{a+h}^{b} f(x)dx$ approaches a limit as $h \to 0$.

Suppose that there is a number μ such that

$$(x - a)^\mu f(x) = \psi(x).$$

Then if $\mu < 1$ and $\psi(x)$ is bounded as $x \to a$ the integral $\int_{a+h}^{b} f(x)dx$ converges. But if $\mu \geq 1$ and $|\psi(x)| > m > 0$ the integral diverges.

If $\psi(x)$ is bounded—that is, if there is a number M such that $|\psi(x)| < M$—we have $|f(x)| < \dfrac{M}{(x-a)^\mu}$. If now $\mu < 1$ the integral $\int_{a+h}^{b} \dfrac{M}{(x-a)^\mu} dx$ converges, and therefore $\int_{a+h}^{b} f(x)dx$ converges.

Since $f(x)$ is continuous for $a < x \leq b$, $\psi(x)$ is also. If now $|\psi(x)| > m > 0$ in the interval (a, c), where $a < c \leq b$, $\psi(x)$ does not change sign in the interval. Let us assume that $\psi(x) > 0$. Then

$$\int_{a+h}^{b} f(x)dx > \int_{a+h}^{b} \frac{m}{(x-a)^\mu} dx.$$

However the second integral diverges as $h \to 0$ if $\mu \geq 1$. Hence under these circumstances $\int_{a+h}^{b} f(x)dx$ diverges.

If $f(x)$ is continuous throughout (a, b), except at $x = b$, where it becomes infinite, we define the improper integral as the limit as $h \to 0$ through positive values of $\int_{a}^{b-h} f(x)dx$, if this limit exists. There are tests for the existence of this limit similar to those given for the existence of a limit of $\int_{a+h}^{b} f(x)dx$. If $f(x)$ becomes infinite for both $x = a$ and

$x = b$ and for no other point of the interval, we define $\int_a^b f(x)dx$ as the sum of the limits of $\int_{a+h}^c f(x)dx$ and of $\int_c^{b-h'} f(x)dx$ as h and h' approach zero independently through positive values, if both of these limits exist. It is to be understood here that c is any point within the interval. If the point of infinite discontinuity is at an intermediate point c, we define the integral $\int_a^b f(x)dx$ as the sum of the limits of $\int_a^{c-h} f(x)dx$ and $\int_{c+h'}^b f(x)dx$ as h and h' approach zero independently through positive values, if both of these limits exist. The procedure in the case of any finite number of discontinuities will be obvious to the reader.

82. Examples.

(a) Let $P(x)$ and $Q(x)$ be two relatively prime polynomials and let $Q(x) = (x-a)^k Q_1(x)$, where $Q_1(x)$ is a polynomial not divisible by $x - a$. Now $(x-a)^k \dfrac{P(x)}{Q(x)} = \dfrac{P(x)}{Q_1(x)}$, and the right member of this equation does not vanish in the neighborhood of $x = a$. If there are no roots of $Q(x)$, except $x = a$, in the interval (a, b), the integral $\int_a^b \dfrac{P(x)}{Q(x)} dx$ diverges, since $k \geqq 1$. More generally, $\int_a^b \dfrac{P(x)}{Q(x)} dx$ diverges if there are any roots of $Q(x)$ in the closed interval (a, b).

(b) $\int_0^a \dfrac{dx}{\sqrt{a^2 - x^2}}$. The integrand is infinite for $x = a$ and $(a-x)^{1/2} \dfrac{1}{\sqrt{a^2-x^2}}$ is bounded in the neighborhood of $x = a$. Moreover $\mu = \tfrac{1}{2}$. Hence the integral converges.

(c) $\int_0^1 \dfrac{\sqrt{1-e^2 t^2}}{\sqrt{1-t^2}} dt$, $(0 < e < 1)$. This integral converges, as may be seen by comparison with (b).

(d) $\int_0^b \dfrac{dx}{\sqrt{(1-x^2)(1-k^2x^2)}}$, $\left(0 < k < 1, b > \dfrac{1}{k}\right)$.

The integrand becomes infinite at two points in the interval of convergence; namely, at $x = 1$ and $x = \dfrac{1}{k}$. It can be shown as in (b) that the integral converges.

(e) It is not easy to see how the rule can be applied to the integral $\int_0^{\pi/2} \log \sin x\, dx$, whose integrand is infinite at $x = 0$. But if we recall that $\lim\limits_{x \to 0} \dfrac{\sin x}{x} = 1$, we see that $\sin x = x\varphi(x)$, where $\lim\limits_{x \to 0} \varphi(x) = 1$. Now $x^\mu \log \sin x = x^\mu \log x + x^\mu \log \varphi(x)$. If we take $\mu > 0$ and let x approach zero, we see that $x^\mu \log \sin x \to 0$, since $x^\mu \log x \to 0$ (see § 50) and $x^\mu \log \varphi(x) \to 0$. Hence the integral converges.

EXERCISES

Pick out the convergent integrals from the following:

1. $\int_{-2}^{2} \dfrac{dx}{x^2}$. 2. $\int_{-1}^{1} \dfrac{dx}{\sqrt[3]{x-1}}$. 3. $\int_0^1 \dfrac{\arcsin x}{\sqrt{1-x^2}}\, dx$.

4. $\int_0^{\pi} \dfrac{d\theta}{1 - \sin \theta}$.

5. Evaluate $\int_0^{\pi/2} \log \sin x\, dx$ by first showing that $\int_0^{\pi/2} \log \sin x\, dx = \int_0^{\pi/2} \log \cos x\, dx$.

6. Show that $\int_0^{\pi/2} \log \tan x\, dx = 0$.

7. Show that $\int_0^a \log x\, dx$ converges $(a > 0)$.

8. Show that $\int_0^a x \log x\, dx$ is a proper integral $(a > 0)$.

Are the following integrals proper or improper?

9. $\int_0^{\pi/2} x \log \sin x\, dx$. 10. $\int_0^{\pi} x \log \sin x\, dx$.

§83] IMPROPER AND INFINITE INTEGRALS

11. Show that the latter integral is equal to $-\dfrac{\pi^2}{2}\log 2$ by relating it to $\displaystyle\int_0^{\pi/2} \log \sin x\, dx$.

12. Show that the improper integral $\displaystyle\int_0^1 x^{m-1}(1-x)^{n-1}dx,\ 0<m<1,\ 0<n<1$, converges. When $m \geqq 1$ and $n \geqq 1$ the integral is a proper one.

13. Discuss the integral $\displaystyle\int_0^{\pi/2} (\sin x)^{m-1}(\cos x)^{n-1}\,dx$ from the same point of view.

14. Show that $\displaystyle\int_0^1 \log \Gamma(x)dx$ converges. Make use of the fact that $\Gamma(x) = \Gamma(x+1)\cdot\dfrac{1}{x}$ and the fact that $\Gamma(x)$ is a continuous function of x for every positive value of x.

15. What is the sign of $\displaystyle\int_0^1 x^2 \log \sin x\,dx$?

16. Does $\displaystyle\int_0^1 \log x \tan \dfrac{\pi x}{2}\,dx$ converge?

17. Show that $\displaystyle\int_0^{\pi/2} \dfrac{x^2}{(\cos x)^n}\,dx$ converges if $n < 1$.

18. Show that $\displaystyle\int_0^{\pi/2} \dfrac{\left(\dfrac{\pi}{2}-x\right)^m}{(\cos x)^n}\,dx$ converges if, and only if, $n < m+1$.

19. Is the integral $\displaystyle\int_0^1 \dfrac{\log x}{\sqrt{1-x^2}}\,dx$ convergent?

83. Infinite integrals. If $f(x)$ is continuous for all values of x greater than a, the integral $\displaystyle\int_a^h f(x)dx$ exists for every h greater than a. It is a function of h that may, or may not, approach a limit as h increases indefinitely. When there is a limit we denote it by the symbol $\displaystyle\int_a^\infty f(x)dx$ and call it an *infinite integral*. These integrals are frequently called improper integrals. It is however desirable to make a distinction between them and the integrals just discussed.

For example, $\int_0^h \dfrac{adx}{x^2 + a^2} = \arctan \dfrac{h}{a} \to \dfrac{\pi}{2}$ as $h \to \infty$ $(a \neq 0)$.

But $\int_0^h \cos x\, dx = \sin h$, and this has no limit as h increases indefinitely.

We can formulate rules for determining the existence or non-existence of a limit in this case without first finding the indefinite integral, just as we did in the case of integrals with infinite integrands. The following tests are useful in many cases, but they are subject to the disadvantage that they are not always applicable:

TEST FOR CONVERGENCE. *The integral* $\int_0^\infty f(x)dx$ *exists if the following conditions are satisfied:*

(a) $f(x)$ *is continuous in the interval* $a \leqq x$,

(b) $|f(x)| \leqq \varphi(x)$ *when* $b \leqq x$, *where b is any number greater than a,*

(c) *The integral* $\int_b^\infty \varphi(x)dx$ *exists.*

TEST FOR DIVERGENCE. *The integral* $\int_a^\infty f(x)dx$ *does not exist if*

(a) $f(x) \geqq \varphi(x) \geqq 0$ *for* $b \leqq x$, *where b is any number greater than a,*

(b) *The integral* $\int_b^\infty \varphi(x)dx$ *does not exist.*

The reader will find it worth his while to work through the details of the proofs. In the light of these tests we can formulate the following rule:

If $x^\mu f(x) = \psi(x)$, the integral $\int_a^\infty f(x)dx$ converges when $|\psi(x)| < M$ for $x > b$ and $\mu > 1$. The integral diverges when $|\psi(x)| > m > 0$ and $\mu \leqq 1$.

But this rule does not cover all cases. For example, if $f(x) = \dfrac{\sin x}{x}$, then $x^\mu f(x) \to 0$ if $\mu < 1$, and is unbounded if $\mu > 1$. If $\mu = 1$ it oscillates between -1 and 1. And yet $\int_0^\infty \frac{\sin x}{x} dx$ converges. This can be seen as follows:

$$\int_0^h \frac{\sin x}{x} dx = u_0 - u_1 + \cdots + u_{n-1} + \theta u_n,$$

where $u_i = \left| \int_{i\pi}^{(i+1)\pi} \frac{\sin x}{x} dx \right|$, $n\pi < h \leq (n+1)\pi$, and $-1 \leq \theta \leq 1$. If we put $x = i\pi + y$, $u_i = \int_0^\pi \frac{\sin y}{y + i\pi} dy$. It follows from this that $u_{i+1} < u_i$ and that $u_i \to 0$ as $i \to \infty$. But $i \to \infty$ as $h \to \infty$. Hence $\int_0^\infty \frac{\sin x}{x} dx$ converges. (See Ch. IX, Th. 10.) It can be shown in a similar way that $\int_0^\infty e^{-\alpha x} \frac{\sin x}{x} dx$ ($\alpha > 0$) converges.

The integral $\int_{-\infty}^a f(x) dx$ is defined as the limit of $\int_h^a f(x) dx$ as $h \to -\infty$, if this limit exists, it being understood that $f(x)$ is continuous for $x \leq a$. The integral $\int_{-\infty}^\infty f(x) dx$ is defined as the sum of the limits of $\int_h^a f(x) dx$ and $\int_a^k f(x) dx$ as $h \to -\infty$ and $k \to \infty$, when both of these limits exist. It is here assumed that $f(x)$ is continuous for every finite value of x. The term *infinite integral* is applied to all these integrals with infinite intervals of integration.

84. EXAMPLES.

(a) Let $P(x)$ and $Q(x)$ be two relatively prime polynomials of degrees p and q respectively. A direct application of the test shows that, if a is greater than any root of $Q(x)$, the integral $\int_a^\infty \frac{P(x)}{Q(x)} dx$ converges or diverges according as $p < q - 1$ or $p \geq q - 1$.

(b) Since $\left| \frac{\cos x}{a^2 + x^2} \right| \leq \frac{1}{a^2 + x^2}$ and $\int_0^\infty \frac{dx}{a^2 + x^2}$ converges we know that $\int_0^\infty \frac{\cos x}{a^2 + x^2} dx$ converges.

(c) $\int_a^\infty \dfrac{dx}{\sqrt{(1-x^2)(1-k^2x^2)}}\left(0 < k < 1, a > \dfrac{1}{k}\right)$. Now

$$\dfrac{1}{\sqrt{(1-x^2)(1-k^2x^2)}} = \dfrac{1}{x^2\sqrt{\left(\dfrac{1}{x^2}-1\right)\left(\dfrac{1}{x^2}-k^2\right)}}.$$ Hence

$$\dfrac{x^2}{\sqrt{(1-x^2)(1-k^2x^2)}} \to \dfrac{1}{k} \quad \text{as} \quad x \to \infty,$$

and the given integral converges.

(d) $\int_0^\infty \dfrac{\sqrt{1-e^2t^2}}{\sqrt{1-t^2}}\,dt$. Here the integrand approaches e as $t \to \infty$. Hence the integral diverges.

(e) It follows as in (d) that $\int_1^\infty \dfrac{x^2\,dx}{\sqrt{ax^4 + bx^2 + c}}$ diverges, if $a > 0$.

85. Integration by parts. We know from the theory of proper integrals that, if $f(x) = u(x)v'(x)$,

$$\int_a^h f(x)\,dx = \int_a^h u(x)v'(x)\,dx = uv\Big|_a^h - \int_a^h v(x)u'(x)\,dx.$$

We have here three functions of h. If they all approach limits as $h \to \infty$ we can apply integration by parts to the infinite integral $\int_a^\infty f(x)\,dx$. Moreover these three functions of h will all approach limits if any two of them do. For example, $\int_0^h x^{\alpha-1}e^{-x}\,dx = -[x^{\alpha-1}e^{-x}]_0^h + (\alpha - 1)\int_0^h x^{\alpha-2}e^{-x}\,dx$. Here all three functions do approach limits as $h \to \infty$ if $\alpha > 1$. (See § 89.) We have therefore the important formula $\int_0^\infty x^{\alpha-1}e^{-x}\,dx = (\alpha - 1)\int_0^\infty x^{\alpha-2}e^{-x}\,dx$.

Similar remarks apply also to improper integrals.

86. Differentiation and integration under the integral sign. It will be shown in § 128 that the integral $\int_a^\infty f(x, \alpha)\,dx$ can be differentiated under the integral sign with respect to α if $f_\alpha(x, \alpha)$ exists and if $\int_a^\infty f_\alpha(x, \alpha)\,dx$

converges uniformly with respect to α in the interval (α_0, α_1), that is, if there is associated to every $\epsilon > 0$ a number N such that $\left|\int_h^\infty f_\alpha(x, \alpha)dx\right| < \epsilon$ whenever $h > N$ for every value of α in the interval (α_0, α_1); and in § 132 that $\int_a^\infty f(x, \alpha)dx$ can be integrated under the integral sign with respect to α if the integral $\int_a^\infty f(x, \alpha)dx$ converges uniformly in the interval (α_0, α_1) and the limits of integration are in this interval.

87. Theorem 2. *If $f(x, \alpha)$ is a continuous function of x and α when $a \leq x$ and $\alpha_0 \leq \alpha \leq \alpha_1$, and if the integral $\int_a^\infty f(x, \alpha)dx$ converges uniformly with respect to the interval (α_0, α_1), the integral is a continuous function of α in this interval.*

For any positive ϵ we can, by virtue of the uniform convergence of the integral, choose an h such that

$$\left|\int_h^\infty f(x, \alpha)dx\right| < \epsilon,$$

$$\left|\int_h^\infty f(x, \alpha + \Delta\alpha)dx\right| < \epsilon$$

if α and $\alpha + \Delta\alpha$ are in the given interval. Having chosen h subject to this condition, we have

$$F(\alpha) = \int_a^\infty f(x, \alpha)dx = \int_a^h f(x, \alpha)dx + \int_h^\infty f(x, \alpha)dx,$$

$$F(\alpha + \Delta\alpha) = \int_a^\infty f(x, \alpha + \Delta\alpha)dx = \int_a^h f(x, \alpha + \Delta\alpha)dx$$

$$+ \int_h^\infty f(x, \alpha + \Delta\alpha)dx,$$

and

$$|F(\alpha + \Delta\alpha) - F(\alpha)| \leq \left|\int_a^h [f(x, \alpha + \Delta\alpha) - f(x, \alpha)dx]\right|$$

$$+ \left|\int_h^\infty f(x, \alpha + \Delta\alpha)dx\right| + \left|\int_h^\infty f(x, \alpha)dx\right|.$$

But we know from the theory of proper integrals (§ 68) that $\int_a^h f(x, \alpha)dx$ is a continuous function of α. Hence if $|\Delta\alpha|$ is sufficiently small

$$\left|\int_a^h [f(x, \alpha + \Delta\alpha) - f(x, \alpha)]dx\right| < \epsilon,$$

and

$$|F(\alpha + \Delta\alpha) - F(\alpha)| < 3\epsilon.$$

The theorem is therefore proved.

88. A similar theorem holds for improper integrals. If $f(x, \alpha)$ is a bounded function of x and α when $a + h \leqq x \leqq b$ and $\alpha_0 \leqq \alpha \leqq \alpha_1$ for every positive h, and if $f(x, \alpha) \to \infty$ as $x \to a + 0$ for every value of α in the interval (α_0, α_1) we say that the integral $\int_a^b f(x, \alpha)dx$ *converges uniformly* with respect to α in the interval (α_0, α_1) if it converges and if to every $\epsilon > 0$ there is associated an $\eta > 0$ such that $\left|\int_a^h f(x, \alpha)dx\right| < \epsilon$ for every α in the interval (α_0, α_1) whenever $0 < h < \eta$.

THEOREM 3. *If $f(x, \alpha)$ is continuous when $a < x \leqq b$ and $\alpha_0 < \alpha < \alpha_1$, and if $f(x, \alpha) \to \infty$ as $x \to a + 0$ for every α in (α_0, α_1), then $F(\alpha) = \int_a^b f(x, \alpha)dx$ is a continuous function of α in this interval provided that the integral converges uniformly.*

For if α and $\alpha + \Delta\alpha$ are both in the interval (α_0, α_1) there is, for every $\epsilon > 0$, an $\eta > 0$ such that

$$\left|\int_a^h f(x, \alpha)dx\right| < \epsilon \quad \text{and} \quad \left|\int_a^h f(x, \alpha + \Delta\alpha)dx\right| < \epsilon$$

whenever $0 < h < \eta$. Having chosen h subject to this condition, we can take $|\Delta\alpha|$ sufficiently small to give us the inequality (§ 68)

$$\left|\int_h^b [f(x, \alpha + \Delta\alpha) - f(x, \alpha)]dx\right| < \epsilon.$$

Hence
$$|F(\alpha + \Delta\alpha) - F(\alpha)| < 3\epsilon$$
and $F(\alpha)$ is continuous in (α_0, α_1).

EXERCISES

Pick out the convergent integrals from the following list:

1. $\int_0^\infty \dfrac{dx}{1+x^4}$. 2. $\int_1^\infty \dfrac{\arctan x}{1+x^2} dx$. 3. $\int_0^\infty \dfrac{\cos x}{1+x} dx$.

4. $\int_1^\infty \dfrac{dx}{x\sqrt{x-1}}$. 5. $\int_1^\infty x^{-1/2} dx$. 6. $\int_0^\infty \sqrt{\sin x}\, dx$.

7. Show that $\int_0^\infty \dfrac{x^{n-1} dx}{1+x}$ converges if, and only if, $0 < n < 1$.

8. Show that $\int_1^\infty \dfrac{x^{n-1} dx}{1+x}$ converges if, and only if, $n < 1$.

Under what circumstances will the following integrals converge:

9. $\int_1^\infty \dfrac{x^{n-1} dx}{1+x^2}$. 10. $\int_0^\infty \dfrac{x^{n-1} dx}{1+x^2}$.

11. Evaluate $\int_0^\infty x^2 e^{-x^2} dx$. Integrate $\int_0^l x^2 e^{-x^2} dx$ by parts and then let $l \to \infty$. Assume that $\int_0^\infty e^{-x^2} dx = \dfrac{\sqrt{\pi}}{2}$ (§ 214).

12. Evaluate $\int_0^\infty e^{-\alpha x^2} dx\ (\alpha > 0)$.

13. Show that the integral in Exercise 12 converges uniformly in the interval $0 < \alpha_0 \leqq \alpha \leqq \alpha_1$.

14. Show that if n is a positive integer $\int_0^\infty x^n e^{-\alpha x} dx = \dfrac{n!}{\alpha^{n+1}}\ (\alpha > 0)$.

15. Show that if n is a positive integer and $\alpha > 0$
$$\int_0^\infty x^{2n} e^{-\alpha x^2} dx = \dfrac{\sqrt{\pi}}{2} \dfrac{1 \cdot 3 \cdot \cdots \cdot (2n-1)}{2^n \cdot \alpha^{n+1/2}}.$$

16. Show that $\int_0^\infty \dfrac{\sin x}{x} dx$ is not absolutely convergent.

17. Show that $\int_{-\infty}^\infty e^{-x^2} \sin x\, dx = 0$.

18. Evaluate $\int_0^\infty \dfrac{\sin^2 x}{x^2} dx$. (Integrate by parts and use the value of $\int_0^\infty \dfrac{\sin x}{x} dx$ that is given in § 214.)

89. The Gamma function. Certain important functions can be defined in terms of improper or infinite integrals whose integrands contain one or more parameters. We shall consider two of these. The first is the *Gamma function*, or *Eulerian integral of the second kind*. By definition

$$\Gamma(\alpha) = \int_0^\infty x^{\alpha-1} e^{-x} dx \quad (\alpha > 0). \tag{1}$$

If $\alpha < 1$ this integral is both an improper and an infinite integral, and if $\alpha \geq 1$ it is an infinite integral. We know from §81 that the integral $\int_h^b x^{\alpha-1} e^{-x} dx$, where $0 < \alpha < 1$ and b is any positive number, approaches a limit as $h \to 0$. For sufficiently great values of x we have $e^{-x} < \dfrac{1}{x^c}$, where c is any positive number. Then $x^{\alpha-1} e^{-x} < \dfrac{1}{x^{c+1-\alpha}}$. If we take $c > \alpha$, $c + 1 - \alpha > 1$ and $\int_b^h \dfrac{dx}{x^{c+1-\alpha}}$ approaches a limit as $h \to \infty$. The symbol $\int_0^\infty x^{\alpha-1} e^{-x} dx$ therefore defines a definite function of α for all positive values of α.

The Gamma function is a continuous function for every positive value of its argument. Since

$$\Gamma(\alpha) = \int_0^b x^{\alpha-1} e^{-x} dx + \int_b^\infty x^{\alpha-1} e^{-x} dx, \tag{2}$$

it is a continuous function of α if each of the integrals in the right member of this equation is. If $\alpha \geq 1$ the first integral is a continuous function of α by virtue of §68. If $0 < \alpha < 1$, we have

$$\int_0^h x^{\alpha-1} e^{-x} dx < \int_0^h \frac{dx}{x^{1-\alpha}} = \frac{h^\alpha}{\alpha} \quad (0 < h < 1).$$

For any $\epsilon > 0$ there is an h such that $\dfrac{h^{\alpha_0}}{\alpha_0} < \epsilon \ (\alpha_0 > 0)$.

Then $\dfrac{h^\alpha}{\alpha} < \epsilon$ for any value of $\alpha > \alpha_0$, since the derivative of $\dfrac{h^\alpha}{\alpha}$ with respect to α is negative if $h < 1$. Hence $\int_0^b x^{\alpha-1} e^{-x} dx$ converges uniformly in the interval (α_0, α_1), $\alpha_1 > \alpha_0$, and is a continuous function of α in this interval (§ 88). The second integral in the right member of (2) converges for any $\alpha_1 > 1$. Hence for $\epsilon > 0$ there is an h such that $\int_h^\infty x^{\alpha_1-1} e^{-x} dx < \epsilon$. Then $\int_h^\infty x^{\alpha-1} e^{-x} dx < \epsilon$ for every positive α less than α_1. That is, $\int_b^\infty x^{\alpha-1} e^{-x} dx$ converges uniformly in (α_0, α_1). This completes the proof of the continuity of $\Gamma(\alpha)$ for any $\alpha > 0$ (§ 87).

It is easy to see that $\Gamma(1) = 1$. It was shown in § 85 that for $\alpha > 1$, $\Gamma(\alpha) = (\alpha - 1)\Gamma(\alpha - 1)$. If α is a positive integer greater than 1, successive applications of this formula show that

$$\Gamma(\alpha) = (\alpha - 1)(\alpha - 2) \cdots 2(1) = (\alpha - 1)!$$

In view of this fact the Gamma function is called the *factorial function*. It has the value of $\alpha!$ when α is a positive integer, and in addition has a value for every positive value of its argument.

If $0 < h < 1$

$$\int_h^1 x^{-1} e^{-x} dx > \frac{1}{e} \int_h^1 x^{-1} dx = -\frac{1}{e} \log h \to \infty \quad \text{as} \quad h \to 0.$$

This means that our definition of the Gamma function does not apply when the argument is zero. But $\Gamma(\alpha) \to +\infty$ as $\alpha \to +0$, since $\Gamma(\alpha) = \dfrac{1}{\alpha} \Gamma(\alpha + 1)$. We can use this last equation to define $\Gamma(\alpha)$ for values of α between -1 and 0, and by successive applications of this process we can define the function for every negative non-integral value of the argument. On the basis of this enlarged definition of $\Gamma(\alpha)$ the function is a continuous function of α, except for α equal to zero or a negative integer. Its graph is shown in the figure.

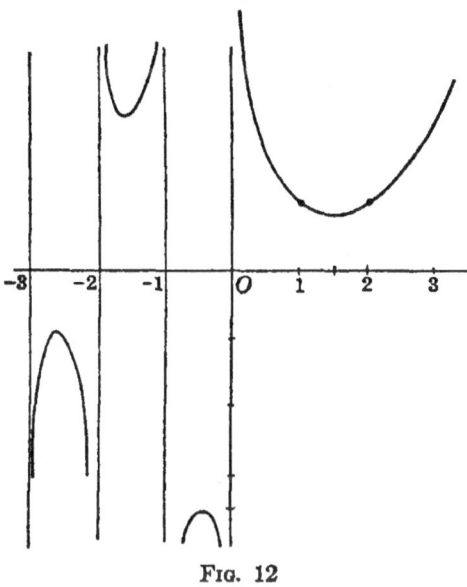

Fig. 12

90. The Beta function. If $m \geqq 1$ and $n \geqq 1$ the integral $\int_0^1 x^{m-1}(1-x)^{n-1}dx$ is a proper integral. If neither one is less than or equal to zero and at least one of them is between zero and 1, we have a convergent improper integral. If at least one of them is equal to or less than zero the integral diverges. The reader will find it a profitable exercise to verify these statements. Assuming then that $m > 0$ and $n > 0$, we have the function of two arguments $B(m, n) = \int_0^1 x^{m-1}(1-x)^{n-1}dx$. This is called the *Beta function*, or the *Eulerian integral of the first kind*.

If we put $x = 1 - y$, we find that

$$B(m, n) = \int_0^1 y^{n-1}(1-y)^{m-1}dy = B(n, m).$$

The Beta function is therefore a symmetric function of its two arguments.

91. The relation between the Beta function and the Gamma function. The Beta function and the Gamma function can be expressed in many forms by the introduction of new variables. For example, we get

$$B(m, n) = 2\int_0^{\pi/2} \sin^{2m-1}\varphi \cos^{2n-1}\varphi \, d\varphi$$

§ 91] IMPROPER AND INFINITE INTEGRALS 149

when $x = \sin^2 \varphi$, and
$$\Gamma(m) = 2\int_0^\infty y^{2m-1}e^{-y^2}dy$$
when $x = y^2$.

If we denote by C_1 the part of the circle $x^2 + y^2 = h^2$ in the first quadrant, by C_2 the part of the circle $x^2 + y^2 = 2h^2$ in this quadrant, and by S the square bounded by the coordinate axes and the lines $x = h$ and $y = h$, we have

$$\iint_{C_1} x^{2m-1}y^{2n-1}e^{-(x^2+y^2)}dxdy < \iint_{S} x^{2m-1}y^{2n-1}e^{-(x^2+y^2)}dxdy$$
$$< \iint_{C_2} x^{2m-1}y^{2n-1}e^{-(x^2+y^2)}dxdy, \quad (3)$$

since $x^{2m-1}y^{2n-1}e^{-(x^2+y^2)} \geqq 0$ in the first quadrant. But we know from § 102 that

$$\iint_{C_1} x^{2m-1}y^{2n-1}e^{-(x^2+y^2)}dxdy$$
$$= \int_0^h r^{2(m+n)-1}e^{-r^2}dr \cdot \int_0^{\pi/2} \sin^{2m-1}\varphi \cos^{2n-1}\varphi d\varphi,$$

$$\iint_{C_2} x^{2m-1}y^{2n-1}e^{-(x^2+y^2)}dxdy$$
$$= \int_0^{h\sqrt{2}} r^{2(m+n)-1}e^{-r^2}dr \cdot \int_0^{\pi/2} \sin^{2m-1}\varphi \cos^{2n-1}\varphi d\varphi,$$

$$\iint_{S} x^{2m-1}y^{2n-1}e^{-(x^2+y^2)}dxdy = \int_0^h x^{2m-1}e^{-x^2}dx \cdot \int_0^h y^{2n-1}e^{-y^2}dy.$$

Since the extreme members of (3) approach the same limit as $h \to \infty$, the middle member also approaches this limit. Hence

$$\int_0^\infty x^{2m-1}e^{-x^2}dx \cdot \int_0^\infty y^{2n-1}e^{-y^2}dy$$
$$= \int_0^{\pi/2} \sin^{2m-1}\varphi \cos^{2n-1}\varphi d\varphi \int_0^\infty r^{2(m+n)-1}e^{-r^2}dr;$$

or
$$\Gamma(m+n) \cdot B(m, n) = \Gamma(m) \cdot \Gamma(n),$$
$$B(m, n) = \frac{\Gamma(m) \cdot \Gamma(n)}{\Gamma(m+n)}. \quad (4)$$

In view of this connection between the two functions it is not necessary to consider both of them in detail.

92. Stirling's formula. By definition

$$\Gamma(x+1) = \int_0^\infty \alpha^x e^{-\alpha} d\alpha \quad (0 < x)$$
$$= x^{x+1} \int_0^\infty t^x e^{-xt} dt \quad (\alpha = xt)$$
$$= x^{x+1} e^{-x} \int_0^\infty e^{-x(t-1-\log t)} dt$$
$$= x^{x+1} e^{-x} \int_0^\infty e^{-xf(t)} dt,$$

where $f(t) = t - 1 - \log t$. Now the important fact here is that $e^{-xf(t)} = 1$ when $t = 1$ regardless of the value of x, and that it approaches zero for a fixed value of t different from 1 as $x \to \infty$. This can be seen as follows: $f'(t) = 1 - \frac{1}{t}$, and is accordingly negative when t is between zero and 1, and positive when t is greater than 1. Hence $f(t)$ is positive for every positive value of t, except 1. The curve $y = e^{-xf(t)}$ for a fixed x is accordingly as shown in the figure. For large positive values of x the curve is close to the t-axis, except in the neighborhood of $t = 1$. This suggests that for fairly small values of ϵ the integrals

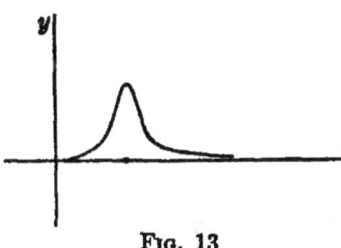

Fig. 13

$$\int_0^\infty e^{-xf(t)} dt \quad \text{and} \quad \int_{1-\epsilon}^{1+\epsilon} e^{-xf(t)} dt$$

will differ by a small amount and that the latter can be taken

§ 92]   IMPROPER AND INFINITE INTEGRALS   151

as an approximation of the former, especially for large values of x. We have now to look into the details in order to see if this surmise is correct, or not. Now

$$\int_0^\infty e^{-xf(t)}dt = \int_0^{1-\epsilon} + \int_{1-\epsilon}^{1+\epsilon} + \int_{1+\epsilon}^4 + \int_4^\infty.$$

We have here written the integrand only once. It is the same in all five of the integrals. For $\frac{1}{2} \leq t \leq 1$ we have

$$t - 1 - \log t = \int_t^1 \left(\frac{1}{u} - 1\right) du > \int_t^1 (1-u) du$$
$$= \frac{1}{2}(t-1)^2 \geq \frac{1}{8}(t-1)^2;$$

and for $1 \leq t \leq 4$

$$t - 1 - \log t = \int_1^t \left(1 - \frac{1}{u}\right) du > \frac{1}{4}\int_1^t (u-1) du$$
$$= \frac{1}{8}(t-1)^2.$$

In the interval $(0, 1-\epsilon)$, where $0 < \epsilon < \frac{1}{2}$, the least value of $t - 1 - \log t$ occurs when $t = 1 - \epsilon$, and the least value in the interval $(1+\epsilon, 4)$ occurs when $t = 1 + \epsilon$. The greatest value of the integrand $e^{-x(t-1-\log t)}$ in each of the intervals is, in view of the preceding inequalities, less than, or equal to, $e^{-x\epsilon^2/8}$. Hence

$$\int_0^{1-\epsilon} e^{-x(t-1-\log t)}dt + \int_{1+\epsilon}^4 e^{-x(t-1-\log t)}dt \leq 4e^{-x\epsilon^2/8}.$$

In the interval $t \geq 4$ we have

$$t - 1 - \log t \geq \frac{3}{4}t - \log t > \frac{t}{4}.$$

If then $x > 4$

$$\int_4^\infty e^{-x(t-1-\log t)}dt < \int_4^\infty e^{-xt/4}dt < e^{-x} < e^{-x\epsilon^2/8}.$$

Now $x^{-2/5} < \frac{1}{2}$ for sufficiently large values of x, and we can therefore put $\epsilon = x^{-2/5}$. Then

$$\int_{1-\epsilon}^{1+\epsilon} e^{-xf(t)}dt < e^x x^{-x-1}\Gamma(x+1) < \int_{1-\epsilon}^{1+\epsilon} e^{-xf(t)}dt + 5e^{-x\epsilon^3/8}.$$

But $e^{-x\epsilon^3/8} \to 0$ as $x \to \infty$. We have therefore

$$e^x x^{-x-1}\Gamma(x+1) = \int_{1-\epsilon}^{1+\epsilon} e^{-xf(t)}dt + \epsilon_1,$$

where $\epsilon_1 \to 0$ as $x \to \infty$.

If we define $\psi(t)$ by the equation

$$f(t) = \frac{(t-1)^2}{2} + (t-1)^3 \psi(t),$$

it is clear that it is bounded in the interval $(\tfrac{1}{2}, \tfrac{3}{2})$ for [1] t. There is then a number M such that for this interval $|\psi(t)| < M$ and

$$e^{-x(t-1)^2/2} \cdot e^{-Mx^{-1/5}} \leq e^{-xf(t)} \leq e^{-x(t-1)^2/2} \cdot e^{Mx^{-1/5}}$$

for $1 - \epsilon \leq t \leq 1 + \epsilon$. But $e^{-Mx^{-1/5}}$ and $e^{Mx^{-1/5}}$ both approach 1 as $x \to \infty$, and therefore for an arbitrary positive δ, $1 - \delta < e^{-Mx^{-1/5}}$ and $1 + \delta > e^{Mx^{-1/5}}$ for sufficiently large values of x. Then

$$(1-\delta)\int_{1-\epsilon}^{1+\epsilon} e^{-x(t-1)^2/2}dt < \int_{1-\epsilon}^{1+\epsilon} e^{-xf(t)}dt$$

$$< (1+\delta)\int_{1-\epsilon}^{1+\epsilon} e^{-x(t-1)^2/2}dt,$$

$$\int_{1-\epsilon}^{1+\epsilon} e^{-xf(t)}dt = (1+\epsilon_2)\int_{1-\epsilon}^{1+\epsilon} e^{-x(t-1)^2/2}dt,$$

where $\epsilon_2 \to 0$ as $x \to \infty$. If we put $\sqrt{\tfrac{x}{2}}(t-1) = u$,

$$\int_{1-\epsilon}^{1+\epsilon} e^{-xf(t)}dt = (1+\epsilon_2)\sqrt{\tfrac{2}{x}}\int_{-\epsilon\sqrt{x/2}}^{\epsilon\sqrt{x/2}} e^{-u^2}du.$$

---

[1] That this is so follows from the fact that $f(t) = \dfrac{(t-1)^2}{2} - \dfrac{(t-1)^3}{3(1-\theta\overline{t-1})^3}$, $0 < \theta < 1$, as may be seen by expanding log t in powers of $(t-1)$ and using Lagrange's form of the remainder.

It will be shown in § 214 that

$$\int_{-\infty}^{\infty} e^{-u^2} du = \sqrt{\pi}.$$

Hence

$$\int_{-\epsilon\sqrt{x/2}}^{\epsilon\sqrt{x/2}} e^{-u^2} du = \int_{-\infty}^{\infty} e^{-u^2} du + \epsilon_3 = \sqrt{\pi} + \epsilon_3,$$

where $\epsilon_3 \to 0$ as $x \to \infty$. We conclude that

$$\int_{1-\epsilon}^{1+\epsilon} e^{-xf(t)} dt = \sqrt{\frac{2\pi}{x}} (1 + \epsilon_2)\left(1 + \frac{\epsilon_3}{\sqrt{\pi}}\right).$$

Hence

$$\Gamma(x+1) = e^{-x} x^{x+1/2} \sqrt{2\pi}(1 + \eta), \tag{5}$$

where $\eta \to 0$ as $x \to \infty$. This relation is described by saying that $\Gamma(x+1)$ is represented asymptotically by the function $e^{-x} x^{x+1/2} \sqrt{2\pi}$, and in symbols thus,

$$\Gamma(x+1) \sim e^{-x} x^{x+1/2} \sqrt{2\pi}. \tag{6}$$

This formula is known as *Stirling's formula*.

93. Stirling's series. Stirling's formula can also be written as follows:

$$\log \Gamma(x+1) = \frac{1}{2} \log (2\pi) - x + \left(x + \frac{1}{2}\right) \log x$$

$$+ \frac{B_2}{1 \cdot 2} \cdot \frac{1}{x} - \frac{B_4}{3 \cdot 4} \frac{1}{x^3} + \cdots$$

$$+ \frac{(-1)^{n-1} B_{2n}}{(2n-1)2n} \frac{1}{x^{2n-1}} + \frac{(-1)^n \theta B_{2n+2}}{(2n+1)(2n+2)} \frac{1}{x^{2n+1}}, \tag{7}$$

where $0 < \theta < 1$, and

$$B_{2n} = \frac{(2n)!}{2^{2n-1} \pi^{2n}} \left(1 + \frac{1}{2^{2n}} + \frac{1}{3^{2n}} + \cdots + \frac{1}{m^{2n}} + \cdots\right),$$

for every positive integral value of n. The proof of this is too long and too complicated to be given here.[1]

If we let n increase indefinitely we get an infinite series in the right member of (7). The absolute value of the

---

[1] See, for example, Serret-Scheffers, *Lehrbuch der Differential- und Integralrechnung*, Vol. 2, 4th and 5th editions, p. 252.

general term of this series is

$$\frac{B_{2n}}{(2n-1)2n} \cdot \frac{1}{x^{2n-1}} = \frac{1}{2\pi x} \cdot \frac{2}{2\pi x} \cdot \ldots \cdot \frac{2n-2}{2\pi x}$$
$$\cdot \frac{1}{2\pi^2 x}\left(1 + \frac{1}{2^{2n}} + \cdots + \frac{1}{m^{2n}} + \cdots\right).$$

This does not approach zero as $n \to \infty$, since the quantity in the parentheses is greater than 1, and only a limited number of the other factors are, for a given value of x, less than 1. All the others exceed 1 and the number of these increases indefinitely with n. Hence the series is divergent for every positive value of x. And yet the series is useful for computation of the numerical value of $\log \Gamma(x+1)$, especially when x is large. The remainder after the term in $\frac{1}{x^{2n-1}}$ is

$$R_n = \frac{(-1)^n \theta \cdot B_{2n+2}}{(2n+1)(2n+2)} \cdot \frac{1}{x^{2n+1}}.$$

It can be shown that [1]

$$|R_n| < \frac{1}{6}\frac{\sqrt{n\pi}}{x}\left(\frac{n}{e\pi x}\right)^{2n} e^{1/24n}. \tag{8}$$

If in the computation the last term used is the one in $\frac{1}{x^{2n-1}}$, the error is in absolute value less than the right member of (8), and also less than the absolute value of the first term neglected. It is therefore desirable in the computation to include all the terms up to the one of least absolute value for the value of x we are using. If we put $u_n = \frac{B_{2n}}{(2n-1)2n} \cdot \frac{1}{x^{2n-1}}$, the absolute value of the ratio of two successive terms in (7) is $\frac{u_{n+1}}{u_n} < \frac{2n(2n-1)}{4\pi^2 x^2} < \frac{n^2}{\pi^2 x^2}$
$< \left(\frac{n}{3x}\right)^2$. If $n < 3x$ the term $\pm u_{n+1}$ is less in absolute value than $|u_n|$. This makes it desirable to proceed with

[1] See Serret-Scheffers, op. cit., p. 255.

the computation until $n = 3x$, if this is possible. And it will be possible if x is a positive integer. For $n = 3x$

$$|R_{3x}| < \frac{1}{2}\left(\frac{3}{\pi}\right)^{6x-1/2}\frac{e^{(1/72x)-6x}}{\sqrt{x}}.$$

If $x = 5$ we have

$$|R_{15}| < \frac{1}{2}\left(\frac{3}{\pi}\right)^{59/2}\frac{e^{(1/360)-30}}{\sqrt{5}} = 0.53829 \times 10^{-14}.$$

Thus we get a good approximation if we use all the terms up to and including the term in $\dfrac{1}{x^{29}}$.

This example gives us a good illustration of the way certain divergent series can be used for numerical computation by observing proper care in determining the number of terms to be used.

EXERCISES

1. Evaluate $B(\frac{1}{2}, \frac{1}{2})$.
2. Show that $\Gamma(\frac{1}{2}) = \sqrt{\pi}$. Use formula (4) with $m = n = \frac{1}{2}$.
3. From the result in Exercise 2 show that $\int_0^\infty e^{-z^2}dz = \dfrac{\sqrt{\pi}}{2}$.
4. Show that the Beta function satisfies the equation

$$B(m+1, n) = \frac{m}{m+n}B(m, n).$$

5. Apply this formula to show that if m and n are positive integers we have $B(m, n) = \dfrac{(m-1)!(n-1)!}{(m+n-1)!}$.

6. Show that $\int_0^\infty e^{-\alpha x}dx$ converges uniformly to $\dfrac{1}{\alpha}$ with respect to the interval $0 < \alpha_0 \leqq \alpha \leqq \alpha_1$.

7. Show that $\int_0^\infty \dfrac{e^{-ax} - e^{-bx}}{x}dx = \log \dfrac{b}{a}$ when $0 < a, 0 < b$. Make use of the result in Exercise 6.

Use Stirling's formula to compute the approximate value of each of the following:

8. $25!$ 9. $31!$ 10. $\Gamma(21)$.   11. $\Gamma(\sqrt{842})$.

In the following exercises use (7) and the fact that $B_2 = \dfrac{1}{6}$:

12. Determine two limits between which $\Gamma(185)$ lies.
13. Determine two limits between which $\Gamma(\sqrt{375})$ lies.

MISCELLANEOUS EXERCISES

1. Does $\int_0^\infty \dfrac{\sin x}{1+x}\, dx$ converge?

2. Show that $\int_0^1 \dfrac{dx}{\sqrt{1-x^{1/4}}} = \dfrac{128}{35}$. (Put $x = y^4$.)

3. Show that $\int_0^{\pi/2} \log \tan x\, dx = 0$.

4. Show that $\int_0^1 \log \sin \pi x\, dx = -\log 2$.

5. Given $u = \int_x^{x+1} \log \Gamma(t)\, dt$. Show that $u = x \log x - x + C$.

First find $\dfrac{du}{dx}$; then apply the formula $\Gamma(x+1) = x\Gamma(x)$ and integrate.

6. It was stated in §92 that $t - 1 - \log t \geqq \dfrac{3}{4} t - \log t > \dfrac{t}{4}$ for $t \geqq 4$. Prove this.

7. The Gamma function satisfies the functional equation $f(x+1) = xf(x)$. Show that in order that a function satisfy this equation it is necessary and sufficient that it be of the form $\Phi(x)\Gamma(x)$, where $\Phi(x)$ is of period 1.

8. Show that $\log \Gamma(x)$ satisfies the equation $f(x+1) - f(x) = \log x$.

9. Prove that if $\int_0^\infty f(x)\, dx$ converges the function $\int_0^x f(x)\, dx$ has no more changes of sign in $0 < x < \infty$ than $f(x)$.

10. Show by direct integration that $\int_0^\infty \dfrac{dx}{(x^2 + a^2)^2} = \dfrac{\pi}{4a^3}$. Then, starting from the formula $\int_0^\infty \dfrac{dx}{x^2 + a^2} = \dfrac{\pi}{2a}$, differentiate each side with respect to a and compare the two results.

11. Show that $I = \int_0^\infty e^{-x^2 - (a^2/x^2)}\, dx = \dfrac{\sqrt{\pi}}{2} e^{-2a}$. (Find the derivative of I with respect to a by differentiating under the integral sign, and in the resulting integral put $\dfrac{a}{x} = t$.)

CHAPTER VIII

DOUBLE AND TRIPLE INTEGRALS

94. The double integral. We consider a region A that is bounded by a closed curve and a function $f(x, y)$ that is one-valued and bounded in A. We subdivide A into partial regions, $\alpha_1, \alpha_2, \cdots, \alpha_n$, whose areas are respectively $\omega_1, \omega_2, \cdots, \omega_n$; and form the sums

$$S = \sum M_i \omega_i \quad \text{and} \quad s = \sum m_i \omega_i,$$

where M_i and m_i are the upper and lower bounds respectively of $f(x, y)$ in the region α_i. Now $S \geq m \sum \omega_i = m\Omega$ and $s \leq M \sum \omega_i = M\Omega$, where M is the upper bound and m is the lower bound of $f(x, y)$ in A and Ω is the area of A. But Ω, M, and m are independent of the particular subdivision of A that was used. Hence no S is less than $m\Omega$ and no s is greater than $M\Omega$. We shall refer to S as an *upper sum* and to s as a *lower sum*.

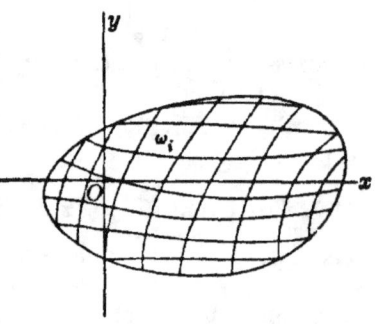

Fig. 14

We can then say that the set of upper sums has a lower limit I and that the set of lower sums has an upper limit I' (§ 3).

INTEGRABLE FUNCTIONS. The bounded function $f(x, y)$ is said to be *integrable* over the finite closed region A if the limits I and I' are equal. The common value of I and I' in the case of an integrable function is represented by the symbol $\iint_A f(x, y) dx dy$ or the symbol $\int_A f(x, y) dA$ and is called the *double integral* of $f(x, y)$ over the region A. The function $f(x, y)$ is called the *integrand* and A the *field* of integration. The word "double" in this designation has reference to the two dimensional set of points that make up the region A.

We have seen (Chap. III, Theorem 1) that if $f(x, y)$ is continuous in A its oscillation in any partial region of A is less than an arbitrary positive number ϵ_1 provided that the diameters of all the partial regions are less than a certain number η. For such a subdivision

$$S - s = \sum(M_i - m_i)\omega_i < \epsilon_1 \sum \omega_i = \epsilon_1 \Omega = \epsilon.$$

But

$$S - s = (S - I) + (I - I') + (I' - s) < \epsilon.$$

We can see as in § 59 that no upper sum is less than any lower sum. Hence $I \geq I'$ and all the quantities in the parentheses on the right are zero or positive. Moreover I and I' are independent of the method of subdivision. Hence $I = I'$ and $S - I < \epsilon$ and $I' - s < \epsilon$.

This means that if a function is continuous in a closed region A it is integrable in A and the upper and lower sums approach the value of the integral as their common limit when the maximum diameter of all the partial regions approaches zero. Of course a function may be integrable even if it is discontinuous. However, a satisfactory discussion of this question is beyond the scope of this book. From now on we shall assume that the integrand is continuous unless it is stated otherwise.

95. Simple properties of the double integral.

(a) Consider the sum $\sum f(x_i, y_i)\omega_i$, where (x_i, y_i) is any point in the partial region α_i. Then

$$S \geq \sum f(x_i, y_i)\omega_i \geq s.$$

Since S and s approach a common limit, $\sum f(x_i, y_i)\omega_i$ approaches the same limit. We see then that in forming the sums whose limit is the definite integral we can multiply the area of each partial region by the value of the integrand at any point inside, or on the boundary, of this region.

We shall take advantage of this fact in the discussion in §§ 96–98.

The reader ought to find no difficulty in supplying proofs of the following statements:

(b) $\iint_A Cf(x, y)\,dxdy = C\iint_A f(x, y)\,dxdy$, if C is a constant.

(c) $\iint_A [f(x, y) \pm \varphi(x, y)]\,dxdy$
$$= \iint_A f(x, y)\,dxdy \pm \iint_A \varphi(x, y)\,dxdy.$$

(d) $\iint_A dxdy = \Omega$.

(e) $\iint_A |f(x, y)|\,dxdy \geqq \left|\iint_A f(x, y)\,dxdy\right|.$

(f) $\iint_A f(x, y)\,dxdy = \iint_{A_1} f(x, y)\,dxdy$
$$+ \iint_{A_2} f(x, y)\,dxdy + \cdots,$$

where $A_1, A_2, \cdots$ are the parts of A.

96. Mean value theorem. Suppose that $f(x, y)$ and $\varphi(x, y)$ are continuous in A and that $\varphi(x, y)$ is positive or zero in this region. If M and m are the maximum and minimum values respectively of $f(x, y)$ in A, we have

$$M\varphi(\xi_i, \eta_i)\omega_i \geqq f(\xi_i, \eta_i)\varphi(\xi_i, \eta_i)\omega_i \geqq m\varphi(\xi_i, \eta_i)\omega_i,$$

where (ξ_i, η_i) is any point in α_i. When we take the sum and pass to the limit we find that

$$M\iint_A \varphi(x, y)\,dxdy \geqq \iint_A f(x, y)\varphi(x, y)\,dxdy$$
$$\geqq m\iint_A \varphi(x, y)\,dxdy.$$

Hence

$$\iint_A f(x, y)\varphi(x, y)dxdy = \mu \iint_A \varphi(x, y)dxdy,$$

where $M \geq \mu \geq m$. But since $f(x, y)$ is continuous, this is equivalent to

$$\iint_A f(x, y)\varphi(x, y)dxdy = f(\xi, \eta) \iint_A \varphi(x, y)dxdy, \quad (1)$$

where (ξ, η) is some point in A (see § 31). The same conclusion can be reached on the supposition that $\varphi(x, y)$ is negative or zero everywhere in A.

If $\varphi(x, y) \equiv 1$, formula (1) becomes

$$\iint_A f(x, y)dxdy = \Omega f(\xi, \eta), \quad (2)$$

since $\iint_A dxdy = \Omega$. Equation (2) is a statement of the

mean value theorem for double integrals.

97. Repeated integrals. By the definition of a double integral the value of such an integral is the result of a single limiting process. This is also true of the value of a simple integral. But whereas we have a means of evaluating the latter through the use of the indefinite integral, there is no corresponding direct way to evaluate a double integral. In order to overcome this difficulty we resort to the expedient of repeated integrals in a way we shall now explain.

We assume first that the field of integration is a rectangle whose sides are $x = a$, $x = b$, $y = c$, $y = d$, and we subdivide it into partial regions by the lines $x = x_1$, $x = x_2$, $\cdots$, $x = x_{m-1}$; and $y = y_1$, $y = y_2$, $\cdots$, $y = y_{n-1}$. The double integral is the limit of the sum

$$S = \sum_{i=1}^{m} \sum_{j=1}^{n} f(\xi_{ij}, \eta_{ij})(x_i - x_{i-1})(y_j - y_{j-1}), \quad (3)$$

§97] DOUBLE AND TRIPLE INTEGRALS

where (ξ_{ij}, η_{ij}) is any point within or upon the boundary of the rectangle bounded by the lines $x = x_{i-1}$, $x = x_i$, $y = y_{j-1}$, and $y = y_j$, where $x_0 = a$, $x_m = b$, $y_0 = c$, and $y_n = d$. The elements of this sum that have the factor $x_i - x_{i-1}$ make a contribution to S that is equal to

$$S_i = (x_i - x_{i-1}) \sum_{j=1}^{n} f(\xi_{ij}, \eta_{ij})(y_j - y_{j-1}).$$

Now $f(x_i, y)$ is a continuous function of y in the interval (c, d), and therefore

$$\int_c^d f(x_i, y) dy = \int_c^{y_1} f(x_i, y) dy + \int_{y_1}^{y_2} f(x_i, y) dy$$
$$+ \cdots + \int_{y_{n-1}}^d f(x_i, y) dy$$
$$= f(x_i, \eta_1)(y_1 - c) + f(x_i, \eta_2)(y_2 - y_1)$$
$$+ \cdots + f(x_i, \eta_n)(d - y_{n-1}) = F(x_i) \quad (4)$$

where $y_{j-1} \leq \eta_j \leq y_j$. If for every j we take $\xi_{ij} = x_i$, and for η_{ij} the number η_j in (4), we have

$$S_i = (x_i - x_{i-1}) F(x_i) \quad (5)$$

and

$$S = \sum_{i=1}^{n} F(x_i)(x_i - x_{i-1}).$$

If we pass to the limit by letting the greatest diagonal of the partial regions approach zero, we find that

$$\iint_A f(x, y) dx dy = \int_a^b \left[\int_c^d f(x, y) dy \right] dx. \quad (6)$$

It should be noted that we did not first let all intervals $y_j - y_{j-1}$ approach zero and then the intervals $x_i - x_{i-1}$; but we let them all approach zero simultaneously. We could do this by virtue of the fact that equation (4) holds for a finite subdivision of the interval (c, d) on the y-axis. We have shown that we can evaluate the double integral by first integrating the integrand with respect to y between the given limits, holding x constant; and then integrating the result, which is a function of x, with respect to x between the given limits for x. Of course we could just as well have performed these integrations in the opposite order.

98. The preceding argument is based on the assumption that the field A of integration is a rectangle with sides parallel to the coordinate axes. We now suppose that the field is bounded by the lines $x = a$, $x = b$ and the curves $Y_1 = \varphi_1(x)$ and $Y_2 = \varphi_2(x)$, where $\varphi_1(x)$ and $\varphi_2(x)$ are continuous in the interval (a, b) and $\varphi_1(x) \leqq \varphi_2(x)$. As before, we shall divide the field of integration by lines parallel to the coordinate axes. In the sum $S = \Sigma f(\xi_i, \eta_i)\omega_i$, whose limit we wish to determine, some of the partial regions are rectangles and some may be more or less irregular figures lying along the boundary of the region.

We first fix our attention on those partial regions that lie between the lines $x = x_{i-1}$ and $x = x_i$. Let $y = y'$ be the lower boundary of the lowest complete rectangle in this strip and let $y = y''$ be the upper boundary of the highest complete rectangle. Then

$$(x_i - x_{i-1}) \int_{y'}^{y''} f(\xi_i, y) dy$$

is the contribution to S made by the complete rectangles in this strip, provided that in the formation of S we take the ξ_i to be the same in the co-factors of the areas of all these rectangles and select the η_i suitably (see § 97).

Since the functions $\varphi_1(x)$ and $\varphi_2(x)$ are continuous in the interval (a, b) we can take $x_i - x_{i-1}$ sufficiently small to insure that the oscillation of each one in the interval $(x_i - x_{i-1})$ is less than a pre-assigned positive number δ. We can also make all the differences $y_j - y_{j-1}$ less than δ. The total area of the irregular partial regions within the strip in question is less than $4\delta(x_i - x_{i-1})$, and the contribution to S of these partial regions is less in absolute value than $4M\delta(x_i - x_{i-1})$, where M is the maximum absolute value of $f(x, y)$ in the whole region of integration.

§98] DOUBLE AND TRIPLE INTEGRALS

Moreover

$$\int_{y'}^{y''} f(\xi_i, y)dy = \int_{Y_1}^{Y_2} f(\xi_i, y)dy + \int_{y'}^{Y_1} f(\xi_i, y)dy + \int_{Y_2}^{y''} f(\xi_i, y)dy,$$

where $Y_1 = \varphi_1(\xi_i)$ and $Y_2 = \varphi_2(\xi_i)$. But $|Y_1 - y'| < 2\delta$ and $|Y_2 - y''| < 2\delta$. Hence

$$\int_{y'}^{y''} f(\xi_i, y)dy = \int_{Y_1}^{Y_2} f(\xi_i, y)dy + 4M\theta_i\delta,$$

where $|\theta_i| < 1$. The contribution to S made by all the partial regions in the strip is therefore

$$(x_i - x_{i-1})\left[\int_{y'}^{y''} f(\xi_i, y)dy + 4M\theta_i'\delta\right]$$
$$= (x_i - x_{i-1})\left[\int_{Y_1}^{Y_2} f(\xi_i, y)dy + 8M\theta_i''\delta\right],$$

where $|\theta_i'| < 1$ and $|\theta_i''| < 1$. Hence

$$S = \sum(x_i - x_{i-1})\left[\int_{Y_1}^{Y_2} f(\xi_i, y)dy + 8M\theta_i''\delta\right].$$

But

$$|8M\delta\sum(x_i - x_{i-1})\theta_i''| < 8M\delta(b - a).$$

This last expression approaches zero as $\delta \to 0$. Hence

$$\iint f(x, y)dxdy = \lim \sum(x_i - x_{i-1})\int_{Y_1}^{Y_2} f(\xi_i, y)dy.$$

If we put

$$\int_{Y_1}^{Y_2} f(x, y)dy = \Phi(x),$$

we can say that

$$\iint_A f(x, y)dxdy = \lim \sum \phi(\xi_i)(x_i - x_{i-1}).$$

That is,

$$\iint_A f(x, y)dxdy = \int_a^b \Phi(x)dx$$
$$= \int_a^b \left[\int_{Y_1}^{Y_2} f(x, y)dy \right] dx. \quad (7)$$

A similar argument shows that also

$$\iint_A f(x, y)dxdy = \int_c^d \left[\int_{X_1}^{X_2} f(x, y)dx \right] dy, \quad (8)$$

where the region A is bounded by the curves $X_1 = \psi_1(y)$ and $X_2 = \psi_2(y)$, and the lines $y = c$ and $y = d$, it being understood that $\psi_1(y)$ and $\psi_2(y)$ are continuous in the interval (c, d), with $\psi_1(y) \leq \psi_2(y)$.

From these results we see that a double integral can be replaced by a properly chosen repeated integral, the integration being with respect to y first and then with respect to x, or vice versa.

We have assumed for simplicity that the integrand is continuous in A, and our conclusion is that in this case both the repeated integrals and the double integral exist and have the same value. This conclusion is still valid if we assume merely that the integrand is integrable in A, but we cannot give the proof here.[1] It is worth noting that if the integrand is discontinuous, one or both of the repeated integrals may exist and the double integral not exist. The following example illustrating this point is due to Thomae.[2] Let the region of integration be the square bounded by the lines $x = 0$, $x = 1$, $y = 0$, $y = 1$; and let $f(x, y) = 1$ for all rational values of x and $f(x, y) = 2y$ for all irrational values of x. Then

$$\int_0^1 \left[\int_0^1 f(x, y)dy \right] dx = 1.$$

[1] See P. duBois Reymond, *Crelle's Journal*, Vol. XCIV, 1883, p. 277. Also Hobson, *Theory of Functions of a Real Variable*, 3d ed., Vol. 1, p. 510.
[2] *Schlömilch's Zeitschrift*, Vol. XXIII, 1878, p. 67.

On the other hand, the double integral does not exist. To see this let I_1 and I_1' be the values of I and I' respectively for the upper quarter of A; and I_2 and I_2' these values for the lower three-quarters. Then $I = I_1 + I_2$ and $I' = I_1' + I_2'$. Now $I_1 > \frac{3}{8}$ and $I_1' = \frac{1}{4}$, and $I_2 \geqq I_2'$. Hence $I > I'$ and the double integral $\iint_A f(x, y)dxdy$ does not exist. Neither does the simple integral $\int_0^1 f(x, y)dx$ for any value of y, except $y = \frac{1}{2}$. If the double integral does not exist, the repeated integrals may exist and be equal.

EXERCISES

1. Verify that $\iint (x^2 + y^2)dxdy = \iint (x^2 + y^2)dydx$ in case the field of integration is the triangle bounded by the lines $x = 0$, $x = y$, and $y = 2$.

2. Show that these two integrals are the same when the field of integration is the circle $x^2 + y^2 = 16$.

Evaluate:

3. $\iint x^2 y\,dxdy$ over the region for which $\dfrac{x^2}{9} + \dfrac{y^2}{4} \leqq 1$.

4. $\iint \dfrac{dxdy}{\sqrt{x^2 + y^2}}$ over the triangle bounded by the lines $y = x$, $y = 0$, and $x = 1$.

When limits are given, it is to be understood that the first set of limits belong to the variable whose differential is written first.

5. $\displaystyle\int_0^\pi \int_0^x \sin(x - y)dxdy$. 6. $\displaystyle\int_0^a \int_0^a (x^3 - y^3)dxdy$.

7. $\displaystyle\int_0^1 \int_0^2 ye^{xy}dxdy$.

8. Verify that $\iint (12 + x - y^2)dxdy = \iint (12 + x - y^2)dydx$ in case the field of integration is the second quadrant of the circle $x^2 + y^2 = 3$.

9. Find the value of the double integral $\iint \dfrac{dxdy}{\sqrt{1 - x^2 - y^2}}$ over the first quadrant of the circle $x^2 + y^2 = 1$. What is the value of this integral when taken over the whole circle?

What is the field of integration determined by the limits in each of the following cases?

10. $\displaystyle\int_0^{\sqrt{2}/2}\int_y^{\sqrt{1-y^2}} f(x,y)\,dy\,dx.$
11. $\displaystyle\int_0^a \int_0^{(b/a)\sqrt{a^2-x^2}} f(x,y)\,dx\,dy.$

12. $\displaystyle\int_0^{2a}\int_{-\sqrt{2ax-x^2}}^{\sqrt{2ax-x^2}} f(x,y)\,dx\,dy.$
13. $\displaystyle\int_0^{4\sqrt{2}/13}\int_{\sqrt{1+y^2}}^{(3/2)\sqrt{4-y^2}} f(x,y)\,dy\,dx.$

14. $\displaystyle\int_0^1 \int_{x^2}^{\sqrt{x}} f(x,y)\,dx\,dy.$
15. $\displaystyle\int_0^{(1+\sqrt{5})/2}\int_{y^2}^{1+y^2} f(x,y)\,dy\,dx.$

16. Integrate $(x+y)^2$ over the circle $x^2 + y^2 = a^2$.

99. Line integrals and Green's theorem. We say that the function $F(x, y)$ is continuous along the arc AB of a curve if to every point (x, y) of the arc and every positive number ϵ there is associated a positive number η such that

$$|F(x, y) - F(x_1, y_1)| < \epsilon$$

whenever $|x - x_1| < \eta$ and $|y - y_1| < \eta$ and (x_1, y_1) is on the arc.

Consider now a function $F(x, y)$ which is continuous along the arc AB of the curve $y = \varphi(x)$, where $A = (a, c)$ and $B = (b, d)$ and $\varphi(x)$ is continuous in the interval (a, b). If the arc be divided into partial arcs by the points (x_1, y_1), $(x_2, y_2), \cdots (x_{n-1}, y_{n-1})$, the sum

$$S = \Sigma F(\xi_i, \eta_i)(x_i - x_{i-1}),$$

where (ξ_i, η_i) is a point on the arc connecting (x_{i-1}, y_{i-1}) and (x_i, y_i), approaches a definite limit as the maximum length of the chords of these arcs approaches zero. For

$$S = \Sigma F[\xi_i, \varphi(\xi_i)](x_i - x_{i-1}),$$

and this has for limit the definite integral $\displaystyle\int_a^b F[x, \varphi(x)]\,dx$, since $F[x, \varphi(x)]$ is a continuous function of x in the interval (a, b). This definite integral is called a *line integral of* $F(x, y)$ *over the arc* AB, and is written thus: $\displaystyle\int_{AB} F(x, y)\,dx$. The limit of the sum $F(\xi_i, \eta_i)(y_i - y_{i-1})$ is the line integral

$\int_{AB} F(x, y)dy$. It is clear from the definition of a line integral that

$$\int_{AB} F(x, y)dx = - \int_{BA} F(x, y)dx.$$

Suppose that we have a region A whose boundary is made up of the lines $x = a$, $x = b$, $y = Y_1(x)$, and $y = Y_2(x)$, where $Y_2(x) \geqq Y_1(x)$ in the interval (a, b). We consider the double integral $\iint_A \frac{\partial P}{\partial y} dxdy$, where $P(x, y)$ and $\frac{\partial P(x, y)}{\partial y}$ are continuous within the region A and on the boundary. It is shown in § 98 that

$$\iint_A \frac{\partial P}{\partial y} dxdy = \int_a^b \left[\int_{Y_1}^{Y_2} \frac{\partial P}{\partial y} dy \right] dx$$
$$= \int_a^b [P(x, Y_2) - P(x, Y_1)]dx.$$

Now $\int_a^b P(x, Y_2)dx$ is the line integral of $P(x, y)$ along the curve $y = Y_2(x)$ from A to B, or minus the line integral of the same function along the same arc from B to A, while $\int_a^b P(x, Y_1)dx$ is the line integral of the same function along the arc $y = Y_1(x)$ from A to B.

DEFINITION. We agree to say that the boundary C is described in the *positive* sense by a man walking along C with the bounded area on his left.

With this understanding we have

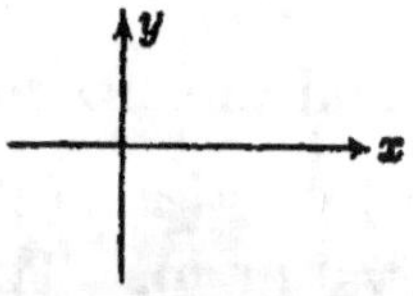

$$\iint_A \frac{\partial P}{\partial y} dxdy = - \int_C P(x, y)dx,$$

FIG. 15

the line integral being taken in the positive sense.[1] Simi-

[1] This assumes that the axes are oriented as shown in the figure.

larly we have

$$\iint_A \frac{\partial Q}{\partial x}dxdy = \int_C Q(x, y)dy,$$

where $Q(x, y)$ and $\dfrac{\partial Q(x, y)}{\partial x}$ are continuous within and on the boundary of A. Hence

$$\iint_A \left(\frac{\partial Q}{\partial x} - \frac{\partial P}{\partial y}\right)dxdy = \int_C Pdx + Qdy. \qquad (9)$$

We have assumed that the boundary C is cut by a line parallel to either axis in not more than two points. We can extend our results to regions not of this type but such that they can be divided up into a finite number of parts each of this type. Consider, for example, the region shown in the figure. Certain lines parallel to the y-axis cut the boundary C in more than two points. But if we draw the line LM we divide A into two regions, A_1 and A_2, neither of whose boundaries C_1 and C_2 is cut by a line parallel to either axis in more than two points. Hence

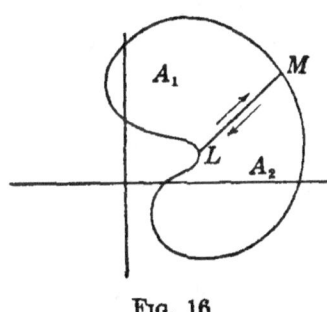

Fig. 16

$$\iint_{A_1} \frac{\partial P}{\partial y}dxdy = -\int_{C_1} Pdx$$

and

$$\iint_{A_2} \frac{\partial P}{\partial y}dxdy = -\int_{C_2} Pdx.$$

If we add the corresponding members of these two equations we get

$$\iint_A \frac{\partial P}{\partial y}dxdy = -\int_{C_1} Pdx - \int_{C_2} Pdx.$$

In the first of the two line integrals on the right we integrate over the line LM from L to M, while in the second one we integrate over the same line from M to L. But

$$\int_{LM} P\,dx = -\int_{ML} P\,dx.$$

Hence
$$\iint_A \frac{\partial P}{\partial y}\,dx\,dy = -\int_C P\,dx.$$

If there were any limited number of these partial regions, we could proceed in a similar way. The same considerations apply whether we are dealing with line integrals with respect to x, or line integrals with respect to y, or both.

The statement that the two members of (9) are equal is called *Green's theorem*.

100. The integral $\int P\,dx + Q\,dy$. We now make use of Green's theorem to prove some important theorems concerning the integral $\int_C P\,dx + Q\,dy$.

THEOREM 1. *If the four functions P, Q, $\dfrac{\partial P}{\partial y}$, $\dfrac{\partial Q}{\partial x}$ are continuous within and upon the boundary C of A, a necessary and sufficient condition that the integral $\int P\,dx + Q\,dy$ should vanish when taken over the boundary of every closed region in A is that*

$$\frac{\partial P}{\partial y} = \frac{\partial Q}{\partial x}$$

for every point of A.

That the condition is sufficient follows immediately from Green's theorem. That it is necessary can be seen as follows: If there were an interior point B of A at which $\dfrac{\partial Q}{\partial x} - \dfrac{\partial P}{\partial y} > 0$, this inequality would hold throughout the interior of a sufficiently small circle Σ with center at B that lies wholly in A. Then

$$\int_C P dx + Q dy = \iint_\Sigma \left(\frac{\partial Q}{\partial x} - \frac{\partial P}{\partial y} \right) dx dy > 0,$$

where C is the circumference of Σ. But this contradicts the hypothesis. A similar contradiction would present itself if we assumed that $\frac{\partial Q}{\partial x} - \frac{\partial P}{\partial y} < 0$ at an interior point of A. If there were a point on the boundary at which $\frac{\partial P}{\partial y} - \frac{\partial Q}{\partial x} \neq 0$, there would be an interior point at which this inequality holds.

THEOREM 2. *If P and Q satisfy the conditions of Theorem 1 and if (a, b) and (x, y) are any two points of A, the integral*

$$\int_{(a,b)}^{(x,y)} P dx + Q dy$$

has the same value when taken over any two paths from (a, b) to (x, y) that lie wholly within A and do not enclose any boundary point of A. If $F(x, y)$ is the common value of these integrals from (a, b) to (x, y), $\frac{\partial F}{\partial x} = P$ and $\frac{\partial F}{\partial y} = Q$.

Consider two such paths C_1 and C_2 between the given points. If they have no common points other than these two, we know from Theorem 1 that

$$\int_{C_1} P dx + Q dy - \int_{C_2} P dx + Q dy = 0,$$

or

$$\int_{C_1} P dx + Q dy = \int_{C_2} P dx + Q dy.$$

If the two paths have common points between the two given ones, we can avoid complications by drawing a third path C_3 between these points that lies wholly in A and has no other points in common with C_1 or C_2. Moreover C_3 can be so drawn that there shall be no boundary points between it and C_1 and none between it and C_2. Then the integral over C_3 is equal to that over either of the other curves, and therefore these latter are equal to each other.

This shows that $F(x, y)$ is a single-valued function throughout the interior of A under the conditions described. We have now to consider whether it has partial derivatives with respect to x and y. We observe in the first place that

$$F(x_0 + \Delta x, y_0) - F(x_0, y_0)$$
$$= \int_{(a,b)}^{(x_0+\Delta x,\, y_0)} Pdx + Qdy - \int_{(a,b)}^{(x_0,\, y_0)} Pdx + Qdy$$

where (x_0, y_0) is any interior point of A. In the first line integral in the right member of this equation we can take the path of integration to be the same as the path of integration in the second integral as far as the point (x_0, y_0); and from here to $(x_0 + \Delta x, y_0)$ we can take it parallel to the x-axis provided that Δx is sufficiently small to keep this latter path wholly in A. This is always possible inasmuch as (x_0, y_0) is by hypothesis an interior point of A. Then

$$F(x_0 + \Delta x, y_0) - F(x_0, y_0)$$
$$= \int_{(x_0,\, y_0)}^{(x_0+\Delta x,\, y_0)} P(x, y_0)dx + Q(x, y_0)dy.$$

But the line integral $\int_{(x_0,\, y_0)}^{(x_0+\Delta x,\, y_0)} Q(x, y_0)dy$ along a line parallel to the x-axis is zero. Hence

$$F(x_0 + \Delta x, y_0) - F(x_0, y_0)$$
$$= \int_{(x_0,\, y_0)}^{(x_0+\Delta x,\, y_0)} P(x, y_0)dx = \Delta x P(\xi, y_0),$$

where ξ lies between x_0 and $x_0 + \Delta x$. Hence

$$\frac{\partial F(x_0, y_0)}{\partial x} = \lim_{\Delta x \to 0} \frac{F(x_0 + \Delta x, y_0) - F(x_0, y_0)}{\Delta x} = P(x_0, y_0).$$

We see in a similar way that

$$\frac{\partial F(x_0, y_0)}{\partial y} = Q(x_0, y_0).$$

If (x_0, y_0) and (x, y) are any two points in A, we have

$$\int_{(a,b)}^{(x,y)} Pdx + Qdy = \int_{(a,b)}^{(x_0,y_0)} + \int_{(x_0,y_0)}^{(x,y)};$$

and therefore

$$\int_{(x_0,y_0)}^{(x,y)} Pdx + Qdy = F(x,y) - F(x_0, y_0).$$

EXERCISES

1. Find the value of the integral $\int_{(0,0)}^{(3,2)} (x^2 + y^2)dx + 2xydy$ when taken along the straight line from $(0, 0)$ to $(3, 2)$.

2. Along the x-axis from the origin to the point $(3, 0)$ and then along the line $x = 3$ to the point $(3, 2)$.

3. Along the parabola $3y^2 = 4x$.

4. Compute $\int_{(0,0)}^{(3,2)} (x^2 + y^2)dx + (x^2 - y^2)dy$ along the same paths as in the preceding exercises.

5. Compute $\int_{(0,0)}^{(2,1)} (3y^2 - x^3)dx + 3y(2x + y)dy$.

6. Compute $\int_{(1,2)}^{(x,y)} (3y^2 - x^3)dx + 3y(2x + y)dy$.

7. What is the function determined by the integral

$$\int_{(0,1)}^{(x,y)} \frac{1+y^2}{y^2} xdx - \frac{1+x^2}{y^3} dy$$

in case the path of integration has no point in common with the x-axis?

8. Differentiate the function of Exercise 7 partially with respect to x and partially with respect to y.

9. Integrate $\dfrac{ydx - xdy}{x^2 + y^2}$ in the positive direction over the circumference of a circle whose center is at the origin.

10. Over the boundary of a circular ring whose center is at the origin.

11. Over the circumference of the circle $(x - 2)^2 + y^2 = 1$.

12. Over the boundary of the ellipse $\dfrac{x^2}{4} + y^2 = 1$.

101. Simply and multiply connected regions. If the region A is the region bounded by two concentric circles

with center at the origin and radii r_1 and r_2 ($r_1 < r_2$), we can connect any two points of A by two curves C_1 and C_2 that enclose boundary points of A.
For two such curves it may be that

$$\int_{C_1} P dx + Q dy \neq \int_{C_2} P dx + Q dy.$$

For example, suppose that $P = \dfrac{y}{x^2 + y^2}$ and $Q = \dfrac{-x}{x^2 + y^2}$. The conditions of the theorem are satisfied at every point except the origin. Take (a, b) to be the point $(-\rho, 0)$ and (x, y) the point $(\rho, 0)$, where $r_1 < \rho < r_2$. If C_1 is the lower half of the circumference $x^2 + y^2 = \rho^2$ and C_2 is the upper half, we have

$$\int_{C_1} P dx + Q dy = -\int_{-\pi}^{0} d\theta = -\pi$$

and

$$\int_{C_2} P dx + Q dy = -\int_{\pi}^{0} d\theta = \pi.$$

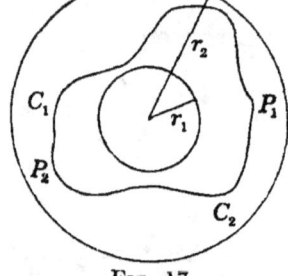

Fig. 17

The discrepancy between this result and the statement of the theorem is due to the fact that the proof of the theorem rested on Green's theorem which was applicable only because the two paths in question formed the complete boundary of the region enclosed by them, whereas in this example the two paths enclose points not in the region A.

SIMPLY CONNECTED REGIONS. In order to have a convenient terminology for such cases we agree to say that a region of the plane is *simply connected* if every closed path in the region encloses only points of the region. The interior of a circle, for example, is a simply connected region, while the circular ring described in the example is not. If we draw a line from the inner boundary of this ring to the outer boundary and make this line part of the boundary, the resulting region will be simply connected. We say that the original region is *doubly connected*. If we can reduce the original region to a simply connected region by inserting two additional boundaries, we say that the original

Fig. 18

region is *triply connected*. The extension of this idea to *n-tuply connected* regions is obvious. These new boundaries which we have introduced are merely lines drawn from one point of the original boundary to another with the understanding that they shall not be crossed in passing from one part of the region to another. They are called "cross cuts" or "barriers." The shaded part of the figure is a triply connected region.

102. Change of variables in a double integral. Suppose we have a one-to-one correspondence between the points of the bounded closed region A of the (x, y)-plane and the points of the bounded closed region A' of the (u, v)-plane defined by the equations

$$x = \varphi(u, v), \qquad y = \psi(u, v),$$

where $\varphi(u, v)$ and $\psi(u, v)$ together with their first partial derivatives are continuous in A'. When the point (u, v) describes the boundary C' of A' in the positive sense, the corresponding point (x, y) will describe the boundary C of A in either the positive or the negative sense. If it reversed its direction at any time, the correspondence would not be one to one.

Now the area Ω of A is given by the line integral

$$\Omega = \int_C x\,dy = \lim \sum x_i(y_i - y_{i-1}).$$

But

$$\begin{aligned}
y_i - y_{i-1} &= \psi(u_i, v_i) - \psi(u_{i-1}, v_{i-1}) \\
&= \psi(u_i, v_i) - \psi(u_{i-1}, v_i) + \psi(u_{i-1}, v_i) \\
&\quad - \psi(u_{i-1}, v_{i-1}) \\
&= \frac{\partial \psi(\xi_i, v_i)}{\partial u}(u_i - u_{i-1}) + \frac{\partial \psi(u_{i-1}, \eta_i)}{\partial v}(v_i - v_{i-1}),
\end{aligned}$$

where ξ_i lies between u_i and u_{i-1}, and η_i lies between v_i and v_{i-1}. Hence

$$\Omega = \lim \sum x_i \frac{\partial \psi(\xi_i, v_i)}{\partial u}(u_i - u_{i-1})$$
$$+ \lim \sum x_i \frac{\partial \psi(u_{i-1}, \eta_i)}{\partial v}(v_i - v_{i-1}),$$

where in the first summation we can take $x_i = \varphi(\xi_i, v_i)$ and in the second $x_i = \varphi(u_{i-1}, \eta_i)$. This is equivalent to the equation

$$\Omega = \pm \int_{C'} \varphi(u, v) \frac{\partial \psi}{\partial u} du + \varphi(u, v) \frac{\partial \psi}{\partial v} dv.$$

We can apply Green's theorem to this line integral by putting $P(u, v) = \varphi \frac{\partial \psi}{\partial u}$ and $Q(u, v) = \varphi \frac{\partial \psi}{\partial v}$. Then

$$\Omega = \pm \iint_A \Delta \, du \, dv,$$

where $\Delta = \begin{vmatrix} \frac{\partial \varphi}{\partial u} & \frac{\partial \varphi}{\partial v} \\ \frac{\partial \psi}{\partial u} & \frac{\partial \psi}{\partial v} \end{vmatrix}$. By the law of the mean

$$\Omega = \pm \Delta(\xi, \eta) \Omega',$$

the point (ξ, η) being somewhere in A'. Since Ω and Ω' are both positive, we have

$$\Omega = |\Delta(\xi, \eta)| \Omega'.$$

It is clear from this discussion that Δ is positive or negative according as the correspondence is direct or inverse—that is, according as the points (x, y) and (u, v) describe their respective paths in the same sense or in opposite senses.

To any subdivision of A' into partial regions α_i' there corresponds a subdivision of A into partial regions α_i and $\omega_i = |\Delta(\xi_i, \eta_i)| \omega_i'$, where (ξ_i, η_i) is some point in α_i'.

Then

$$\iint_A f(x, y)dxdy = \lim \Sigma f(x_i, y_i)\omega_i$$
$$= \lim \Sigma f[\varphi(\xi_i, \eta_i), \psi(\xi_i, \eta_i)] \times |\Delta(\xi_i, \eta_i)|\omega_i'$$
$$= \iint_{A'} f[\varphi(u, v), \psi(u, v)] \times |\Delta(u, v)|dudv, \quad (10)$$

where $x_i = \varphi(\xi_i, \eta_i)$ and $y_i = \psi(\xi_i, \eta_i)$.

In evaluating this last integral by means of repeated integrals, the limits of integration are determined from C' in the same way that the limits of integration for the original repeated integrals are determined from C.

EXAMPLE. The substitution of polar coordinates for rectangular affords frequent occasion for the application of formula (10). In this case we have

$$x = \rho \cos \theta, \quad y = \rho \sin \theta,$$

with $\Delta = \rho$. If we wish to evaluate

$$\iint e^{-x^2-y^2}dxdy$$

over the circle $x^2 + y^2 = R^2$, we have

$$\iint_C e^{-x^2-y^2}dxdy = \int_0^{2\pi}\int_0^R e^{-\rho^2}\rho d\rho d\theta$$
$$= \int_0^{2\pi} -\frac{1}{2}e^{-\rho^2}\bigg]_0^R d\theta = \pi(1 - e^{-R^2}). \quad (11)$$

If the reader will attempt to evaluate the original integral directly, he will see the importance of the introduction of new variables.

103. Geometric applications of double integrals. In § 64 it was pointed out that the area bounded by the curve $y = f(x)$, the x-axis, and the ordinates $x = a$, $y = b$ as defined in § 57 is the value of the definite integral $\int_a^b f(x)dx$.

§ 103] DOUBLE AND TRIPLE INTEGRALS 177

In the same order of ideas we can attach a geometric meaning to the double integral $\iint_A f(x, y)\,dx\,dy$.

The upper and lower sums $S = \sum M_i \omega_i$ and $s = \sum m_i \omega_i$ which entered into the definition of the double integral are the sums of the volumes of two sets of cylinders whose bases are the partial regions α_i. No definition of the volume bounded by the surface $z = f(x, y)$, the (x, y)-plane, and the cylindrical surface whose elements are parallel to the z-axis and pass through the boundary of A would be satisfactory if it assigned to this portion of space a volume greater than any upper sum S or less than any lower sum s. Now if $f(x, y)$ is continuous over the region A, there is only one number that satisfies these requirements, and that is the number represented by the definite integral $\iint_A f(x, y)\,dx\,dy$.

We accordingly take this definite integral as our definition of the volume in question. As a consequence of this definition that part of the surface that is below the (x, y)-plane will correspond to a negative volume.

We have now to define what we mean by the area of a limited part of a curved surface and to derive a formula for the area so defined.

Let Γ be a bounded portion of the surface $z = f(x, y)$ and suppose that a line parallel to the z-axis does not cut it in more than one point. Denote by A the orthogonal projection of Γ onto the (x, y)-plane.

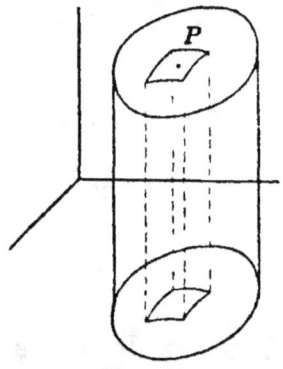

Fig. 19

We have already considered the division of such a bounded region A into partial regions α_i. We draw the tangent plane to the surface at any point (ξ_i, η_i, ζ_i) of the surface within the cylinder whose base is α_i and whose elements are parallel to the z-axis. If the sum $\sum \omega_i'$ of all the plane areas α_i' cut out from these tangent planes by the respective cylinders approaches a unique limit as the maximum diameter of the regions α_i approaches zero, we shall call this unique limit the *area of the curved surface* Γ. Now

$$\omega_i' = \sec \gamma_i \omega_i,$$

where ω_i is the area of α_i and γ_i is the acute angle between the normal to the surface at the point (ξ_i, η_i, ζ_i) and the z-axis. If $z = f(x, y)$ is a continuous function of x and y throughout A, with continuous first partial derivatives, then

$$\lim \sum \omega_i' = \lim \sum \sec \gamma_i \cdot \omega_i = \iint_A \sec \gamma \cdot dx dy$$
$$= \iint_A \sqrt{1 + p^2 + q^2}\, dx dy,$$

where $p = \dfrac{\partial z}{\partial x}$ and $q = \dfrac{\partial z}{\partial y}$. This definite integral then is by definition the area S of the portion Γ of the surface $z = f(x, y)$. That is,

$$S = \iint_A \sec \gamma \cdot dx dy \tag{12}$$

is the formula which gives us the area. We have assumed that no line parallel to the z-axis cuts the area Γ in more than one point. If such a line cuts Γ in more than one point, and we can divide Γ into a finite number of parts each one of which meets the requirements of the assumption, we shall take the sum of the areas of all these parts as the area of the whole.

If we apply the mean value theorem to the double integral in (12), we obtain the formula

$$S = \sec \gamma_i \iint_A dx dy = \sec \gamma_i \cdot \Omega,$$

where γ_i is the angle made with the z-axis by the normal to the surface at some point of Γ, and Ω is the area of A. In particular we shall call σ_i the *element of area* of the surface if $\sigma_i = \sec \gamma_i \cdot \omega_i$.

If no line parallel to the y-axis cuts Γ in more than one point and we project Γ into the region B of the (z, x)-plane, similar considerations lead us to take the value of the integral $\iint_B \sec \beta \cdot dz dx$ to be the measure of the area of Γ, β being the angle between the normal to the surface and the y-axis. It is important therefore to know that these two expressions for this area are equal. Now the partial regions α_i into which A is divided project into the partial regions σ_i on the respective tangent planes, and these in turn project into a set of partial regions β_i of B whose respective areas we denote by ω_i'. Then $\iint_A \sec \gamma \, dx dy = \lim \sum \sec \gamma_i \cdot \omega_i$
$= \lim \sum \sigma_i = \lim \sum \sec \beta_i \cdot \omega_i' = \iint_B \sec \beta \, dz dx$. Under corresponding conditions the area of Γ is also equal to $\iint_C \sec \alpha \cdot dy dz$, where α is the angle between the normal to the surface and the x-axis.

Our definition of the area of Γ seems to make this area dependent upon a particular orientation of the xy-plane, whereas we intuitively feel that area should be an intrinsic property of the surface. We could avoid this difficulty by dividing Γ into partial regions σ_i, projecting these orthogonally upon the respective tangent planes, and taking the limit of the sum of all these projections for our area. It turns out however that this limit is the same as the other one, and that therefore the difficulty was only an apparent one. We omit the details of this alternative proof because they are somewhat complicated.[1]

[1] See, for example, Goursat-Hedrick, *Mathematical Analysis*, Vol. I, p. 272.

If we are dealing with a surface of revolution we can determine the area of that part of it that is contained between two planes perpendicular to the axis by evaluating a certain simple integral. Suppose, for example, that the surface is generated by revolving the curve $y = f(x)$ around the x-axis. Its equation is $y^2 + z^2 = f^2(x)$ and the area between the planes $x = a$ and $x = b$ is

$$S = 4 \int_a^b \int_0^{f(x)} \sec \gamma \cdot dx dy,$$

where

$$\sec \gamma = \frac{f(x)\sqrt{1 + f'^2(x)}}{\sqrt{f^2(x) - y^2}}.$$

Hence

$$S = 4 \int_a^b \int_0^{f(x)} \frac{f(x)\sqrt{1 + f'^2(x)}}{\sqrt{f^2(x) - y^2}} dx dy$$

$$= 2\pi \int_a^b f(x)\sqrt{1 + f'^2(x)}\, dx.$$

EXERCISES

Compute the following areas by double integration:

1. The ellipse $\dfrac{x^2}{a^2} + \dfrac{y^2}{b^2} = 1$. 2. The astroid $x^{2/3} + y^{2/3} = 1$.
3. That bounded by the curves $y = x^2$ and $y^2 = x$.
4. That between the axes and $x^{1/2} + y^{1/2} = a^{1/2}$.
5. That common to the circles $x^2 + y^2 = 4$ and $(x - 2)^2 + y^2 = 1$.
6. Express the double integral $\iint dx dy$ in terms of polar coordinates ρ and θ, with $x = \rho \cos \theta$ and $y = \rho \sin \theta$. This formula is useful in finding an area bounded by two lines through the origin and the arc of a curve.
7. Find the area of the cardioid $\rho = 2a(1 - \cos \theta)$.
8. Find the area of one loop of the curve $\rho = \sin 2\theta$.
9. Find the area of the curve $\rho^2 = \cos 2\theta$.
10. Find the area bounded by the curves $y = \sin x$, $y = \cos x$, and $x = 0$.
11. Find the volume between the xy-plane and the paraboloid $z = 1 - x^2 - \dfrac{y^2}{16}$.

12. Find the volume enclosed by the surfaces $xy = 0$, $z = 0$, and $2x - 3y + z = 1$.

13. Find the volume enclosed by the cylindrical surface $y^2 + z^2 - 2ay - 2bz = 0$, the plane $x = 0$, and the paraboloid $x = 2yz$.

14. Find the volume common to the sphere $x^2 + y^2 + z^2 = a^2$ and the cylinder $y^2 + z^2 - ay = 0$.

15. Find the area of the surface of the sphere $x^2 + y^2 + z^2 = a^2$ between the planes $z = 0$ and $z = 1$.

16. Find the area of that part of the surface of the paraboloid $az = x^2 + y^2$ that lies between the planes $z = 1$ and $z = 2$.

17. Find the volume and the surface area of the solid generated by revolving the circle $x^2 + (y - 3)^2 = 1$ about the x-axis.

18. Find the moment of inertia of a circular disc about an axis through its center and perpendicular to its plane.

19. Of a ring bounded by two concentric circles about an axis through their common center and perpendicular to their plane.

20. Of a circle about a diameter.

21. Find the volume between the xz-plane and the surface $y = 4 - x^2 - 3z^2$.

22. Find the volume common to the cylinders $x^2 + y^2 = 1$ and $y^2 + z^2 = 1$.

23. Find the moment of inertia about the x-axis of the area bounded by the x-axis and one arch of the cycloid $x = a(\theta - \sin \theta)$, $y = a(1 - \cos \theta)$.

24. Find the area of the surface generated by revolving the astroid about one of its axes.

25. Find the area of that part of the surface of a sphere of radius 5 inches that is included between two planes 3 and 4 inches from the center respectively and on the same side of the center.

26. Find the area of that part of the surface generated by revolving the catenary $y = \dfrac{a}{2}(e^{x/a} + e^{-x/a})$ about its directrix that is contained between the planes $x = 0$ and $x = a$.

27. A circular hole one inch in diameter is bored through a sphere four inches in diameter, the axis of the hole coinciding with a diameter of the sphere. Find the volume cut out.

104. Triple integrals. Let the function $f(x, y, z)$ be single valued and bounded throughout a certain limited portion A of space whose volume is V. If we divide A into partial regions $\alpha_1, \alpha_2, \cdots, \alpha_n$ whose volumes are respectively $\omega_1, \omega_2, \cdots, \omega_n$, the function $f(x, y, z)$ will be bounded in each of these regions. We can accordingly form the upper and lower sums

$$S = \sum M_i \omega_i \quad \text{and} \quad s = \sum m_i \omega_i,$$

where M_i and m_i are the upper and lower bounds respectively of $f(x, y, z)$ in α_i. No upper sum is less than $\sum m\omega_i = mV$, where m is the lower bound of $f(x, y, z)$ in A, and no lower sum is greater than $\sum M\omega_i = MV$, where M is the upper bound of $f(x, y, z)$ in A. The set of all upper sums has a lower limit I, and the set of all lower sums has an upper limit I'. Moreover

$$I \geq I'.$$

The proof of this is similar to the proof given in § 59 of the corresponding statement in connection with simple integrals.

INTEGRABLE FUNCTIONS. The bounded function $f(x, y, z)$ is said to be *integrable* over a finite closed region A in case $I = I'$. The common value of I and I' is represented by the symbol $\iiint_A f(x, y, z) dx dy dz$, or the symbol $\int_A f(x, y, z) dV$, and is called the *triple integral* of $f(x, y, z)$ over A. The function $f(x, y, z)$ is called the *integrand* and A is called the *field of integration*. If $f(x, y, z)$ is continuous in the closed region A, it is integrable over A (see § 94).

105. Repeated integrals. In order to show that a triple integral is equal to a repeated integral we assume first that the field of integration is a rectangular parallelopiped whose faces are parallel to the coordinate planes.

Let the bounding planes be $x = x_0$, $x = X$, $y = y_0$, $y = Y$, $z = z_0$, and $z = Z$; and let the partial regions be bounded by the planes $x = x_i$, $y = y_j$, $z = z_k$. Then we have

$$\iiint_A f(x, y, z) dx dy dz$$
$$= \lim \sum f(\xi_{i,j,k}, \eta_{i,j,k}, \zeta_{i,j,k})(x_i - x_{i-1})$$
$$\times (y_j - y_{j-1})(z_k - z_{k-1}),$$

where $\xi_{i,j,k}$, $\eta_{i,j,k}$, and $\zeta_{i,j,k}$ are the coordinates of any point in the partial region bounded by the planes $x = x_{i-1}$, $x = x_i$, $y = y_{j-1}$, $y = y_j$, $z = z_{k-1}$, and $z = z_k$. (Cf. § 95 (a).) The contribution to the integral by those regions which lie in the column bounded by the planes $x = x_{i-1}$, $x = x_i$, $y = y_{j-1}$, $y = y_j$, is

$$(x_i - x_{i-1})(y_j - y_{j-1})[f(x_{i-1}, y_{j-1}, \zeta_1)(z_1 - z_0) + \cdots],$$

if we take $\xi_{i,j,k} = x_{i-1}$ and $\eta_{i,j,k} = y_{j-1}$ for all values of k. Now the ζ's may be so chosen that the part in the square brackets is equal to

$$\int_{z_0}^{Z} f(x_{i-1}, y_{j-1}, z) dz = \Phi(x_{i-1}, y_{j-1}) \qquad \text{(see § 97)}.$$

The triple integral in question is therefore equal to the limit of the sum

$$\sum (x_i - x_{i-1})(y_j - y_{j-1}) \Phi(x_{i-1}, y_{j-1}),$$

and this in turn is equal to the double integral

$$\iint \Phi(x, y) dx dy$$

over the region of the (x, y)-plane bounded by the lines $x = x_0$, $x = X$, $y = y_0$, and $y = Y$.

This conclusion can be extended to the case in which A is not a rectangular parallelopiped with faces parallel to the coordinate planes, provided its bounding surface is not cut by any line parallel to one of the axes in more than two points, and provided the integrand is continuous throughout A. The argument is as follows: The points of the bounding surface of A project into the points of a region R of the (x, y)-plane bounded by a curve C. Every point (x, y) inside C is the projection of two points (x, y, Z_1) and (x, y, Z_2) of the bounding surface. The z-coordinates of these points are functions of x and y. We shall denote them by $\varphi_1(x, y)$ and $\varphi_2(x, y)$ respectively with the understanding that $\varphi_1(z) \leq \varphi_2(z)$, and that both are continuous within and upon C.

If we divide the region of integration into partial regions by planes parallel to the coordinate planes, the part of the sum $\Sigma f(\xi_i, \eta_i, \zeta_i)v_i$, where the v_i are parallelopipeds, or portions of parallelopipeds, which are made up of the elements bounded by the planes $x = x_{i-1}$, $x = x_i$, $y = y_{j-1}$, $y = y_j$, is given by the sum

$$\Sigma(x_i - x_{i-1})(y_j - y_{j-1})\left[\int_{Z_1}^{Z_2} f(x_{i-1}, y_{j-1}, z)dz + \epsilon_{ij}\right].$$

The reader should refer to § 98 for indications as to how this result is reached. Since $\varphi_1(x, y)$ and $\varphi_2(x, y)$ are uniformly continuous in the closed region bounded by C, the numbers $\epsilon_{i,j}$ can all be made less in absolute value than an arbitrary positive number ϵ for every choice of i and j by making all the differences $x_i - x_{i-1}$ and $y_j - y_{j-1}$ sufficiently small. The sum

$$\Sigma\Sigma(x_i - x_{i-1})(y_j - y_{j-1})\epsilon_{ij}$$

can therefore be made arbitrarily small in absolute value, and the original triple integral is equal to the double integral

$$\iint_R \psi(x, y)dxdy,$$

where $\psi(x, y) = \int_{Z_1}^{Z_2} f(x, y, z)dz$.

If a line in the (x, y)-plane parallel to the y-axis meets C in the two points whose coordinates are $Y_1 = \theta_1(x)$ and $Y_2 = \theta_2(x)$, we have

$$\iiint_A f(x, y, z)dxdydz = \int_a^b dx \int_{Y_1}^{Y_2} dy \int_{Z_1}^{Z_2} f(x, y, z)dz. \quad (13)$$

It is to be understood that in the repeated integral in the right member of this formula the integration with respect to z is to be performed first and that the limits Z_1 and Z_2 are functions of x and y. Then the integration with respect to y is to be performed with limits Y_1 and Y_2 which are functions of x. In the last integration the limits are constants. This establishes the fact that a triple integral can be evaluated by means of repeated integrals. The order of the integrations in the latter is of course arbitrary, provided that proper regard is had for the limits in each case. The restriction that a line parallel to one of the coordinate axes shall not meet the bounding surface in more than two points can be replaced by a less restrictive one.

106. Change of variables in a triple integral. The formula for a change of variables in a triple integral is similar to the formula for a change of variables in a double integral. If the new variables u, v, and w are connected with the original ones by the equations

$$x = \varphi(u, v, w), \qquad y = \psi(u, v, w), \qquad z = \theta(u, v, w), \quad (14)$$

which establish a one-to-one correspondence between the original region A of integration and a new one A_1, this formula is

$$\iiint_A f(x, y, z)\,dx\,dy\,dz$$
$$= \iiint_{A_1} f[\varphi(u, v, w), \psi(u, v, w), \theta(u, v, w)]$$
$$\times \frac{D(\varphi, \psi, \theta)}{D(u, v, w)}\,du\,dv\,dw, \quad (15)$$

where $\dfrac{D(\varphi, \psi, \theta)}{D(u, v, w)}$ represents the functional determinant

$$\begin{vmatrix} \dfrac{\partial \varphi}{\partial u} & \dfrac{\partial \varphi}{\partial v} & \dfrac{\partial \varphi}{\partial w} \\ \dfrac{\partial \psi}{\partial u} & \dfrac{\partial \psi}{\partial v} & \dfrac{\partial \psi}{\partial w} \\ \dfrac{\partial \theta}{\partial u} & \dfrac{\partial \theta}{\partial v} & \dfrac{\partial \theta}{\partial w} \end{vmatrix}.$$ We omit the proof of this formula.

The variables u, v, and w may be considered as the coordinates of a point with reference to a rectangular system of coordinates. If

$$u = \Phi(x, y, z), \qquad v = \Psi(x, y, z), \qquad w = \Theta(x, y, z) \quad (16)$$

are the solutions of equations (14), it is clear that the equations $u = c_1$, $v = c_2$, $w = c_3$ represent three families of surfaces and that the constants determine a point in A. Conversely, any point in A determines the values of the three constants. We can therefore think of these surfaces as determining a curvilinear system of coordinates.

107. Surface integrals. If A is the orthogonal projection of a certain portion Γ of the surface $z = \varphi(x, y)$ upon the (x, y)-plane the double integral

$$\iint_A f[x, y, \varphi(x, y)]dxdy$$

is called the *surface integral* of $f(x, y, z)$ over Γ. We shall assume that $f(x, y, z)$ and $\varphi(x, y)$ are continuous throughout Γ and A respectively, and that Γ is cut in only one point by a line parallel to the z-axis, or at least that it can be divided up into a finite number of parts, each of which has this property.

It will be observed that a surface integral is closely analogous to a line integral, with the difference that there appears to be nothing in the former to correspond to the reversal of sign in the latter when the direction of integration is reversed. There is however a property of surface integrals that can with propriety be regarded as supplying this lack. In order to describe this property it is necessary to distinguish between *unilateral* and *bilateral* surfaces.

DEFINITION. At any point of a surface at which there is a normal we can consider that part of the normal that extends in either of two opposite directions. If now starting at any point of the surface with a definite direction of the normal we can describe a continuous closed curve on the surface with the direction of the normal changing continuously in such a way that we return to the starting point with the direction of the normal opposite to its original direction, we say that the surface is *unilateral*.[1] Otherwise, we say that it is *bilateral*. We shall confine our discussion to bilateral surfaces.

If at any point we take that direction of the normal that makes an acute angle with the positive direction of the z-axis, we shall say

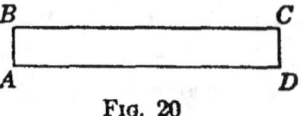

Fig. 20

that we are on the upper side of the surface. If we take the opposite direction of the normal we shall say that we are on the lower side. Of course this is a mere convention— a surface does not have sides.

If now we write the surface integral $\iint_\Gamma f(x, y, z) dx dy$ in the form $\iint_\Gamma f(x, y, z) \cos \gamma\, d\sigma$, where γ is the angle between the normal to the surface and the z-axis, we see that the integral has one sign or the other according as we integrate over the upper side or the lower side of the surface. In certain cases it might of course be zero.

108. Green's theorem in space. Let S be a closed surface which is not cut by any line parallel to an axis in more than two points. If $R(x, y, z)$ is a function which together with $\dfrac{\partial R}{\partial z}$ is continuous on S and in the region V bounded by S, we take advantage of the fact that the triple integral $\iiint_V \dfrac{\partial R}{\partial z} dx dy dz$ can be replaced by a certain iterated integral:

$$\iiint_V \frac{\partial R}{\partial z} dx dy dz = \iint_A dx dy \int_{\varphi_1(x, y)}^{\varphi_2(x, y)} \frac{\partial R}{\partial z} dz$$

$$= \iint_A \{R[x, y, \varphi_2(x, y)] - R[x, y, \varphi_1(x, y)]\} dx dy,$$

---

[1] If a narrow strip of paper such as is shown in the figure is formed into a band by making C coincide with A and D with B, a unilateral surface will be formed.

where $z = \varphi_2(x, y)$ is the equation of the part S_2 of S that forms the upper boundary of V, $z = \varphi_1(x, y)$ is the equation of the lower boundary S_1, and A is the projection of S on the (x, y)-plane. Now $\iint_A R[x, y, \varphi_2(x, y)]dxdy$ is the surface integral of $R(x, y, z)$ taken over the upper side of S_2 and $-\iint_A R[x, y, \varphi_1(x, y)]dxdy$ is the surface integral of $R(x, y, z)$ taken over the lower side of S_1. Hence

$$\iiint_V \frac{\partial R}{\partial z} dxdydz = \iint_S R(x, y, z)dxdy$$

taken over the exterior of S. If we make similar assumptions concerning $P(x, y, z)$, $\frac{\partial P}{\partial x}$, $Q(x, y, z)$, and $\frac{\partial Q}{\partial y}$, then also

$$\iiint_V \frac{\partial P}{\partial x} dxdydz = \iint_S P(x, y, z)dydz$$

and

$$\iiint_V \frac{\partial Q}{\partial y} dxdydz = \iint_S Q(x, y, z)dzdx.$$

If we combine these three integrals we obtain the formula

$$\iiint_V \left(\frac{\partial P}{\partial x} + \frac{\partial Q}{\partial y} + \frac{\partial R}{\partial z}\right) dxdydz$$
$$= \iint_S Pdydz + Qdzdx + Rdxdy.$$

This is *Green's theorem* for space. It gives us a certain volume integral in terms of a surface integral over the bounding surface of the volume just as Green's theorem for the plane gives us a certain double integral over a closed area in terms of a line integral over the bounding curve.

109. Stokes' theorem. Consider a closed twisted curve L and a bilateral surface Γ bounded by L and not cut by

§ 109] DOUBLE AND TRIPLE INTEGRALS

any line parallel to one of the coordinate axes in more than one point. If $P(x, y, z)$ is continuous along L we define the line integral $\int_L P(x, y, z)dx$ in a way suggested by the definition of a plane line integral in § 99; that is, by definition,

$$\int_L P(x, y, z)dx = \int_L P[x, \varphi(x), \psi(x)]dx,$$

where $y = \varphi(x)$ and $z = \psi(x)$ are the equations of L. If we denote by A the orthogonal projection of Γ upon the (x, y)-plane, we have

$$\iint_\Gamma P(x, y, z)d\sigma = \iint_A P[x, y, F(x, y)]dxdy,$$

where $z = F(x, y)$ is the equation of the surface Γ, and $P(x, y, z)$ is now assumed to be continuous, together with its first partial derivatives, on the surface Γ. Now the line integral

$$\int_L P(x, y, z)dx$$

is the same as the line integral

$$\int_l P[x, y, F(x, y)]dx,$$

l being the orthogonal projection of L upon the (x, y)-plane. If we put

$$P_1(x, y) = P[x, y, F(x, y)]$$

we have

$$\frac{\partial P_1}{\partial y} = \frac{\partial P}{\partial y} + \frac{\partial P}{\partial z}\frac{\partial F}{\partial y}.$$

But

$$\cos \mu : \cos \nu = \frac{\partial F}{\partial y} : -1,$$

or

$$\frac{\partial F}{\partial y} = -\frac{\cos \mu}{\cos \nu}.$$

Here $\cos \lambda$, $\cos \mu$, and $\cos \nu$ are the direction cosines of the normal to the surface. Hence

$$\frac{\partial P_1}{\partial y} = \frac{\partial P}{\partial y} - \frac{\partial P}{\partial z}\frac{\cos \mu}{\cos \nu}.$$

By § 99

$$\int_l P_1(x, y)dx = -\iint_A \frac{\partial P_1}{\partial y} dxdy$$

$$= \iint_A \left[\frac{\partial P}{\partial z}\cos \mu - \frac{\partial P}{\partial y}\cos \nu\right]\sec \nu \, dxdy$$

$$= \lim \sum \left[\frac{\partial P}{\partial z}\cos \mu_i - \frac{\partial P}{\partial y}\cos \nu_i\right]\sec \nu_i \cdot \omega_i$$

$$= \lim \sum \left[\frac{\partial P}{\partial z}\cos \mu_i - \frac{\partial P}{\partial y}\cos \nu_i\right]\sigma_i$$

$$= \lim \sum \left(\frac{\partial P}{\partial z}\omega_i'' - \frac{\partial P}{\partial y}\omega_i\right)$$

$$= \iint_\Gamma \left(\frac{\partial P}{\partial z} dzdx - \frac{\partial P}{\partial y} dxdy\right),$$

where ω_i'' is the projection of σ_i upon the (z, x)-plane. If we take all the σ_i as positive, the ω_i'' will be positive or negative according as μ_i is acute or obtuse. We agree to take the direction of the normal such that ν_i is acute. If then the μ_i are obtuse we must determine the limits of integration in the repeated integral $\iint_\Gamma \frac{\partial P}{\partial z} dzdx$ in a way that would make the integral $\iint dzdx$ negative. With this understanding we have

$$\int_L P(x, y, z)dx = \iint_\Gamma \left(\frac{\partial P}{\partial z} dzdx - \frac{\partial P}{\partial y}\cdot dxdy\right), \quad (17)$$

where the direction of the line integral corresponds to the positive direction around l.

This direction around L corresponds to the positive or the negative direction around the projection l' of L upon the (y, z)-plane according as the normals to Γ make acute or obtuse angles with the positive direction of the x-axis. If then these angles are acute we have

$$\int_L Q(x, y, z) dy = \iint_\Gamma \left(\frac{\partial Q}{\partial x} dx dy - \frac{\partial Q}{\partial z} dy dz \right), \quad (17')$$

where the direction of integration around L is the same as in (17). But if these angles are obtuse this direction around L is the opposite to what it is in (17). A change of sign of the integrand brings the direction of integration around L into agreement with the direction in (17). Similar remarks apply to

$$\int_L R(x, y, z) dz = \iint_\Gamma \left(\frac{\partial R}{\partial y} dy dz - \frac{\partial R}{\partial x} dz dx \right). \quad (17'')$$

If now we have proper regard for the limits of integration in the repeated integrals we can combine (17), (17'), and (17''). This gives us

$$\int_L P(x, y, z) dx + Q(x, y, z) dy + R(x, y, z) dz$$
$$= \iint_\Gamma \left(\frac{\partial Q}{\partial x} - \frac{\partial P}{\partial y} \right) dx dy + \left(\frac{\partial R}{\partial y} - \frac{\partial Q}{\partial z} \right) dy dz$$
$$+ \left(\frac{\partial P}{\partial z} - \frac{\partial R}{\partial x} \right) dz dx. \quad (18)$$

This is *Stokes' theorem*. It gives a certain surface integral over a limited portion of a surface in terms of a line integral over the bounding curve.

The rule to be followed can be formulated as follows: Select that direction of the normal that makes an acute angle with the positive direction of one of the axes, say with the z-axis. Then adjust the limits in the second and third repeated integrals in the right member of (18) in the way described if this direction of the normal makes an obtuse angle with the corresponding axis. In view of the restriction that no line parallel to one of the coordinate axes shall meet Γ in more than one point, the angles λ are either all acute or all obtuse. This is also true of the angles μ and ν.

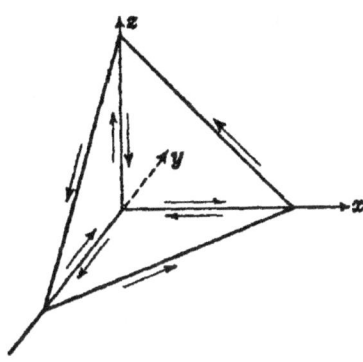

Fig. 21

EXAMPLE. Verify Stokes' theorem in case $P = x - y$, $Q = x + y$, $R = x + y + z$, and Γ is that part of the plane $x - y + z = 1$ that is cut out by the coordinate planes.

$$\int P dx + Q dy + R dz = \int_1^0 x dx + \int_0^1 dx + \int_{-1}^0 (1 + 2y) dy$$
$$+ \int_0^{-1} y dy + \int_0^1 dz + \int_1^0 (2z - 1) dz = 2.$$

$$\iint_\Gamma \left(\frac{\partial Q}{\partial x} - \frac{\partial P}{\partial y} \right) dx dy + \iint_\Gamma \left(\frac{\partial R}{\partial y} - \frac{\partial Q}{\partial z} \right) dy dz$$
$$+ \iint_\Gamma \left(\frac{\partial P}{\partial z} - \frac{\partial R}{\partial x} \right) dz dx$$

$$= \int_0^1 \int_{x-1}^0 2 dx dy + \int_{-1}^0 \int_0^{1+y} dy dz - \int_0^1 \int_{1-z}^0 dz dx$$

$$= \int_0^1 (2 - 2x) dx + \int_{-1}^0 (1 + y) dy - \int_0^1 (z - 1) dz = 2.$$

Here the direction of the normals that makes an acute angle with the positive direction of the z-axis makes an acute angle with the positive direction of the x-axis and an obtuse angle with the positive direction of the y-axis. Note the corresponding adjustment in the limits for the double integrals.

EXERCISES

1. Evaluate $\iiint_V x^2 dx\,dy\,dz$, where V is the volume common to the sphere $x^2 + y^2 + z^2 = a^2$ and the cylinder $y^2 + z^2 = ay$.

2. Evaluate $\iiint_V y^2 dx\,dy\,dz$, where V is the volume common to the sphere $x^2 + y^2 + z^2 = a^2$ and the cylinder $x^2 + z^2 = az$.

3. Evaluate $\iiint_V (x - 2y + 5z)xyz\,dx\,dy\,dz$, V being the region defined by the inequality $x^2 + y^2 + z^2 \leq 4$.

4. Evaluate $\iiint (x + y + z)(3x - 5y^2 + 2z)dx\,dy\,dz$ throughout the region defined by the inequalities $x \geq 0$, $y \geq 0$, $z \geq 0$, $x + y + z \leq 2$.

5. Evaluate $\iiint \dfrac{dx\,dy\,dz}{x^2 + y^2 + z^2}$ subject to the condition $x^2 + y^2 + z^2 \leq 1$. Use spherical coordinates.

6. Find the moment of inertia of a homogeneous right circular cylinder about its axis. Compare this with the kinetic energy of the cylinder when it is revolving about its axis with the angular velocity ω.

7. Find the mass and the moment of inertia about a diameter of a sphere of radius 1 whose density at any point is a linear function of the distance of this point from the center.

8. Prove that if A, B, and C are the moments of inertia of a solid with respect to the coordinate axes, then $A + B > C$.

9. Find the moment of inertia of the homogeneous ellipsoid $\dfrac{x^2}{a^2} + \dfrac{y^2}{b^2} + \dfrac{z^2}{c^2} = 1$ about the y-axis.

10. Find the volume of the solid bounded by the surfaces $z^2 + x^2 = 4y$, $x^2 = y$, and $y = 5$.

11. Evaluate $\iint_S (lx^2 + my^2 + nz^2)d\sigma$, where S is the surface of the sphere $x^2 + y^2 + z^2 = r^2$, and l, m, and n are the direction cosines of the normals to the sphere.

12. We have seen that the triple integral $\iiint_V dx\,dy\,dz$ taken between suitable limits is equal to the volume of a given bounded region. Express this in terms of cylindrical coordinates, and also in terms of spherical coordinates.

13. Show directly that $\iint\limits_S (xzdydz + yzdzdx + z^2dxdy) = 0$ in case S is the surface of the sphere $x^2 + y^2 + z^2 = a^2$ and the integration is over the exterior of the surface.

14. Verify this result by applying Green's theorem and evaluating the triple integral that appears.

15. Apply Green's theorem to show that

$$\iint\limits_S (adydz + bdzdx + cdxdy)$$

taken over any closed surface is zero in case a, b, and c are constants.

16. Evaluate $\iint\limits_S (xdydz - ydzdx - zdxdy)$ in case S is the part of the cylindrical surface $x^2 + y^2 = r^2$ between the planes $z = \pm a$.

17. If $P = u\dfrac{\partial v}{\partial x}, Q = u\dfrac{\partial v}{\partial y}, R = u\dfrac{\partial v}{\partial z}$, we have from Green's theorem

$$\iiint\limits_V u\left(\frac{\partial^2 v}{\partial x^2} + \frac{\partial^2 v}{\partial y^2} + \frac{\partial^2 v}{\partial z^2}\right) dxdydz$$

$$+ \iiint\limits_V \left(\frac{\partial u}{\partial x}\frac{\partial v}{\partial x} + \frac{\partial u}{\partial y}\frac{\partial v}{\partial y} + \frac{\partial u}{\partial z}\frac{\partial v}{\partial z}\right) dxdydz$$

$$= \iint\limits_S \left(u\frac{\partial v}{\partial x}dydz + u\frac{\partial v}{\partial y}dzdx + u\frac{\partial v}{\partial z}dxdy\right).$$

Show that the right member of this equation can be written in the form $\iint\limits_S u\dfrac{\partial v}{\partial n}d\sigma$, where $\dfrac{\partial v}{\partial n}$ is the derivative of v in the direction of the exterior normal.

18. Show that

$$\iiint\limits_V (u\Delta v - v\Delta u)dxdydz = \iint\limits_S \left(u\frac{\partial v}{\partial n} - v\frac{\partial u}{\partial n}\right)d\sigma,$$

where $\Delta u = \dfrac{\partial^2 u}{\partial x^2} + \dfrac{\partial^2 u}{\partial y^2} + \dfrac{\partial^2 u}{\partial z^2}$.

19. What does this equation become if we put $v = 1$?

20. What is the value of $\iint\limits_S \dfrac{\partial u}{\partial n}d\sigma$ in case u is a solution of Laplace's equation $\Delta u = 0$?

21. What results from (Ex. 18) if we put $v = u$?

22. Show that the volume enclosed by a closed surface S is given by the formula $V = \dfrac{1}{3}\displaystyle\iint_S r\cos\theta\, d\sigma$, where r is the radius vector and θ is the angle between it and the exterior normal.

23. Compute $\displaystyle\int_{(0,0,0)}^{(2,-1,3)} z(x^2+y^2)dx + 2xyz\,dy + xy^2 dz$ over two paths and compare the results.

24. Compute $\displaystyle\int_{(1,2,0)}^{(2,3,4)} (dx + dy + dz)$ along the paths: (a) the line from $(1, 2, 0)$ to $(2, 3, 4)$; (b) a broken line whose parts are parallel to the axes.

25. Show that the two line integrals $\displaystyle\int P\,dx + Q\,dy + R\,dz$ and $\displaystyle\int \sqrt{P^2+Q^2+R^2}\cos\theta\, ds$ are equal in case θ is the angle between the curve and the line whose direction cosines are proportional to P, Q, and R.

110. Potential.

The fraction $\dfrac{m}{r}$ is called the *potential* at a point P due to a particle of mass m whose distance from P is r. If there are n particles of masses $m_1, m_2, \cdots, m_n$ at the respective distances $r_1, r_2, \cdots, r_n$ from P, the sum

$$\frac{m_1}{r_1} + \frac{m_2}{r_2} + \cdots + \frac{m_n}{r_n}$$

is called the *potential* of the system at P. The potential at P due to a continuous distribution of matter over a finite region is by definition the value u of the triple integral $\displaystyle\int_V \frac{\rho\, dv}{r}$, where $\rho(\xi, \eta, \zeta)$ is the density of the matter at the point (ξ, η, ζ), and r is the distance between this point and the point $P = (x, y, z)$. The integral is to be extended over the region occupied by the matter in question. The variables of integration are ξ, η, ζ and x, y, z are parameters. The integral is therefore a function $V(x, y, z)$ of x, y, and z.

The reader can convince himself of the naturalness of this definition by the following considerations: If we conceive of the total mass divided up into n partial masses $\Delta_1 m$, $\Delta_2 m$, $\cdots$, $\Delta_n m$, and if r_i is the distance of some point in $\Delta_i m$ from P, the sum $\sum \dfrac{\Delta_i m}{r_i}$ is approximately what we would naturally want to call the potential of the total mass. Now $\Delta_i m = \bar{\rho}_i \Delta_i V$, where $\bar{\rho}_i$ is the average density[1] of the mass $\Delta_i m$ and $\Delta_i V$ is the volume occupied by it. We assume that the density is a continuous function of position. Then $\bar{\rho}_i$ is the actual density at some point of $\Delta_i V$. If we take r_i to be the distance of this point from P, we see immediately that

$$\lim \sum \frac{\Delta_i m}{r_i} = \int_V \frac{\rho\, dV}{r}.$$

It is therefore natural to take this integral as our definition of potential.

If the point P is outside the region occupied by the matter, the integrand is finite throughout the field of integration and the integral is a proper one. But if P is in this region $r = 0$ at one point of the field and the integrand is infinite here. The improper integral nevertheless converges. For we can take the element of volume dV to be $r^2 dr d\omega$, where $d\omega$ is the element of surface on a unit sphere. Then

$$u = \int_V \frac{\rho r^2 dr d\omega}{r} = \int_V \rho r\, dr\, d\omega.$$

Now

$$r^2 = (\xi - x)^2 + (\eta - y)^2 + (\zeta - z)^2,$$

and hence

$$\frac{\partial \left(\dfrac{1}{r}\right)}{\partial x} = \frac{\xi - x}{r^3} = \frac{\cos \alpha}{r^2}$$

---

[1] That is, $\bar{\rho}_i = \dfrac{\int_{\Delta_i V} \rho\, dV}{\Delta_i V}$. Then $\bar{\rho}_i \Delta_i V = \int_{\Delta_i V} \rho\, dV$. But for any mass M distributed throughout the volume V with the density $\rho(x, y, z)$ we have $\sum \rho_i'' \omega_i \leqq M \leqq \sum \rho_i' \omega_i$, where ρ_i' and ρ_i'' are the maximum and minimum values of ρ in the elementary volume ω_i. Hence $M = \int_V \rho\, dV$, and therefore $\bar{\rho}_i \Delta_i V = \int_{\Delta_i V} \rho\, dV = \Delta_i m$.

§ 110]   DOUBLE AND TRIPLE INTEGRALS   197

if α is the angle between the positive x-axis and the line from P to the point (ξ, η, ζ). If we represent the triple integral $\int \rho \dfrac{\partial \left(\frac{1}{r}\right)}{\partial x} dV$ by X, we have

$$ X = \int \frac{\rho \cos \alpha}{r^2} dV = \int \frac{\cos \alpha \, dm}{r^2}. $$

By properly selecting the element of volume we see that this integral also converges when P is in the region occupied by the matter.

Now [1]

$$ \int_{x_0}^{x} X \, dx = \int_{x_0}^{x} \int_{V} \rho \frac{\partial \left(\frac{1}{r}\right)}{\partial x} dV \, dx = \int_{V} \int_{x_0}^{x} \rho \frac{\partial \left(\frac{1}{r}\right)}{\partial x} dx \, dV $$

$$ = \int_{V} \rho \left(\frac{1}{r} - \frac{1}{r_0} \right) dV = u - u_0, $$

where[2] u_0 is the value of u at the point (x_0, y_0, z_0). It follows from this that

$$ X = \frac{\partial u}{\partial x}. \qquad (19) $$

The x-component of the attraction due to the matter in the volume $\Delta_i V$ upon a unit mass at P lies between the product $\operatorname{Max} \left(\dfrac{\rho \cos \alpha}{r^2} \right) \Delta_i V$ and the product $\operatorname{Min} \left(\dfrac{\rho \cos \alpha}{r^2} \right) \Delta_i V$, where the coefficients of $\Delta_i V$ denote the maximum and minimum values respectively of $\dfrac{\rho \cos \alpha}{r^2}$ in $\Delta_i V$. This component is therefore equal to the value $\dfrac{\rho_i \cos \alpha_i}{r_i^2} \Delta_i V$ of $\dfrac{\rho \cos \alpha}{r^2} \Delta_i V$ at some point of $\Delta_i V$. The sum $\sum \dfrac{\rho_i \cos \alpha_i}{r_i^2} \Delta_i V$ is the x-component of the attraction of the whole mass upon

---

[1] We assume that we can integrate with respect to x under the integral sign.
[2] We should keep in mind that ρ is independent of x, y, z; it is a function of ξ, η, ζ.

a unit mass at P. The limit of this sum, that is, $\int \frac{\rho \cos \alpha}{r^2} dV$, or X, is therefore also equal to this component. Formula (19) shows that this component is equal to the partial derivative of the potential u with respect to x. Similarly

$$Y = \frac{\partial u}{\partial y} \quad \text{and} \quad Z = \frac{\partial u}{\partial z}$$

where Y and Z are the y- and z-components of the attraction.

The component of this attraction along a line whose direction cosines are $\cos \alpha$, $\cos \beta$, and $\cos \gamma$ is

$$X \cos \alpha + Y \cos \beta + Z \cos \gamma,$$

or

$$\frac{\partial u}{\partial x} \cos \alpha + \frac{\partial u}{\partial y} \cos \beta + \frac{\partial u}{\partial z} \cos \gamma.$$

That is, the component of the attraction in a given direction is the corresponding directional derivative of the potential. Since

$$du = X dx + Y dy + Z dz$$

the right member of this equation is a complete differential.

111. Work. If a constant force F is directed along a straight line and moves a given mass a distance l on this line, the product Fl is taken as a measure of the work done. If the force is directed at a fixed angle θ to the line we consider the components along the line and perpendicular to it. These components are $F \cos \theta$ and $F \sin \theta$. But if the mass is constrained to remain on the line the latter component produces no motion. We therefore agree to say that it does no work, and that the other one does it all. That is, the work done is $W = Fl \cos \theta$.

If X and Y are the components of F along the x- and y-axes respectively, and the given line makes the angle α with the positive direction of the x-axis, we have

$$F \cos \theta = X \cos \alpha + Y \sin \alpha,$$

§ 111]   DOUBLE AND TRIPLE INTEGRALS

and therefore

$W = Xl\cos\alpha + Yl\sin\alpha.$

But

$l\cos\alpha = x_2 - x_1$

and

$l\sin\alpha = y_2 - y_1,$

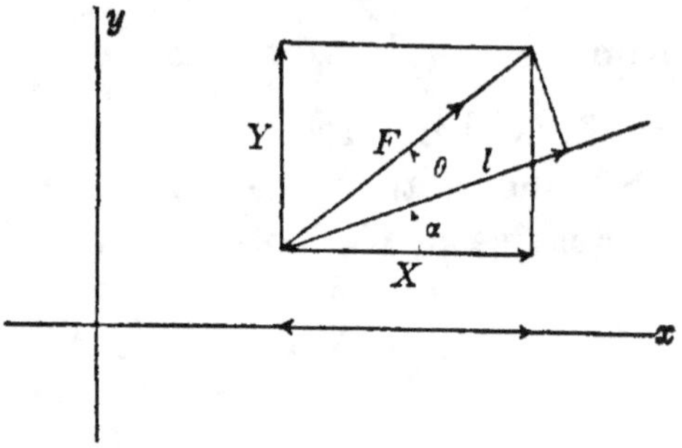

where (x_1, y_1) and (x_2, y_2) are the ends of the line. Hence

$$W = X(x_2 - x_1) + Y(y_2 - y_1). \tag{20}$$

Fig. 22

Suppose now that the force varies continuously in magnitude and position, and that it moves a given mass along a curve C from A to B. We have to consider how we shall define the work done. We shall want a definition that is consistent with that just given for a special case. In order to arrive at a decision on this point we divide the arc into a number of partial arcs and draw the chords to these. If the force along each chord were the same in magnitude and direction as at the beginning of the chord, the total work done in moving the mass from A to B along these chords would be

$$\sum F_i l_i \cos\theta_i,$$

where l_i is the length of the ith chord, F_i is the magnitude of the force at the beginning of this chord, and θ_i is the angle between the direction of this force and the chord. If the chords are all short, this sum will be a good approximation to what we must define as the work done in moving the mass along C from A to B, if we want to be consistent with the special definition already given. This consideration leaves us no alternative—we must define the work done as the limit of $\sum F_i l_i \cos\theta_i$ as we vary the arcs in such a way as to make the maximum length of the chords approach zero. Now if $\Delta_i s$ is the length of the arc whose chord is l_i, it follows from the discussion in §§ 186 and 187

that the ratios $\frac{\Delta_i s}{l_i}$ approach one uniformly. Therefore
$$\lim \sum F_i l_i \cos \theta_i = \lim \sum F_i \cos \theta_i \Delta_i s = \int_{s_1}^{s_2} F \cos \theta \, ds = W.$$
Here F and θ are functions of s. In view of (20) it amounts to the same thing to say that
$$W = \lim \sum (X_i \Delta_i x + Y_i \Delta_i y) = \int_{(x_1, y_1)}^{(x_2, y_2)} X dx + Y dy,$$
where X_i and Y_i are the components of the force at the initial point of the ith chord, and $\Delta_i x$ and $\Delta_i y$ are the projections of this chord on the axes. We see then that the work done is given by a line integral.

In space similar considerations lead us to the definition
$$W = \int_{(x_1, y_1, z_1)}^{(x_2, y_2, z_2)} X dx + Y dx + Z dz.$$

112. Total differentials. It follows from (18) that if
$$\frac{\partial Q}{\partial x} - \frac{\partial P}{\partial y} = 0, \quad \frac{\partial R}{\partial y} - \frac{\partial Q}{\partial z} = 0, \quad \text{and} \quad \frac{\partial P}{\partial z} - \frac{\partial R}{\partial x} = 0 \quad (21)$$
everywhere on Γ, we have
$$\int P dx + Q dy + R dz = 0.$$
If now this integral is zero when taken over any closed curve in a given region, equations (21) must be satisfied in this region. For if there were a point at which, for example,
$$\frac{\partial Q}{\partial x} - \frac{\partial P}{\partial y} > 0$$
and we draw through this point a plane perpendicular to the z-axis, we could surround the point by a closed curve in this plane of sufficiently small diameter to insure that $\frac{\partial Q}{\partial x} - \frac{\partial P}{\partial y}$ be positive throughout the region of the plane enclosed by this curve. This is due to the fact that this function is continuous within this region. Then the surface

§ 112] DOUBLE AND TRIPLE INTEGRALS

integral in the right member of (18) when applied to this region would not be zero and Stokes' theorem would be contradicted.

If A is a region throughout which equations (21) are satisfied, the line integrals

$$\int P\,dx + Q\,dy + R\,dz$$

along two curves connecting two points M and N of the region are equal, provided that there is a surface Γ passing through the two curves and lying wholly in A. Under these restrictions the line integral from M to N is independent of the path of integration. If we keep M fixed and let N vary, the integral in question is then a function of the coordinates of N.

$$\int_M^N P\,dx + Q\,dy + R\,dz = u(x, y, z).$$

Now

$$u(x + \Delta x, y, z) - u(x, y, z) = \int_{(x, y, z)}^{(x+\Delta x, y, z)} P\,dx + Q\,dy + R\,dz,$$

since the line integral that defines $u(x + \Delta x, y, z)$ can be taken in part along the path from M to N that was used in determining $u(x, y, z)$. Moreover the integral in the right member of this equation can be taken along a path parallel to the x-axis, since for sufficiently small absolute values of Δx all points of the straight line from (x, y, z) to $(x + \Delta x, y, z)$ lie in the given region. That is,

$$\int_{(x, y, z)}^{(x+\Delta x, y, z)} P\,dx + Q\,dy + R\,dz = \int_x^{x+\Delta x} P\,dx.$$

But

$$\int_x^{x+\Delta x} P\,dx = P(x + \theta \Delta x, y, z)\Delta x,$$

where $0 < \theta < 1$. Hence

$$\frac{\partial u}{\partial x} = \lim_{\Delta x \to 0} \frac{u(x + \Delta x, y, z) - u(x, y, z)}{\Delta x}$$

$$= \lim P(x + \theta \Delta x, y, z) = P(x, y, z).$$

Similarly $\dfrac{\partial u}{\partial y} = Q(x, y, z)$ and $\dfrac{\partial u}{\partial z} = R(x, y, z)$. This is equivalent to saying that

$$Pdx + Qdy + Rdz = \frac{\partial u}{\partial x}dx + \frac{\partial u}{\partial y}dy + \frac{\partial u}{\partial z}dz.$$

Hence when equations (21) are satisfied

$$Pdx + Qdy + Rdz$$

is the total differential of some function. Conversely, if this function is the differential of a function u, then $P = \dfrac{\partial u}{\partial x}$, $Q = \dfrac{\partial u}{\partial y}$, and $R = \dfrac{\partial u}{\partial z}$, and equations (21) are satisfied.

As in the case of two variables, we see that

$$\int_{(x_0, y_0, z_0)}^{(x, y, z)} (Pdx + Qdy + Rdz) = u(x, y, z) - u(x_0, y_0, z_0).$$

EXERCISES

1. Find the potential and attraction of a thin uniform rod at a point on its perpendicular bisector.
2. At a point on the perpendicular to the rod at one end.
3. Of a circular disc at a point on its axis.
4. Of a spherical shell at an exterior point.
5. At an interior point.
6. Find the attraction of a homogeneous sphere of radius r upon a unit mass at a distance d from the center $(d > r)$. Suppose the unit mass to be on the x-axis. The resultant attraction is then along the x-axis and amounts to

$$A = \rho \iiint_V \frac{(d - x)dxdydz}{[(d - x)^2 + y^2 + z^2]^{3/2}}.$$

The integral is taken over the sphere and ρ is the density. (Cf. Ex. 4.)

7. Show that the potential V at (ξ, η, ζ) due to a homogeneous volume satisfies the partial differential equation

$$\frac{\partial^2 V}{\partial \xi^2} + \frac{\partial^2 V}{\partial \eta^2} + \frac{\partial^2 V^2}{\partial \zeta} = 0.$$

Assume that (ξ, η, ζ) is outside the volume.

8. Compare the work done against gravity in moving a particle from the point (x_0, y_0, z_0) to the point (x_1, y_1, z_1) along two different paths.

9. A certain force at any point of the xy-plane is proportional to the distance of this point from the origin and is directed toward the origin. Compare the work done in moving a particle against this force from the point $(1, 0)$ to the point $(1, 4)$ along the two arcs of the circle $(x - 1)^2 + (y - 2)^2 = 4$.

10. Show that the work done in moving a unit mass from the surface of the earth to infinity is equal to the potential at a point on the surface due to the earth's attraction.

MISCELLANEOUS EXERCISES

1. Prove that, if an arc of a plane curve revolve about an axis in its plane, the area of the surface formed is equal to the length of the arc multiplied by the length of the path described by the center of mass of the arc. (Assume that the arc does not cut the axis.)

2. Prove that, if a plane area revolve about an axis in its own plane, the volume of the solid generated is equal to the area multiplied by the length of the path described by the center of mass of the area. (Assume that the axis does not cut the area.)

The two theorems in Exercises 1 and 2 are due to Pappus of Alexandria, about 300 A.D.

3. Prove that the sum of the moments of inertia of a plane area about two perpendicular axes in its plane is equal to its moment about an axis through their intersection and perpendicular to the plane.

4. Find the difference between the moment of inertia of a plane area about a given axis and its moment of inertia about a parallel axis through its center of mass.

5. Solve the same problem with respect to the moment of inertia of a solid about a given axis.

6. Show that the definition of volume given in § 103 is consistent with the formula $V = \iiint_V dx\,dy\,dz$.

7. Prove that the double integral $\iint_C f(x, y)dx\,dy$ approaches a limit as C becomes infinite in case $\iint_C |f(x, y)|dx\,dy$ does.

8. Show that the double integral $\int_0^a \int_0^a \sin(x^2 + y^2)dx\,dy \to \dfrac{\pi}{4}$ as $a \to \infty$ but that the same integral over the quadrant bounded by the axes and the arc $x^2 + y^2 = R^2$ does not approach any limit as $R \to \infty$. See § 214.

9. Prove that if $F(\alpha) = \iint_S f(x, y, \alpha)dxdy$, where $f(x, y, \alpha)$ and $f_\alpha(x, y, \alpha)$ are continuous when the point (x, y) is in the closed region S and $\alpha_0 \leqq \alpha \leqq \alpha_1$, then $F'(\alpha) = \iint_S f_\alpha(x, y, \alpha)dxdy$. It is assumed that S is independent of α.

10. Show that any force in space that is directed to a fixed point and is equal to a continuous function of the distance from this point has its component in any direction equal to the derivative of a certain function in this direction.

11. Find the moment of inertia of a rectangle about a vertex in case the density varies as the square of the distance from the center.

12. Describe the region over which the integral

$$\int_{-r}^{r} \int_{-\sqrt{r^2-x^2}}^{\sqrt{r^2-x^2}} \int_{-\sqrt{r^2-x^2-y^2}}^{\sqrt{r^2-x^2-y^2}} f(x, y, z)dxdydz$$

is taken.

13. Is the integral $\displaystyle\int_{-r}^{r} \int_{-\sqrt{r^2-x^2}}^{\sqrt{r^2-x^2}} \int_{0}^{x^2+y^2} dxdydz$ equal to

$$4\int_{0}^{r} \int_{0}^{\sqrt{r^2-x^2}} \int_{0}^{x^2+y^2} dxdydz?$$

14. Write the integral in Exercise 12 in five other ways.

CHAPTER IX

INFINITE SERIES

113. Definition. A set of numbers arranged in a one-to-one correspondence with the positive integers is called a *sequence*. The numbers (which need not be distinct) are called the terms of the sequence, but they are not sufficient to determine the sequence. It is necessary that they be arranged in a definite order. There may, or may not, be a last term. If there is, the sequence is said to be finite and is without importance in this discussion. If there is no last term the sequence is said to be an infinite sequence.

Examples.

(a) The set of all the positive integers arranged in the order of magnitude.

(b) The negative integers arranged in the order of their absolute values.

(c) The numbers $1, \frac{1}{2}, \frac{1}{3}, \cdots, \frac{1}{n}, \cdots$, where n is a positive integer.

(d) The minor approximations to the square root of 5; that is, $2, 2.2, 2.23, \cdots$. The identity of the terms of this sequence is not immediately obvious, but can readily be determined.

(e) It is possible to put the rational numbers from 0 to 1 into one-to-one correspondence with the positive integers, and thus to form a sequence from them.

We shall use the symbol (a_n) to denote the sequence $a_1, a_2, \cdots, a_n, \cdots$. In many cases it is important to know whether a_n approaches a finite limit as n increases without limit. If it does, the sequence is said to be *convergent*; otherwise it is said to be *divergent*.

114. The greatest limit of a bounded sequence. If the sequence is bounded—that is, if there is a number M such that $|a_n| < M$ for all values of n—we divide all real numbers into two classes by assigning to class A every real number that is exceeded by an infinite number of terms of the sequence, and to class B all other real numbers. This classification determines a number λ (see Chapter I, Theorem 4). Every number less than λ is in A and every number greater than λ is in B. The number itself may be in either class. It is called the *greatest limit* of the sequence. Similarly there is a number λ' which is called the *least limit* of the sequence. We shall use the following notation: $\lambda = \overline{\lim} \, a_n$; $\lambda' = \underline{\lim} \, a_n$. The reader should not confuse λ with the upper limit of the sequence or λ' with its lower limit. In the sequence $2, 1\frac{1}{2}, 1\frac{1}{4}, \cdots 1 + \frac{1}{2^n}, \cdots$, for example, the greatest limit is 1, while the upper limit is 2. On the other hand, for the sequence in § 113(c) the greatest limit and the upper limit are the same.

Let $\epsilon_1, \epsilon_2, \epsilon_3, \cdots, \epsilon_n, \cdots$, be an infinite sequence of positive numbers such that $\epsilon_{n-1} > \epsilon_n$ and $\epsilon_n \to 0$ as $n \to \infty$. If (a_n) is a bounded sequence for which the greatest limit is λ, there is an unlimited number of terms of the sequence in the interval $(\lambda - \epsilon_1, \lambda + \epsilon_1)$. Let a_1' be such a term distinct from λ; and in general let a_n' be a term of the sequence that lies in the interval $(\lambda - \epsilon_n, \lambda + \epsilon_n)$ and is distinct from a_i' when $i < n$. Then (a_n') is an infinite partial sequence of (a_n) that converges to λ. If there existed an infinite partial sequence that converged to a limit greater than λ, there would be an unlimited number of members of the original sequence greater than $\lambda + \epsilon$, where ϵ is a suitably chosen positive number. But this is impossible. These facts justify the name given to λ; it is the greatest limit of all possible infinite partial sequences. That there may be infinite partial sequences that converge to limits less than λ is clear from the following example:

$$\frac{1}{2}, \frac{1}{4}, \frac{3}{4}, \frac{1}{8}, \frac{7}{8}, \cdots, \frac{1}{2^n}, 1 - \frac{1}{2^n}, \cdots.$$

§ 114] INFINITE SERIES 207

In this case 0 is the limit of the partial sequence $\frac{1}{2}, \frac{1}{4}, \cdots,$ $\frac{1}{2^n}, \cdots$, while 1 is the limit of the partial sequence $\frac{1}{2}, \frac{3}{4}, \cdots, 1 - \frac{1}{2^n}, \cdots$, and $\lambda = 1$.

THEOREM 1. *In order that the sequence* (a_n) *be convergent it is necessary and sufficient that for any preassigned positive number ϵ there be a number N such that $|a_n - a_{n+p}| < \epsilon$ for any positive integer n greater than N and every positive integer p.*

This is merely a special case of Theorem 6, Chapter I.

THEOREM 2. *A monotonic bounded sequence is convergent.*
This is a special case of Theorem 7, Chapter I.

EXAMPLE. Consider the sequence in which

$$a_n = \left(1 + \frac{1}{n}\right)^{n+1}.$$

For $n > 1$

$$\frac{\left(1 + \frac{1}{n-1}\right)^n}{\left(1 + \frac{1}{n}\right)^n} = \left(\frac{n^2}{n^2 - 1}\right)^n$$

$$= \left(1 + \frac{1}{n^2 - 1}\right)^n > 1 + \frac{n}{n^2 - 1},$$

as may be seen from the expansion of $\left(1 + \frac{1}{n^2 - 1}\right)^n$ by the binomial theorem. Hence

$$\frac{\left(1 + \frac{1}{n-1}\right)^n}{\left(1 + \frac{1}{n}\right)^n} > 1 + \frac{n}{n^2} = 1 + \frac{1}{n}.$$

That is, $\left(1 + \frac{1}{n-1}\right)^n > \left(1 + \frac{1}{n}\right)^{n+1}$, and the sequence is monotonically decreasing. Moreover it has a lower bound since $\left(1+\dfrac{1}{n}\right)^{n+1} > 1$ for all positive integral values of n. The sequence therefore converges. Its limit is the same as the limit of $\left(1+\dfrac{1}{n}\right)^{n}$. Cauchy represented this limit by e as far back as 1731.

EXERCISES

Is the sequence (a_n) in each of the following cases bounded or unbounded; convergent or divergent; monotonic?

1. $a_n = \dfrac{1}{n}$. 2. $a_n = \dfrac{1}{n^2}$. 3. $a_n = (-1)^n \dfrac{1}{n}$. 4. $a_n = (-1)^n$.

5. $a_n = \dfrac{n-2}{n+1}$. 6. $a_n = \log n$. 7. $a_n = \dfrac{\sin \theta_n}{n}$.

8. $1, 2, \dfrac{1}{2}, -2, -\dfrac{1}{2}, 3, \dfrac{1}{3}, -3, -\dfrac{1}{3}, \cdots, n, \dfrac{1}{n}, -n, -\dfrac{1}{n}, \cdots$.

9. $a_n = \dfrac{1}{2}[1 - (-1)^n]$. 10. $a_n =$ the nth prime number.

11. A sequence that converges to zero is called a "null sequence." Prove that every infinite partial sequence of a null sequence is a null sequence.

12. Show that (a_n) is a null sequence if $a_n = \dfrac{1 + \dfrac{1}{2} + \cdots + \dfrac{1}{n}}{n}$.

13. Show that if $a_n = \dfrac{1}{n+1} + \dfrac{1}{n+2} + \cdots + \dfrac{1}{2n}$, then (a_n) is monotonic.

14. What is the limit of a_n as $n \to \infty$ if $a_{n+2} = \frac{1}{2}(a_n + a_{n+1})$?

15. Show that $a_n \to (a_1 \cdot a_2{}^2)^{1/3}$ as $n \to \infty$ in case $a_n > 0$ and $a_{n+2} = \sqrt{a_n \cdot a_{n+1}}$.

115. Infinite series. Associated with any sequence (a_n) is another sequence $S_0 = a_0, S_1 = a_0 + a_1, S_2 = a_0 + a_1 + a_2, \cdots, S_n = \sum_{\nu=0}^{n} a_\nu$. It is customary to represent this second sequence by one or the other of the symbols $a_0 + a_1 + \cdots + a_n + \cdots$ and $\sum_{\nu=0}^{\infty} a_\nu$. We call either of these symbols an *infinite series* with the *terms* $a_0, a_1, \cdots, a_n, \cdots$. The numbers S_n are called the *partial sums* of the series. We say that the series converges or diverges according as the sequence (S_n) converges or diverges. If the series converges, the limit of the sequence is called the *sum* of the series. It should be remembered however that this sum is not obtained by the mere process of summation, but by such a process followed by a passage to a limit.

It is important to observe that the series $\sum_{\nu=k}^{\infty} a_\nu$ is convergent or divergent according as $\sum_{\nu=0}^{\infty} a_\nu$ is convergent or divergent. For if $\sum_{n-k} = a_k + a_{k+1} + \cdots + a_n$, then $S_n - \sum_{n-k} = S_{k-1}$. But S_{k-1} is independent of n. Hence S_n and $\sum_{n-k}$ are both convergent or both divergent. If then we are interested in determining merely whether a series is convergent or not, we may neglect as many terms at the beginning of the series as seems convenient.

116. General test for convergence. We have directly from Theorem 1 the following general test for convergence:

In order that the series $\sum a_n$ converge it is necessary and sufficient that for an arbitrary positive ϵ there be a positive number N such that when $n > N$ we have $|S_{n+p} - S_n| < \epsilon$, for every positive integral value of p.

If this condition is satisfied, then for an arbitrary positive ϵ and any sufficiently large n we have

$$|S_{n+1} - S_n| < \epsilon,$$

or

$$|a_{n+1}| < \epsilon.$$

It follows that in order for a series to be convergent it is necessary that $a_n \to 0$ as $n \to \infty$. But this condition is not sufficient to insure convergence, as may be seen from a consideration of the harmonic series $\sum \dfrac{1}{n}$, which is divergent. (See § 117.)

117. Series with positive terms. The simplest series are those whose terms are real constants of the same sign. We shall assume that they are all positive. The sequence of partial sums of such a series is a monotonically increasing sequence. The series is therefore convergent if this se-

quence is bounded (Theorem 2). Consider for example the series $\sum \frac{1}{(n+1)!}$. Here
$$S_n = 1 + \frac{1}{2!} + \cdots + \frac{1}{n!}.$$
Now compare S_n with Σ_n, where
$$\Sigma_n = 1 + \frac{1}{2} + \cdots + \frac{1}{2^{n-1}}.$$
We have $3! = 3 \cdot 2 > 2^2$, $4! > 2^3$, and, in general, $n! > 2^{n-1}$. Thus beginning with the third term on, every term of S_n is less than the corresponding term of Σ_n, and the first two terms of the two sums are the same. Hence $S_n < \Sigma_n$ for $n > 2$. But $\Sigma_n = \dfrac{1 - \frac{1}{2^n}}{1 - \frac{1}{2}} = 2 - \frac{1}{2^{n-1}} < 2$. The original series is therefore convergent.

THEOREM 3. *If each of the terms of a series of positive terms is less than, or equal to, the corresponding term of a convergent series, the series is convergent.*

For all the partial sums of the given series are, for all values of n, less than, or equal to, the sum of the second series.

THEOREM 4. *If each of the terms of a series of positive terms is greater than, or equal to, the corresponding term of a divergent series of positive terms, the series is divergent.*

For the sum of the first n terms of the first series is not less than the sum of the first n terms of the second series, and this latter increases without limit with n.

EXAMPLE. If the series is $\sum_{n=0}^{\infty} \frac{1}{n+1}$, we have
$$S_n = 1 + \frac{1}{2} + \cdots + \frac{1}{n+1}.$$
$$S_{2^m-1} = \left(1 + \frac{1}{2}\right) + \left(\frac{1}{3} + \frac{1}{4}\right) + \cdots$$
$$+ \left(\frac{1}{2^{m-1}+1} + \cdots + \frac{1}{2^m}\right).$$

If we put

$$\Sigma_{2^m-1} = \left(\frac{1}{2}+\frac{1}{2}\right) + \left(\frac{1}{4}+\frac{1}{4}\right) + \cdots$$
$$+ \left(\frac{1}{2^m} + \cdots + \frac{1}{2^m}\right)$$

it is clear that $S_{2^m-1} > \Sigma_{2^m-1}$. But the first group in Σ_{2^m-1} is equal to 1, and each of the remaining groups is equal to $\frac{1}{2}$. Hence

$$\Sigma_{2^m-1} = 1 + \frac{1}{2}(m-1) = \frac{1}{2}(m+1).$$

It follows that if N is an arbitrarily large positive number we have

$$S_{2^m-1} > \Sigma_{2^m-1} > N$$

for all positive integral values of m such that $m > 2N - 1$. The series therefore diverges, as does also the series $\sum_{n=0}^{\infty} \frac{1}{(n+1)^\alpha}$, when $\alpha < 1$. A slight modification of the discussion shows that the latter series converges when $\alpha > 1$.

THEOREM 5. *If the series of positive terms Σa_n converges, the series of positive terms Σb_n also converges if there is a number m such that when $n > m$ we have $\frac{b_{n+1}}{b_n} \leq \frac{a_{n+1}}{a_n}$. If the first series diverges and for $n > m$ we have $\frac{b_{n+1}}{b_n} \geq \frac{a_{n+1}}{a_n}$, the second series diverges.*

Since the convergence or divergence of a series is not affected by multiplying every term by the same number different from zero, and since the ratio of two terms is not affected by this multiplication, we can assume that $b_{m+1} < a_{m+1}$. If now

$$\frac{b_{m+2}}{b_{m+1}} \leq \frac{a_{m+2}}{a_{m+1}}, \quad \text{then} \quad b_{m+2} \leq a_{m+2} \cdot \frac{b_{m+1}}{a_{m+1}} < a_{m+2};$$

and in general $b_{m+p} < a_{m+p}$ (p a positive integer). It follows that the series $\sum b_n$ is convergent.

The proof of the second part of the theorem is similar to this.

The last three theorems are of the nature of comparison tests—they depend upon particular series whose convergence behavior is known. The most useful standard for this purpose is the geometric series since we know that it converges when $|r| < 1$ and diverges when $|r| \geq 1$, where r is the common ratio.

118. Cauchy's root criterion. We shall make use of the comparison test to establish the following important theorem:

THEOREM 6. *If all the terms of the series $\sum a_n$ are positive and if $\sqrt[n]{a_n} < k < 1$ for all sufficiently large values of n, the series converges. If $\sqrt[n]{a_n} \geq 1$ for all sufficiently large values of n, the series diverges.*

For if $\sqrt[n]{a_n} < k$, $a_n < k^n$ and from a certain point on the terms of the given series are less than the corresponding terms of a convergent series. The given series is therefore convergent. If on the other hand $\sqrt[n]{a_n} \geq 1$ for all sufficiently large values of n, then a_n does not approach zero as n increases indefinitely, and the series diverges.

This criterion is known as *Cauchy's root criterion*.[1] It may be stated in somewhat less general form as follows:

If $\sqrt[n]{a_n} \to k$ the series converges when $k < 1$ and diverges when $k > 1$. When $k = 1$ the series may be either convergent or divergent. But if $\sqrt[n]{a_n} \to 1$ and is always greater than 1, the series diverges.

119. D'Alembert's test. *If in a given series of positive terms the ratio of any term after a given one to the preceding term is less than a fixed number that is itself less than one, the series converges. If this ratio after a certain term remains greater than, or equal to, one, the series diverges.*

The proof is left to the reader.

[1] Sometimes referred to as *Cauchy's quotient criterion*.

120. Application of the greatest limit. If a_n is the general term of a series of positive terms, the sequence

$$a_1, a_2^{1/2}, \cdots, a_n^{1/n}, \cdots$$

may, or may not, be bounded. If it is not, a_n does not approach zero as n increases indefinitely, and the series diverges. If the sequence is bounded and λ is its greatest limit, the series converges when $\lambda < 1$ and diverges when $\lambda > 1$.

If $\lambda < 1$, let $1 - \alpha$ be a number between λ and 1. Then $a_n^{1/n} < 1 - \alpha$ for all sufficiently large values of n, and the series converges. In case $\lambda > 1$ we choose a number $1 + \alpha$ between 1 and λ. Then $a_n^{1/n} > 1 + \alpha$ for all sufficiently large values of n, and the series diverges. This leaves the case $\lambda = 1$ undecided.

121. Series with arbitrary terms. We have been considering series all of whose terms are of one sign. But what we have said applies also with obvious modifications to series with only a finite number of terms of a given sign. When we remove all restrictions upon the signs of the terms an important difference in series presents itself. Consider, for example, the two series

$$1 - \frac{1}{2} + \frac{1}{3} + \cdots + (-1)^{n-1}\frac{1}{n} + \cdots$$

and

$$1 - \frac{1}{2^2} + \frac{1}{3^2} + \cdots + (-1)^{n-1}\frac{1}{n^2} + \cdots.$$

It will be shown later (Theorem 10) that they are both convergent. But the series whose terms are the absolute values of the terms of the first series is divergent, while the series formed in the same way from the second series is convergent (see § 117). This leads to the important notion of absolute convergence.

Absolute convergence. If the series $\sum |a_n|$ is convergent the series $\sum a_n$ is said to be *absolutely convergent*. If $\sum a_n$ converges and $\sum |a_n|$ does not, the former is said to be *conditionally convergent*.

Theorem 7. *If a series is absolutely convergent it is convergent.*

Since $\sum |a_n|$ is convergent, there is associated to every positive ϵ an N such that when $n > N$

$$|a_n| + |a_{n+1}| + \cdots + |a_{n+p}| < \epsilon$$

for every positive integral value of p. But

$$|a_n + \cdots + a_{n+p}| \leqq |a_n| + \cdots + |a_{n+p}|,$$

and therefore $\sum a_n$ is convergent.

122. Let $b_0, b_1, \cdots$ denote the positive terms of $\sum a_n$ in the order of their occurrence, and $-c_0, -c_1, \cdots$ the negative terms in the order of their occurrence. Put $\sum_n = b_0 + b_1 + \cdots + b_n$ and $\sum_n' = c_0 + c_1 + \cdots + c_n$. If $\sum a_n$ converges, either $\sum b_n$ and $\sum c_n$ both converge or both diverge. For $S_n = \sum_m - \sum'_{m'}$, if $n = m + m'$ and if either $\sum_m$ or $\sum'_{m'}$ converged and the other diverged S_n would increase indefinitely in absolute value.

The series $\sum a_n$ converges absolutely if $\sum b_n$ and $\sum c_n$ both converge, and conversely. For $|a_0| + |a_1| + \cdots + |a_n| = \sum_m + \sum'_{m'}$.

Theorem 8. *The order of the terms in a convergent series of positive terms may be changed in any way without altering the sum of the series.*

If the given series is

$$S = a_0 + a_1 + \cdots + a_n + \cdots,$$

let

$$\Sigma = b_0 + b_1 + \cdots + b_n + \cdots$$

be a series every one of whose terms is in the original series and which contains every term of the original series without repetition. Let $S_m' = \sum_0^m b_i$. Then since every b_i occurs in the first series and all the terms are positive we have for sufficiently large values of n

$$S_m' < S_n < S.$$

It follows that S_m' approaches a limit $S' \leqq S$ as $n \to \infty$. On the other hand, for a given n and sufficiently large values of m
$$S_n < S_m' < S'.$$
Hence $S \leqq S'$ and therefore $S = S'$.

THEOREM 9. *The terms of a convergent series of positive terms may be grouped together in any manner without altering the sum of the series.*

In view of Theorem 8 we may assume that the terms included within any group are consecutive terms. Suppose then that
$$b_0 = a_0 + a_1 + \cdots + a_p, \qquad b_1 = a_{p+1} + \cdots + a_q, \cdots.$$
Now $S_m' = \sum_0^m b_i = S_N$ for some $N \geqq m$. Moreover $N \to \infty$ as $m \to \infty$, and therefore $\lim_{m \to \infty} S_m' = \lim_{N \to \infty} S_N = S$.

Suppose now that $S = \sum a_n$ is any absolutely convergent series. Then $\sum(a_n + |a_n|)$ is convergent since its terms are all zero or positive and the sum of any number of them does not exceed $2S$. Denote its sum by S_1 and the sum $\sum|a_n|$ by S_2. Then $S_1 = S + S_2$. If now $S' = \sum a_n'$ is obtained from the original series by any rearrangement of terms, and $S_1' = \sum(a_n' + |a_n'|)$ and $S_2' = \sum|a_n'|$ then $S_1' = S_1$ and $S_2' = S_2$. Hence $S' = S$. A similar argument shows that no regrouping of the terms of an absolutely convergent series affects the sum. Hence from the point of view of numerical computation an absolutely convergent series may be treated as if it were the sum of a finite number of terms. But a conditionally convergent series may not be so treated. As a matter of fact it is possible so to arrange the terms of a conditionally convergent series as to get a convergent series whose sum is any preassigned number A. For if $\sum a_n$ is a conditionally convergent series, the series $\sum b_n$ and $\sum c_n$ are both divergent. If now A is an arbitrary number (say positive), we select positive terms from the given series in the order in which they occur in it until the sum exceeds A. This is possible since the series of the positive terms is divergent. Then we add to these in the order in which they occur in the given series just enough negative terms to make the total sum less than A. We now add positive terms, beginning where we left off until the sum is greater than A. We then return to the negative terms, and so on. The partial sums of the new series differ from A by amounts whose absolute values do not exceed the last term of the original series used. But this approaches zero as we go further and further out. Hence the new series converges to A. If A were negative, the proof would be similar. We could even make the new series divergent.

THEOREM 10. *The series $\sum a_n$ converges if the following conditions are satisfied:*

(1) *The terms are alternately positive and negative.*
(2) $|a_{n+1}| < |a_n|$.
(3) $a_n \to 0$.

If we put $|a_n| = b_n$, then $b_n > 0$ and $b_{n+1} < b_n$. Hence
$$S_{2k} = (b_0 - b_1) + (b_2 - b_3) + \cdots \\ + (b_{2k-2} - b_{2k-1}) + b_{2k} > 0.$$
Moreover
$$S_{2k} = b_0 - (b_1 - b_2) - \cdots - (b_{2k-1} - b_{2k}) < b_0.$$
From this it follows that (S_{2k}) is a bounded monotonically decreasing sequence. It therefore converges. But $S_{2k+1} = S_{2k} + a_{2k+1}$, and $a_{2k+1} \to 0$. That is, S_{2k+1} approaches the same limit as S_{2k}, and the series converges.

It follows immediately from this theorem that the series $\sum (-1)^{n-1} \frac{1}{n}$ and $\sum (-1)^{n-1} \frac{1}{n^2}$ to which reference was made in § 121 are convergent.

EXERCISES

Which of the following series are convergent?

1. $\sum (-1)^n$. 2. $\sum \frac{1}{1+n^2}$. 3. $\sum \frac{n!}{n^a} (a > 1)$. 4. $\sum \frac{1}{(\log n)^3}$.
5. $\sum \frac{1}{(\log n)^n}$. 6. $\sum \left(\frac{2n}{2n+1} - \frac{2n-1}{2n} \right)$.

7. $\sum \dfrac{1}{\sqrt{(n+1)(n+2)}}$. 8. $\sum a^n \ (0 < a < 1)$. 9. $\sum \dfrac{2^{n-1}}{n!}$.

10. Show that $\dfrac{n}{2} < 1 + \dfrac{1}{2} + \cdots + \dfrac{1}{2^n - 1} < n$.

11. Make a rough estimate of the value of n in order that $S = \sum\limits_{\nu=1}^{n} \dfrac{1}{\nu}$ shall exceed 16.

12. Show how to rearrange and group the terms of the series $\sum (-1)^{n-1} \dfrac{1}{n}$ in order that the resulting series shall converge to $\dfrac{3}{2}$. Give the first three groups.

123. Series with variable terms. We pass now to the consideration of series $\sum f_n(x)$ whose terms are functions of a single real variable. If the series converges for every value of the variable in a given interval, we say that *it converges in the interval.*

If the series converges to $S(x)$ in the interval (a, b), then for any given x in the interval and any $\epsilon > 0$ there exists a positive number m such that $|R_n(x)| < \epsilon$ for every $n > m$, where $\sum\limits_{\nu=0}^{n} f_\nu(x) + R_n(x) = S(x)$. There is considerable latitude in the choice of m, since if the stated conditions are satisfied for one value of m they are satisfied for every greater value. We shall in each case deal with the least possible value. It is obvious that this value of m depends in general upon both ϵ and x. We shall at times emphasize its dependence upon x by writing it m_x. For a given ϵ this function m_x may, or may not, have an upper bound in the interval (a, b). Consider, for example, the series

$$1 + x + x^2 + \cdots + x^n + \cdots$$

in the interval $\left(-\dfrac{1}{2}, \dfrac{1}{2}\right)$. Now $S_n(x) = \dfrac{1 - x^{n+1}}{1 - x}$ and $S(x) = \dfrac{1}{1 - x}$. Hence

$$R_n(x) = \dfrac{1}{1 - x} - \dfrac{1 - x^{n+1}}{1 - x} = \dfrac{x^{n+1}}{1 - x}$$

and $|R_n(x)| \leq (\tfrac{1}{2})^n$ for every value of x in the interval. If now after ϵ has been chosen we so select m that $(\tfrac{1}{2})^m < \epsilon$, we shall have $|R_n(x)| < \epsilon$ for every $n > m$ irrespective of the value of x in the given interval. That is, m_x is bounded in the interval.

Consider now on the other hand the series defined by $S_n(x) = nxe^{-nx^2}$ in the interval $(0, 1)$. The terms of the series are

$$f_0(x) = S_0(x) = 0, f_1(x) = S_1(x) - S_0(x) = xe^{-x^2}, \cdots,$$
$$f_n(x) = S_n(x) - S_{n-1}(x) = nxe^{-nx^2} - (n-1)xe^{-(n-1)x^2}, \cdots.$$

Since $\lim S_n(x) = 0$ for all finite values of x, we have $R_n(x) = 0 - nxe^{-nx^2}$, and $|R_n(x)| = e^{-1/n}$ for $x = \dfrac{1}{n}$. But $e^{-1/n} > \dfrac{1}{e}$ for $n > 1$. Hence if ϵ is taken less than $\dfrac{1}{e}$, there is for an arbitrarily large number N, an x in the interval $(0, 1)$, namely $x = \dfrac{1}{n}$ (n a positive integer greater than N), for which $m_x > N$. That there is a definite value of m_x for this value of x follows from the convergence of the series. In this case m_x is unbounded in the interval $(0, 1)$.

DEFINITION. If the series $\sum f_n(x)$ converges in the interval (a, b), and if m_x is bounded in this interval, the series is said to converge *uniformly* in the interval. If m_x is unbounded in the interval, the series is said to converge *nonuniformly* in the interval.

It is to be noted that uniform convergence is always associated with an interval. It means nothing to say that a series converges uniformly at a point.

The distinction between these two kinds of convergence can be made clear geometrically. We draw the three parallel curves $y = S(x)$ and $y = S(x) \pm \epsilon$. If now there is a positive number m such that when $n > m$ all the approximation curves $y = S_n(x)$ lie between the curves $y = S(x) \pm \epsilon$ throughout the interval, the series is uniformly convergent in this interval. If however, no matter how great m is, there is an $n > m$ such that the curve $y = S_n(x)$ does not lie within this band throughout the whole interval, the series does not converge uniformly in the interval. Figures 23 and 24 illustrate this distinction with respect to the two series just discussed.

In Figure 23 the curve $y = S_n(x)$ has been sketched only roughly. In this case $S(x) = \dfrac{1}{1-x}$ and $S_n(x) = \dfrac{1 - x^{n+1}}{1 - x}$.

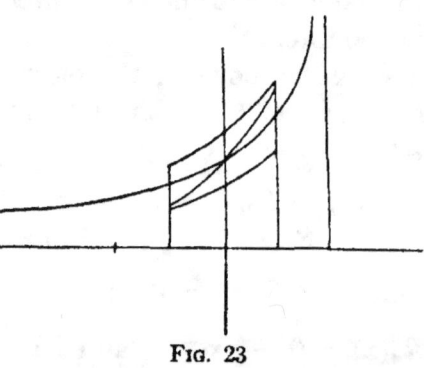

Fig. 23

For the other series we have $S(x) = 0$ and $S_n(x) = nxe^{-nx^2}$. This last curve has a maximum at the point $\left(\dfrac{1}{\sqrt{2n}}, \sqrt{\dfrac{n}{2e}}\right)$.

Hence if $\sqrt{\dfrac{n}{2e}} \not< \epsilon$ the curve $y = nxe^{-nx^2}$ will not lie within the required band throughout the interval, although for any particular value of x in this interval, as x_1, there is an m such that for $n > m$ the curve $y = nxe^{-nx^2}$ cuts the line $x = x_1$ between the x-axis and the line $y = \epsilon$.

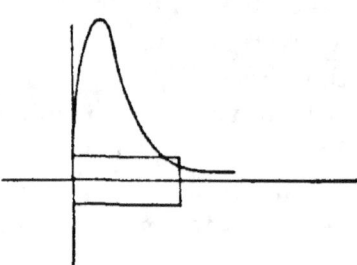

Fig. 24

THEOREM 11. *If the series $\Sigma f_n(x)$ converges uniformly to $S(x)$ in the interval (a, b) and the terms of the series are all continuous in this interval, then $S(x)$ is continuous in this interval.*

Let x_1 and $x_1 + h$ be two points in (a, b). By virtue of the uniform convergence of the series there is, for any $\epsilon > 0$, an n such that $|R_n(x)| < \dfrac{\epsilon}{3}$ throughout the interval. Using this value of n, we have $S(x) = \sum\limits_{\nu=0}^{n} f_\nu(x) + R_n(x)$.

This holds for any value of x in the interval. Hence

$$S(x_1) = \sum_{\nu=0}^{n} f_\nu(x_1) + R_n(x_1),$$

$$S(x_1 + h) = \sum_{\nu=0}^{n} f_\nu(x_1 + h) + R_n(x_1 + h).$$

Therefore

$$S(x_1 + h) - S(x_1) = \left[\sum_{0}^{n} f_\nu(x_1 + h) - \sum_{0}^{n} f_\nu(x_1)\right] \\ + R_n(x_1 + h) - R_n(x_1).$$

But

$$|R_n(x_1 + h)| < \frac{\epsilon}{3} \quad \text{and} \quad |R_n(x_1)| < \frac{\epsilon}{3}.$$

Moreover, since $\sum_{\nu=0}^{n} f_\nu(x)$ is the sum of a finite number of functions each continuous in (a, b), it is itself continuous in this interval, and we can take $|h|$ sufficiently small to insure that

$$\left|\sum_{\nu=0}^{n} f_\nu(x_1 + h) - \sum_{\nu=0}^{n} f_\nu(x_1)\right| < \frac{\epsilon}{3}.$$

For such a choice of h we have

$$|S(x_1 + h) - S(x_1)| < \frac{\epsilon}{3} + \frac{\epsilon}{3} + \frac{\epsilon}{3} = \epsilon.$$

This completes the proof of the theorem.

THEOREM 12. *If the series $\sum f_n(x)$ converges uniformly to $S(x)$ in the finite interval (a, b), and if its terms are continuous in this interval, the series*

$$\sum \int_\alpha^\beta f_n(x) dx,$$

where $a \leq \alpha < \beta \leq b$, converges to $\int_\alpha^\beta S(x) dx$.

Since the given series converges uniformly in the interval, there is associated to every positive ϵ a positive number m such that $|R_n(x)| < \epsilon$ for every x in (a, b) when $n > m$.

Moreover $S(x)$ is continuous in (a, b) and therefore integrable in (α, β). Now

$$\int_\alpha^\beta S(x)dx - \int_\alpha^\beta f_0(x)dx - \int_\alpha^\beta f_1(x)dx \cdots$$
$$\int_\alpha^\beta f_n(x)dx = \int_\alpha^\beta R_n(x)dx.$$

But $\left|\int_\alpha^\beta R_n(x)dx\right| < \epsilon(\beta - \alpha)$ for $n > m$. This is equivalent to saying that

$$\int_\alpha^\beta S(x)dx = \sum_{n=0}^\infty \int_\alpha^\beta f_n(x)dx.$$

If we keep α fixed and put x for β, we have

$$\int_\alpha^x S(x)dx = \sum \int_\alpha^x f_n(x)dx.$$

The argument shows that the series in the right member of this equation converges uniformly in (a, b) to $\int_\alpha^x S(x)dx$.

If the series converges to an integrable function, but the convergence is not uniform, it may, or may not, be integrable term by term. Consider first the series for which $S_n(x) = \dfrac{nx}{1 + n^2x^2}$. For every finite value of x we have $S(x) = 0$. Hence $\int_0^1 S(x)dx = 0$. On the other hand,

$$\sum_{\nu=1}^n \int_0^1 f_\nu(x)dx = \int_0^1 \frac{nx}{1 + n^2x^2}dx$$
$$= \frac{1}{2n}\log(1 + n^2x^2)\Big]_0^1 = \frac{1}{2n}\log(1 + n^2).$$

But $\dfrac{1}{2n}\log(1 + n^2) \to 0$ as $n \to \infty$. Hence

$$\sum_1^\infty \int_0^1 f_n(x)dx = 0 = \int_0^1 S(x)dx = 0.$$

The series can therefore be integrated term by term in the

interval $(0, 1)$. Is the convergence uniform in this interval? Is it uniform in the interval, (α, β), where $0 < \alpha < \beta$?

As a second example, consider the series for which $S_n(x) = nxe^{-nx^2}$. For any finite value of x we have $S(x) = 0$, and therefore $\int_0^1 S(x)dx = 0$. But

$$\int_0^1 nxe^{-nx^2}dx = -\frac{1}{2}e^{-nx^2}\Big|_0^1 = \frac{1}{2}(1 - e^{-n}) \to \frac{1}{2} \text{ as } n \to \infty.$$

Hence this series cannot be integrated term by term in the interval $(0, 1)$.

When we come to term-by-term differentiation of a series, the situation is somewhat different inasmuch as not every uniformly convergent series of differentiable functions can be differentiated term by term. Consider, for example, the series

$$\sum \frac{\sin nx}{n^2}.$$

It follows from § 124 that this series converges uniformly in every interval. But term-by-term differentiation gives the series $\sum \frac{\cos nx}{n}$, which diverges for $x = 0$. Under suitable conditions however a series of differentiable functions can be differentiated term by term. A set of such conditions is given in the following theorem:

THEOREM 13. *If the series $\sum f_n(x)$ converges for at least one value, c, of x in the interval (a, b) within which all its terms have continuous derivatives and if the series $\sum f_n'(x)$ converges uniformly in this interval to $\varphi(x)$, then $\sum f_n(x)$ converges uniformly in the interval to a function $S(x)$ and $S'(x) = \varphi(x)$.*

By hypothesis

$$\varphi(x) = f_0'(x) + f_1'(x) + \cdots + f_n'(x) + \cdots$$

and the series on the right is a uniformly convergent series of continuous functions in the interval (a, b). Hence $\varphi(x)$

is continuous and therefore integrable in this interval.

$$\int_c^x \varphi(x)dx = [f_0(x) - f_0(c)] + [f_1(x) - f_1(c)]$$
$$+ \cdots + [f_n(x) - f_n(c)] + \cdots.$$

The series on the right converges uniformly in (a, b) and the series $\Sigma f_n(c)$ converges by hypothesis. Hence $\Sigma f_n(x)$ converges uniformly to a function $S(x)$ and

$$\int_c^x \varphi(x)dx = S(x) - S(c).$$

From this we get by differentiation $\varphi(x) = S'(x)$.

124. The Weierstrass test for uniform convergence. It is in general a difficult matter to determine whether a given series converges uniformly in an interval, or not, even when we know that it converges. The following test is useful in many cases. It is known as the *Weierstrass M-Test*.

If there is a convergent series of positive constants M_0, M_1, $\cdots$, M_n, $\cdots$ such that $|f_n(x)| \leq M_n$ for every n and every x in the interval (a, b), the series $\Sigma f_n(x)$ converges absolutely and uniformly in this interval.

Since the series ΣM_n converges, there is an N for every positive ϵ such that when $n > N$

$$\sum_{\nu=n}^{n+k} M_\nu < \epsilon,$$

for every k. But $\left|\sum_{\nu=n}^{n+k} f_\nu(x)\right| \leq \sum_{\nu=n}^{n+k} M_\nu < \epsilon$, for every x in (a, b). That is, the series $\Sigma f_n(x)$ converges absolutely and uniformly in (a, b).

The only series that can be shown by this test to be uniformly convergent in an interval are absolutely convergent in this interval. But a series may be uniformly convergent in an interval without being absolutely convergent everywhere in the interval, and a series may be absolutely convergent throughout an interval without being uniformly convergent there. For example, the series

$\sum \dfrac{x^2}{(1+x^2)^n}$ is absolutely convergent in the interval (0, 1), but not uniformly convergent. On the other hand, the series $\sum \dfrac{\sin(2n+1)x}{2n+1}$ is uniformly convergent in the interval $\left(\dfrac{\pi}{4}, \dfrac{3\pi}{4}\right)$ (see Exercise 1, § 160 and Theorem 2, § 156), but is not absolutely convergent at the point $x = \dfrac{\pi}{2}$.

125. Abel's lemma. In order to establish a more delicate test for uniform convergence we first prove a set of inequalities the statement of which constitutes what is known as *Abel's lemma*.

If the set of positive numbers u_n ($n = 1, 2, \cdots, p$) is such that $u_n \leqq u_{n-1}$, and a_n is another set of p numbers, not necessarily positive, then

$$mu_1 < \sum_{n=1}^{p} a_n u_n < Mu_1,$$

where M and m are the greatest and the least respectively of the sums $a_1, a_1 + a_2, \cdots, a_1 + a_2 + \cdots + a_p$.

If we put $S_n = a_1 + a_2 + \cdots + a_n$ ($n = 1, 2, \cdots, p$), we have

$$a_1 = S_1,\ a_2 = S_2 - S_1,\ \cdots,\ a_p = S_p - S_{p-1}.$$

Then

$$\sum_{n=1}^{p} a_n u_n = S_1 u_1 + (S_2 - S_1)u_2 + \cdots + (S_p - S_{p-1})u_p$$
$$= S_1(u_1 - u_2) + S_2(u_2 - u_3) + \cdots$$
$$+ S_{p-1}(u_{p-1} - u_p) + S_p u_p.$$

But no one of the factors $u_{n-1} - u_n$ is negative, and therefore

$$S_n(u_n - u_{n+1}) \leqq M(u_n - u_{n+1}) \quad (n = 1, 2, \cdots, p-1).$$
$$S_p u_p \leqq M u_p.$$

Hence

$$\sum_1^p a_n u_n \leq M[(u_1 - u_2) + (u_2 - u_3) + \cdots \\ + (u_{p-1} - u_p) + u_p] = Mu_1.$$

It can be shown in a similar way that

$$\sum a_n u_n \geq m u_1.$$

126. Abel's test for uniform convergence. *The series $\sum a_n u_n(x)$ is uniformly convergent in the interval (a, b) provided*

(1) $\sum a_n$ *is convergent.*

(2) $u_n(x)$ *is positive throughout the interval and does not increase with n for a fixed value of x.*

(3) $u_1(x) < k$ *for all values of x in the interval.*

By virtue of the convergence of $\sum a_n$ there is associated to every positive ϵ a number m such that

$$\left| \sum_{n=m+1}^{m+p} a_n \right| < \frac{\epsilon}{k}$$

for any positive integral value of p. By Abel's lemma, for a fixed p,

$$\left| \sum_{m+1}^{m+p} a_n u_n(x) \right| < \frac{\epsilon}{k} u_{m+1}(x) < \epsilon,$$

since $u_{m+1}(x) \leq u_1(x) < k$. This proves the uniform convergence of the series $\sum a_n u_n(x)$ in the interval (a, b), since m is independent of x.

127. THEOREM 14. *If the functions $f_n(x)$ are all continuous in the closed interval (a, b) and if the series $\sum f_n(x)$ is uniformly convergent in the corresponding open interval, it is uniformly convergent in the closed interval.*

For, a positive ϵ having been chosen arbitrarily, there is an m such that

$$|S_n(x) - S_m(x)| < \epsilon$$

for $n > m$ and $a < x < b$. But $S_n(x)$ and $S_m(x)$ are continuous functions of x at the point $x = a$. Then there is a

$\delta > 0$ such that when $|x - a| < \delta$
$$|S_n(a) - S_n(x)| < \epsilon,$$
$$|S_m(x) - S_m(a)| < \epsilon.$$

Hence $|S_n(a) - S_m(a)| < 3\epsilon$. It follows that $S_n(a)$ converges and that a can be included in the interval of uniform convergence. A similar argument applies to the point $x = b$.

EXERCISES

1. Does the series $\sum \log(nx^2)$ converge for any value of x?

2. For what values of x does the series $\sum \dfrac{x^n}{n}$ converge?

3. For what values of x does the series $\sum \dfrac{\cos nx}{n}$ converge absolutely?

4. Prove that the series $\sum (-1)^{n-1} \dfrac{\sin nx}{n^\alpha}$ $(\alpha > 1)$ converges absolutely for every value of x. Does it converge uniformly in any interval?

5. Can the series $\sum \dfrac{1}{n} \sin \dfrac{x}{n}$ be differentiated term by term?

6. Show that the series in Exercise 5 converges uniformly in any finite interval.

7. Give an example of a series that converges uniformly and absolutely in an interval.

8. Give an example of a series that converges uniformly, but not absolutely, in an interval.

9. Give an example of a series that converges absolutely, but not uniformly, in an interval.

10. Discuss the convergence of the series for which $S_n(x) = \dfrac{nx}{1 + n^2 x^2}$ in the interval $(0, 1)$.

11. Do the same for the interval $(2, 3)$.

12. Draw the curve $y = \dfrac{nx}{1 + n^2 x^2}$ for $n = 3$.

13. Does the series for which $S_n(x) = x^n$ converge uniformly in the open interval $(0, 1)$?

14. Prove that the series $\sum c_n \cos nx$ and $\sum c_n \sin nx$ converge absolutely and uniformly in any interval in case $\sum c_n$ converges absolutely.

15. Show that the series $\sum \dfrac{1}{n^x}$ converges uniformly in the interval $1 + a < x$ $(a > 0)$.

16. Does the series for which $S_n(x) = \arctan(nx)$ converge uniformly in the interval $(0, 1)$?

Show that if $\sum a_n$ converges, the series in each of the next two exercises converges uniformly in the interval $(0, 1)$.

17. $\sum a_n \dfrac{x^n}{1 + x^n}.$ **18.** $\sum a_n \dfrac{x^n}{1 + x^{2n}}.$

19. Find $\displaystyle\int_0^1 f(x)dx$ in case $f(x) = \sum a_n x^n$, where $\sum a_n$ is convergent.

20. Find the derivative of $f(x) = \sum \dfrac{x^n}{n^2}$ for $|x| < 1$.

21. Integrate $f(x) = \sum \dfrac{a_n}{n^x}$ between the limits 0 and x ($0 < x \leq 1$), it being understood that $\sum a_n$ is convergent.

22. Differentiate $f(x) = \sum \dfrac{\cos nx}{n^3}.$

23. Obtain a power series for $\dfrac{1}{(1-x)^2}$. First expand $\dfrac{1}{1-x}$ and then apply the theorem concerning the Cauchy product. (See § 141.)

128. Differentiation of an infinite integral. It was shown in § 68 that if $f(x, \alpha)$ is a continuous function of x and α in the region $a \leq x \leq b$, $\alpha_1 \leq \alpha \leq \alpha_2$, and

$$F(\alpha) = \int_a^b f(x, \alpha) dx,$$

then

$$F'(\alpha) = \int_a^b f_\alpha(x, \alpha) dx,$$

provided that $f(x, \alpha)$ has a derivative with respect to α which is continuous in this region, and the limits a and b are independent of α. The proof depends essentially upon the fact that the limits a and b are both finite, and as a matter of fact the conclusion does not always hold when this condition is not satisfied. If, for example,

$$F(\alpha) = \int_0^\infty \frac{\sin \alpha x}{x} dx,$$

the integrand is continuous in the region $0 \leq x$ and $1 \leq \alpha$, and has a derivative with respect to α which is also continuous in this region. But $\int_0^\infty \cos \alpha x\, dx$ has no meaning, whereas, on the other hand,

$$F(\alpha) = \int_0^\infty \frac{\sin \alpha x}{x} dx = \int_0^\infty \frac{\sin y}{y} dy = \frac{\pi}{2}$$

(§ 129), and therefore $F'(\alpha) = 0$. We can however be sure that the function of α represented by the improper integral will be differentiable under the integral sign if we impose certain further conditions upon $f(x, \alpha)$. The first of these is that the improper integral

$$\int_a^\infty f(x, \alpha) dx$$

shall converge *uniformly* with respect to a certain interval for α (see § 86). It is understood that $f(x, \alpha)$ shall be continuous for $x \geqq a$ and $\alpha_1 \leqq \alpha \leqq \alpha_2$.

We are considering this question at this point because it is closely related to the theorems concerning infinite series that we have been considering. Take an infinite sequence of numbers $a = a_0, a_1, a_2, \cdots, a_n, \cdots$ such that $a_{n-1} < a_n$ and $a_n \to \infty$ as $n \to \infty$; and put

$$F_0(\alpha) = \int_a^{a_1} f(x, \alpha) dx, \quad F_1(\alpha) = \int_{a_1}^{a_2} f(x, \alpha) dx, \quad \cdots,$$

$$F_n(\alpha) = \int_{a_n}^{a_{n+1}} f(x, \alpha) dx, \quad \cdots.$$

Now the series

$$\sum_{n=0}^\infty F_n(\alpha)$$

converges to the value $F(\alpha)$ of $\int_a^\infty f(x, \alpha) dx$ in case the latter converges. Moreover if this infinite integral converges uniformly in (α_1, α_2), the series converges uniformly in this interval, since

$$|R_n(\alpha)| = \left|\int_{a_n}^\infty f(x, \alpha) dx\right|,$$

and this is by hypothesis less than ϵ for all values of α in the interval and every $a_n > N$. Now $F_n(\alpha)$ is continuous in the interval, and therefore also $F(\alpha)$ is continuous there (Theorem 11).

We now impose the following additional conditions:

(1) $\dfrac{\partial f}{\partial \alpha}$ is a continuous function of x and α when $a \leqq x$ and $\alpha_1 \leqq \alpha \leqq \alpha_2$.

(2) The integral $\displaystyle\int_a^\infty \dfrac{\partial f}{\partial \alpha} dx$ converges uniformly in the interval (α_1, α_2).

Now if $G_n(\alpha) = \displaystyle\int_{a_n}^{a_{n+1}} \dfrac{\partial f}{\partial \alpha} dx$, then

$$\int_a^\infty \frac{\partial f}{\partial \alpha} dx = G_0(\alpha) + G_1(\alpha) + \cdots + G_n(\alpha) + \cdots,$$

and the series on the right converges uniformly. Moreover

$$F_n'(\alpha) = \int_{a_n}^{a_{n+1}} \frac{\partial f}{\partial \alpha} dx = G_n(\alpha).$$

Hence the series $\sum F_n'(\alpha)$ converges uniformly, and therefore (Theorem 13)

$$F'(\alpha) = \int_a^\infty \frac{\partial f}{\partial \alpha} dx.$$

In the same order of ideas we can define uniform convergence of the improper integral $F(\alpha) = \displaystyle\int_a^b f(x, \alpha) dx$ with finite limits but with the integrand infinite at one or both of the limits or at some intermediate point. And we can prove a similar theorem concerning $F'(\alpha)$. The reader should fill in the details.

129. Illustrative Example. We have seen (§ 83) that the infinite integral $\displaystyle\int_0^\infty e^{-\alpha x} \dfrac{\sin x}{x} dx$ converges when $\alpha > 0$. We represent its value by $F(\alpha)$. Now $-e^{-\alpha x} \sin x$ is the partial derivative of the integrand with respect to α, and the integral $\int_0^\infty e^{-\alpha x} \sin x \, dx$ converges uniformly for $0 < \alpha_1 \leq \alpha$. This can be seen as follows:

$$\left| \int_l^\infty e^{-\alpha x} \sin x \, dx \right| < \int_l^\infty e^{-\alpha x} dx = -\frac{1}{\alpha} e^{-\alpha x} \Big]_l^\infty = \frac{1}{\alpha e^{\alpha l}}.$$

If we select N in such a way that $\alpha_1 e^{\alpha_1 N} > \frac{1}{\epsilon}$, where ϵ is an arbitrary positive number, then for all values of $l > N$ and every $\alpha \geq \alpha_1$

$$\left| \int_l^\infty e^{-\alpha x} \sin x \, dx \right| < \epsilon.$$

Hence

$$F'(\alpha) = -\int_0^\infty e^{-\alpha x} \sin x \, dx$$

$$= \left[\frac{e^{-\alpha x}(\cos x + \alpha \sin x)}{1 + \alpha^2} \right]_{x=0}^{x=\infty} = -\frac{1}{1 + \alpha^2}$$

and $F(\alpha) = C - \arctan \alpha$. Now the absolute value of $\frac{\sin x}{x}$ is never greater than 1; therefore

$$|F(\alpha)| < \int_0^\infty e^{-\alpha x} dx = \frac{1}{\alpha}.$$

Hence $F(\alpha) \to 0$ as $\alpha \to \infty$. Also $\arctan \alpha \to \frac{\pi}{2}$ under the same conditions. From this it follows that $C = \frac{\pi}{2}$, and

$$\int_0^\infty e^{-\alpha x} \frac{\sin x}{x} dx = \frac{\pi}{2} - \arctan \alpha.$$

The infinite series corresponding to this infinite integral (see § 128) converges uniformly in the closed interval $(0, \alpha)$, where α is any positive number. Hence $F(\alpha) \to F(0)$ as $\alpha \to 0$ through positive values, and therefore

$$\int_0^\infty \frac{\sin x}{x} dx = \frac{\pi}{2}.$$

130. Test for uniform convergence. *If the function $f(x, \alpha)$ is continuous for $x \geq a$ and $\alpha_1 \leq \alpha \leq \alpha_2$, and if*

$|f(x, \alpha)| \leq \varphi(x)$ for $a \leq x$ and $\alpha_1 \leq \alpha \leq \alpha_2$, where $\int_a^\infty \varphi(x)dx$ converges, then $\int_a^\infty f(x, \alpha)dx$ converges uniformly.

The proof is omitted since it is similar to the proof given in § 124 for the uniform convergence of a series.

Consider, for example, the integral $\int_0^\infty \frac{\cos \alpha x}{1+x^2} dx$. For all values of x and α we have

$$\left|\frac{\cos \alpha x}{1+x^2}\right| \leq \frac{1}{1+x^2}.$$

Moreover $\int_0^\infty \frac{dx}{1+x^2} = \frac{\pi}{2}$. Hence the original integral converges uniformly to a function $F(\alpha)$. In this case $f_\alpha(x, \alpha) = -\frac{x \sin \alpha x}{1+x^2}$ and the integral

$$\int_0^\infty f_\alpha(x, \alpha)dx = -\int_0^\infty \frac{x \sin \alpha x}{1+x^2} dx$$

converges uniformly for $0 < \alpha_1 \leq \alpha \leq \alpha_2$, as may be seen by identifying it with the series

$$\sum_{n=0}^\infty \int_{n\pi/\alpha}^{(n+1)\pi/\alpha} \frac{x \sin \alpha x}{1+x^2} dx. \tag{1}$$

Now

$$\left|\int_{n\pi/\alpha}^{(n+1)\pi/\alpha} \frac{x \sin \alpha x}{1+x^2} dx\right| = \int_0^{\pi} \frac{\left(y + \frac{n\pi}{\alpha}\right) \sin \alpha y}{1 + \left(y + \frac{n\pi}{\alpha}\right)^2} dy,$$

if $x = y + \frac{n\pi}{\alpha}$; and the reader will have no difficulty in showing that, starting with a sufficiently large value of n, this latter integral decreases as n increases. Moreover, it approaches zero as $n \to \infty$. Since then (1) is a convergent alternating series with each term after a certain one less than the preceding one in absolute value, the remainder $R_n(x)$ is less in absolute value than the first term omitted if

n is sufficiently great. That is,

$$|R_n(x)| < \left|\int_{(n+1)\pi/\alpha}^{(n+2)\pi/\alpha} \frac{x \sin \alpha x}{1 + x^2} dx\right|$$

$$< \frac{(n + 2)\dfrac{\pi}{\alpha}}{1 + (n + 1)^2 \dfrac{\pi^2}{\alpha^2}} \left|\int_{(n+1)\pi/\alpha}^{(n+2)\pi/\alpha} \sin \alpha x\, dx\right|$$

$$= \frac{2(n + 2)\pi}{\alpha^2 + (n + 1)^2\pi^2} \leqq \frac{2(n + 2)\pi}{\alpha_1^2 + (n + 1)^2\pi^2} < \epsilon.$$

for sufficiently large values of n and any α in the interval (α_1, α_2). Hence

$$\int_0^\infty \frac{x \sin \alpha x}{1 + x^2} dx$$

converges uniformly in (α_1, α_2). We can then differentiate

$$F(\alpha) = \int_0^\infty \frac{\cos \alpha x}{1 + x^2} dx$$

under the integral sign. This gives us

$$F'(\alpha) = -\int_0^\infty \frac{x \sin \alpha x}{1 + x^2} dx.$$

But [1]

$$F(\alpha) = \frac{\pi}{2} e^{-\alpha}$$

and

$$F'(\alpha) = -\frac{\pi}{2} e^{-\alpha}.$$

Hence

$$\int_0^\infty \frac{\cos \alpha x}{1 + x^2} dx = -\int_0^\infty \frac{x \sin \alpha x}{1 + x^2} dx.$$

131. Infinite integrals of infinite series. In Theorem 12 we saw that in case $u_n(x)$ is continuous in (a, b)

$$\int_a^b \sum_{n=0}^\infty u_n(x)\,dx = \sum_{n=0}^\infty \int_a^b u_n(x)\,dx$$

[1] See Serret-Scheffers, *Lehrbuch der Differential- und Integralrechnung*, Vol. 2, 4th and 5th ed., p. 192.

§ 131]   INFINITE SERIES   233

provided that the series $\sum u_n(x)$ converges uniformly in (a, b). It is essential in the proof of this formula that the interval of integration be finite, and the formula does not always hold when we are dealing with infinite integrals. Consider, for example, the series the sum of the first n terms of which is $S_n(x) = \dfrac{2x}{n^2} e^{-x^2/n^2}$. This converges to zero for every finite value of x. Moreover, the maximum value of $S_n(x)$ is $\dfrac{1}{n}\sqrt{\dfrac{2}{e}}$, and occurs when $x = \dfrac{n}{\sqrt{2}}$. Hence the series converges uniformly in the interval $(0, h)$, where h is an arbitrary positive number. On the other hand

$$\int_0^\infty S_n(x)dx = -e^{-x^2/n^2}\Big]_0^\infty = 1,$$

and therefore $\lim\limits_{n \to \infty} \int_0^\infty S_n(x)dx = 1$, while $\int_0^\infty S(x)dx = 0$, where $S(x) = \lim\limits_{n \to \infty} S_n(x) = 0$.

Various sets of conditions can be given which are sufficient to insure the validity of the formula

$$\int_a^\infty \sum_{n=0}^\infty u_n(x)dx = \sum_{n=0}^\infty \int_a^\infty u_n(x)dx.$$

We shall confine our attention to one rather special set of such conditions.

THEOREM 15. *If the series $\sum |u_n(x)|$ is uniformly convergent in the interval (a, h), where h is an arbitrary number greater than a, and if either*

$$\int_a^\infty [\sum |u_n(x)|]dx \quad or \quad \sum \int_a^\infty |u_n(x)|dx$$

is convergent, then $\int_a^\infty [\sum u_n(x)]dx = \sum \int_a^\infty u_n(x)dx.$

We shall assume in the first place that $u_n(x) \geq 0$ for every $x \geq a$ and for every positive integer n. Then the

function
$$F(\lambda, \mu) = \int_a^\lambda \left[\sum_{n=0}^\mu u_n(x) \right] dx = \sum_{n=0}^\mu \int_a^\lambda u_n(x) dx$$

does not decrease as λ and μ increase. It follows as in § 138 that if one of the repeated limits $\lim_{\lambda \to \infty} [\lim_{\mu \to \infty} F(\lambda, \mu)]$ and $\lim_{\mu \to \infty} [\lim_{\lambda \to \infty} F(\lambda, \mu)]$ exists, the other one does also and the two limits are equal.

Since the series $\sum u_n(x)$ is uniformly convergent in the interval (a, λ),
$$\lim_{\mu \to \infty} F(\lambda, \mu) = \int_a^\lambda \left[\sum_{n=0}^\infty u_n(x) \right] dx.$$
Hence
$$\lim_{\lambda \to \infty} [\lim_{\mu \to \infty} F(\lambda, \mu)] = \int_a^\infty \left[\sum_{n=0}^\infty u_n(x) \right] dx \qquad (2)$$

if the improper integral in the right member converges. On the other hand
$$\lim_{\lambda \to \infty} F(\lambda, \mu) = \sum_{n=0}^\mu \int_a^\infty u_n(x) dx$$
and
$$\lim_{\mu \to \infty} [\lim_{\lambda \to \infty} F(\lambda, \mu)] = \sum_{n=0}^\infty \int_a^\infty u_n(x) dx \qquad (3)$$

if the series in the right member converges. But by hypothesis either this series converges or the infinite integral in the right member of (2) converges. Hence the right members of (2) and (3) both converge and their limits are equal. That is,
$$\int_a^\infty \left[\sum_0^\infty u_n(x) \right] dx = \sum_0^\infty \int_a^\infty u_n(x) dx.$$

If $u_n(x)$ is not always positive or zero we can apply the argument just given to $u_n(x) + |u_n(x)|$ which does satisfy this condition. Moreover
$$0 < \sum [u_n(x) + |u_n(x)|] \leq 2 \sum |u_n(x)|$$
and $\int_a^\infty \sum_0^\infty [u_n(x) + |u_n(x)|] dx$ converges if $\int_a^\infty \sum_0^\infty |u_n(x)| dx$

does. Also $0 \leq \int_a^\infty [u_n(x) + |u_n(x)|] dx \leq 2 \int_a^\infty |u_n(x)| dx$
and $\sum_0^\infty \int_a^\infty [u_n(x) + |u_n(x)|] dx$ converges if $\sum_0^\infty \int_a^\infty |u_n(x)| dx$ does. Hence

$$\int_a^\infty \left[\sum_0^\infty \{u_n(x) + |u_n(x)|\}\right] dx = \sum_0^\infty \int_a^\infty [u_n(x) + |u_n(x)|] dx$$

and

$$\int_a^\infty \left[\sum_0^\infty |u_n(x)|\right] dx = \sum_0^\infty \int_a^\infty |u_n(x)| dx.$$

From these it follows that

$$\int_a^\infty \left[\sum_0^\infty u_n(x)\right] dx = \sum_0^\infty \int_a^\infty u_n(x) dx.$$

132. Integration of an infinite integral. It was shown in § 97, that if $f(x, \alpha)$ is continuous in the finite region $a \leq x \leq b$, $\alpha_1 \leq \alpha \leq \alpha_2$, then

$$\int_{\alpha_1}^{\alpha_2} d\alpha \int_a^b f(x, \alpha) dx = \int_a^b dx \int_{\alpha_1}^{\alpha_2} f(x, \alpha) d\alpha.$$

That this formula does not hold without further restrictions when $b = \infty$ can be seen from the following considerations:

$$\int_0^\infty \frac{\sin \alpha x}{x} dx = \int_0^\infty \frac{\sin x}{x} dx = \frac{\pi}{2} \quad (\S\ 129).$$

Hence

$$\int_0^\infty dx \int_{\alpha_1}^{\alpha_2} \cos \alpha x\, d\alpha = \int_0^\infty \frac{\sin \alpha_2 x - \sin \alpha_1 x}{x} dx = 0.$$

But we cannot say that

$$\int_0^\infty dx \int_{\alpha_1}^{\alpha_2} \cos \alpha x\, d\alpha = \int_{\alpha_1}^{\alpha_2} d\alpha \int_0^\infty \cos \alpha x\, dx$$

since the infinite integral on the right does not converge, although the integrand is continuous in the region $0 \leq x$, $\alpha_1 \leq \alpha \leq \alpha_2$.

But if $\int_a^\infty f(x, \alpha)dx$ converges uniformly to $F(\alpha)$ in the interval (α_1, α_2) there is associated to an arbitrary positive ϵ a number N such that when $l > N$

$$\left| \int_l^\infty f(x, \alpha)dx \right| < \epsilon,$$

and

$$\int_{\beta_1}^{\beta_2} F(\alpha)d\alpha = \int_{\beta_1}^{\beta_2} d\alpha \int_a^l f(x, \alpha)dx + \int_{\beta_1}^{\beta_2} d\alpha \int_l^\infty f(x, \alpha)dx,$$

where $\alpha_1 \leq \beta_1 < \beta_2 \leq \alpha_2$. But

$$\int_{\beta_1}^{\beta_2} d\alpha \int_a^l f(x, \alpha)dx = \int_a^l dx \int_{\beta_1}^{\beta_2} f(x, \alpha)d\alpha$$

by the theorem referred to at the beginning of this paragraph and

$$\left| \int_{\beta_1}^{\beta_2} d\alpha \int_l^\infty f(x, \alpha)dx \right| < (\beta_2 - \beta_1)\epsilon \leq (\alpha_2 - \alpha_1)\epsilon.$$

Hence

$$\int_{\beta_1}^{\beta_2} F(\alpha)d\alpha = \int_{\beta_1}^{\beta_2} d\alpha \int_a^\infty f(x, \alpha)dx = \int_a^\infty dx \int_{\beta_1}^{\beta_2} f(x, \alpha)d\alpha.$$

This shows that if the stated conditions are satisfied we can integrate under the integral sign. We leave it to the reader to justify the assumption that $F(\alpha)$ is integrable in the interval (α_1, α_2).

EXERCISES

Establish the uniform convergence of each of the following infinite integrals for $0 \leq \alpha \leq \alpha_1$:

1. $\int_1^\infty \dfrac{\cos \alpha x}{x^{3/2}} dx.$ 2. $\int_1^\infty \dfrac{\sin \alpha x}{x^{4/3}} dx.$ 3. $\int_0^\infty \dfrac{\cos \alpha x}{1 + x^3} dx.$

4. $\int_1^\infty e^{-\alpha x} \dfrac{1}{x^2} dx \ (\alpha > 0).$ 5. $\int_0^\infty \dfrac{e^{-\alpha x}}{1 + x^2} dx \ (\alpha > 0).$

6. Show that $\Gamma'(x) = \int_0^\infty t^{x-1} \log t \, e^{-t} dt.$

7. Show that $\int_0^\infty t^{x-1}e^{-t}dt$ converges uniformly in the interval $0 < \alpha \leqq \alpha_1$.

8. Given $F(\alpha) = \int_\pi^\infty \frac{\cos \alpha x}{x^{5/2}}dx$. Find $F'(\alpha)$.

9. Given $F(\alpha) = \int_\pi^\infty e^{-\alpha x}\frac{\cos x}{x^2}dx$. Find $F'(\alpha)$ for $\alpha > 0$.

133. A continuous function without a derivative. If a is a positive number less than 1, the series $\sum_0^\infty a^n \cos b^n \pi x$ converges uniformly in any interval since $|a^n \cos b^n \pi x| \leqq a^n$. We shall represent the continuous function to which it converges by $F(x)$. In case $ab < 1$

$$F'(x) = -\sum a^n b^n \pi \sin b^n \pi x$$

since the series on the right converges uniformly in any interval.

We proceed to show that on the other hand if b is a positive odd integer greater than 1 and ab is sufficiently great the function $F(x)$ has no derivative anywhere.

For a fixed x let β_m be the integer nearest to $b^m x$, or the smaller of two equally near integers. Then

$$b^m x = \beta_m + \xi_m,$$

where $-\frac{1}{2} < \xi_m \leqq \frac{1}{2}$. If we put $h = \frac{e_m - \xi_m}{b^m}$, where $e_m = \pm 1$, then

$$b^m(x + h) = \beta_m + e_m.$$

The sign of h is the same as the sign of e_m and $\frac{1}{2b^m} \leqq |h| \leqq \frac{3}{2b^m}$. We see then that $h \to 0$ as $m \to \infty$.

Now

$$\frac{F(x + h) - F(x)}{h} = S_m + R_m,$$

where
$$S_m = \frac{1}{h} \sum_0^{m-1} a^n [\cos b^n \pi (x+h) - \cos b^n \pi x]$$
and
$$R_m = \frac{1}{h} \sum_m^{\infty} a^n [\cos b^n \pi (x+h) - \cos b^n \pi x].$$

By the law of the mean
$$\cos b^n \pi (x+h) - \cos b^n \pi x = -b^n \pi \sin (b^n \pi \eta) h,$$
where η lies between x and $x+h$. Hence
$$|\cos b^n \pi (x+h) - \cos b^n \pi x| \leq b^n \pi |h|,$$
and
$$|S_m| < \pi \sum_0^{m-1} a^n b^n = \pi \frac{a^m b^m - 1}{ab - 1} < \frac{\pi a^m b^m}{ab - 1},$$
in case $ab > 1$. Furthermore,
$$b^n \pi (x+h) = b^{n-m} b^m \pi (x+h) = b^{n-m} \pi (\beta_m + e_m).$$
The product $b^{n-m}(\beta_m + e_m)$ is even or odd with $\beta_m + 1$, since b is odd and $e_m = \pm 1$. Hence
$$\cos b^n \pi (x+h) = (-1)^{\beta_m+1}.$$
Also
$$\cos b^n \pi x = \cos (b^{n-m} b^m \pi x) = \cos b^{n-m} \pi (\beta_m + \xi_m)$$
$$= \cos b^{n-m} \pi \beta_m \cos b^{n-m} \pi \xi_m.$$
But $b^{n-m}\beta_m$ and β_m are either both even or both odd. Therefore
$$\cos b^n \pi x = (-1)^{\beta_m} \cos b^{n-m} \pi \xi_m$$
and
$$R_m = \frac{(-1)^{\beta_m+1}}{h} \sum_m^{\infty} a^n (1 + \cos b^{n-m} \pi \xi_m).$$

There is no negative term in the series on the right. The sum is therefore greater than the first term, which in turn is not less than a^m, since $\cos b^{n-m} \xi_m \pi = \cos \xi_m \pi$ for $n = m$, and $-\frac{1}{2} < \xi_m \leq \frac{1}{2}$. Then
$$|R_m| \geq \frac{a^m}{|h|}.$$

But
$$|h| \leq \frac{3}{2b^m},$$
and therefore
$$|R_m| \geq \frac{2}{3}(ab)^m.$$

We can so choose the positive odd integer b as to make
$$ab > 1 + \frac{3\pi}{2}.$$
Then
$$\frac{2}{3} > \frac{\pi}{ab-1}$$
and
$$\frac{2}{3}(ab)^m > \frac{\pi(ab)^m}{ab-1}.$$
Then
$$|R_m| > |S_m|.$$
It follows that
$$\left|\frac{F(x+h) - F(x)}{h}\right| \geq |R_m| - |S_m| > \frac{2}{3}(ab)^m \frac{ab - 1 - \frac{3\pi}{2}}{ab - 1}.$$

As m increases without limit, $h \to 0$, and the expression on the right of this inequality increases without limit. Then for any positive value of ϵ there is an h less than ϵ in absolute value for which
$$\left|\frac{F(x+h) - F(x)}{h}\right|$$
exceeds any pre-assigned number. The function $F(x)$ is then a continuous function without a derivative anywhere in case $ab > 1 + \frac{3\pi}{2}$. This example is due to Weierstrass.

134. Quasi-uniform convergence. If the series $\sum f_n(x)$ converges in the interval (a, b) and if to any two positive numbers ϵ and N there is associated a third number $N' > N$ such that for every value of x in the interval there is an integer n_x between N and N' for which

$$|R_{n_x}(x)| < \epsilon$$

the series is said to converge *quasi-uniformly* in the interval.

In connection with quasi-uniform convergence we have the following theorem which definitely settles the question as to the continuity of the sum of a convergent series of continuous functions of x.

THEOREM 16. *In order that the sum of a series of functions of x each continuous in the interval (a, b) be itself a continuous function of x in this interval it is necessary and sufficient that the series converge quasi-uniformly in the interval.*

We omit the proof.[1]

The series for which

$$S_n(x) = nxe^{-nx^2}$$

is quasi-uniformly convergent in the interval $(0, 1)$, while the series for which

$$S_n(x) = \frac{x^2}{(1+x^2)^n}$$

is not quasi-uniformly convergent in this interval. The reader can convince himself of the truth of this by drawing the curves $y = S_n(x)$ and $y = S(x)$ for each of the series.

A series that is uniformly convergent in an interval is quasi-uniformly convergent there. The converse of this statement is however not true.

135. Double series. The symbol

$$\begin{vmatrix} a_{0,0} + a_{0,1} + \cdots + a_{0,n} + \cdots \\ + a_{1,0} + a_{1,1} + \cdots + a_{1,n} + \cdots \\ + \cdots\cdots\cdots\cdots\cdots\cdots\cdots\cdots\cdots \\ + \cdots\cdots\cdots\cdots\cdots\cdots\cdots\cdots\cdots \\ + a_{m,0} + a_{m,1} + \cdots + a_{m,n} + \cdots \\ + \cdots\cdots\cdots\cdots\cdots\cdots\cdots\cdots\cdots \\ + \cdots\cdots\cdots\cdots\cdots\cdots\cdots\cdots\cdots \end{vmatrix}$$

which is constructed from the doubly infinite sequence of

[1] See, for example, Borel, *Leçons sur les fonctions de variables réelles*, p. 42.

§ 135]   INFINITE SERIES   241

terms $a_{m,n}$ ($m = 0, 1, 2, \cdots$; $n = 0, 1, 2, \cdots$) is called an *infinite double series*. The sums $S_{m,n} = \sum_{\mu=0}^{m} \sum_{\nu=0}^{n} a_{\mu,\nu}$ are called the *partial sums* of the double series, which is also represented by the symbol $\sum_{m,n=0}^{\infty} a_{m,n}$.

We say that the series $\sum a_{m,n}$ converges to the limit S if there is associated to every positive number ϵ a positive number N such that $|S - S_{m,n}| < \epsilon$ whenever $m > N$ and $n > N$. Now

$$a_{m,n} = S_{m,n} - S_{m,n-1} + S_{m-1,n-1} - S_{m-1,n}.$$

If therefore the series converges, $|a_{m,n}| < \epsilon$ for all sufficiently large values of m and n. This means that if the series converges, it is necessary that $a_{m,n} \to 0$ as m and n both tend to infinity. It does not mean that $a_{m,n}$ necessarily approaches zero if one of its subscripts increases indefinitely while the other one remains finite. Suppose, for example, that the series is such that

$$S_{m,n} = \frac{(-1)^{m+n}}{a+1}\left(\frac{1}{a^m} + \frac{1}{a^n}\right), \quad a > 1.$$

Then

$$a_{0,0} = \frac{2}{a+1}, \quad a_{m,0} = S_{m,0} - S_{m-1,0}$$
$$= \frac{(-1)^m}{a+1}\left(\frac{1}{a^m} + 1\right) - \frac{(-1)^{m-1}}{a+1}\left(\frac{1}{a^{m-1}} + 1\right)$$
$$= (-1)^m\left(\frac{1}{a^m} + \frac{2}{a+1}\right).$$

In a similar way we find that

$$a_{0,n} = (-1)^n\left(\frac{2}{a+1} + \frac{1}{a^n}\right),$$

and, in general, $a_{m,n} = (-1)^{m+n}\left(\frac{2}{a^m} + \frac{2}{a^n}\right)$. From this it follows that

$$\lim_{m \to \infty} |a_{m,0}| = \lim_{n \to \infty} |a_{0,n}| = \frac{2}{a+1}$$

and

$$\lim_{m \to \infty} |a_{m,n}| = \frac{2}{a^n}, \quad \text{for a fixed } n,$$

$$\lim_{n \to \infty} |a_{m,n}| = \frac{2}{a^m}, \quad \text{for a fixed } m.$$

On the other hand it is easy to see that the series converges to zero.

136. General condition for convergence. *In order that the double series $\sum a_{m,n}$ converge, it is necessary and sufficient that there be associated to every positive number ϵ a positive number N such that $|S_{m+p, n+q} - S_{m,n}| < \epsilon$ whenever $m > N, n > N, p > 0, q > 0$.*

The proof is similar to that of the analogous theorem for simple series.

137. Double series whose terms are all positive or zero. We make the following observations in connection with double series, all of whose terms are zero or positive:

(a) *In testing for convergence it is sufficient to consider the sequence $S_{m,m}$.*

In order to show this we put $S_{m,m} = \sigma_m$. Then, in the first place, $\sigma_m \to S$ if $S_{m,n} \to S$. Furthermore $\sigma_{m+n} \geqq S_{m,n} \geqq \sigma_\mu$, in case $m > \mu, n > \mu$. If now $\sigma_m \to S$ we can, for an arbitrarily chosen positive ϵ, take μ sufficiently great to bring about the inequalities

$$S - \epsilon < \sigma_\mu \leqq S,$$
$$S - \epsilon < \sigma_{m+n} \leqq S.$$

m and n being both greater than μ. It follows that

$$S - \epsilon < S_{m,n} \leqq S;$$

or, in other words, that $S_{m,n}$ also approaches S.

(b) Consider the sequence of curves, $C_0, C_1, \cdots$ drawn as follows:

(1) Any one of them forms a closed area with the two straight lines that bound the array $(a_{m,n})$ above and on the left, and entirely encloses the preceding area.

(2) The distance from the intersection of the two straight lines to any point of C_n becomes infinite with n.

We denote the sum of the elements inside the curve C_n by S_n. For all sufficiently great values of n, $\sigma_m \leq S_n$. Moreover for any n there is an $m' > n$ such that $S_n \leq \sigma_{m'}$. Hence if either σ_m or S_n approaches a limit, the other one approaches the same limit. It thus appears that the limit of S_n, when it exists, is independent of the system of curves used, and any system of the kind described can replace the system of rectangles.

138. Theorem 17. *If a double series of positive or zero terms converges, the series formed from the elements of any row or any column converges, and the series formed from the sums of the respective rows or columns converges to the sum of the double series.*

If the double series converges to S, $S_{m,n} \leq S$ for every m and n. Hence
$$\lim_{n \to \infty} S_{m,n} \quad (m \text{ fixed})$$
and
$$\lim_{m \to \infty} S_{m,n} \quad (n \text{ fixed})$$

exist and do not exceed S. Now for any positive ϵ there is an N such that for $m > N$, $n > N$, we have $S_{m,n} > S - \epsilon$. Hence
$$\lim_{m \to \infty} [\lim_{n \to \infty} S_{m,n}] = S$$
and
$$\lim_{n \to \infty} [\lim_{m \to \infty} S_{m,n}] = S.$$

139. Test for convergence. *If the terms of a double series are either zero or positive and are equal to, or less than, the corresponding terms of a convergent series, the series converges.*

The proof is similar to the proof of an analogous test concerning simple series whose terms are positive.

140. Absolute convergence. We pass now to the consideration of series whose terms are not restricted as to sign.

Definition. *The double series $\sum a_{m,n}$ is said to be absolutely convergent if the series $\sum |a_{m,n}|$ is convergent.*

Theorem 18. *If a double series is absolutely convergent, it is convergent.*

In order to see this, form a new series by replacing each negative term of $\sum a_{m,n}$ by zero and retaining the positive and zero terms as they stand. Then form a second new series by replacing each positive term of $\sum a_{m,n}$ by zero and changing the sign of each negative term, leaving the zero terms unchanged. These new series are convergent since the original series is absolutely convergent. Let their respective sums be S' and S''. If we denote the sum of the terms in the first m rows and the first n columns of these new series by $S'_{m,n}$ and $S''_{m,n}$ respectively, we have

$$S_{m,n} = S'_{m,n} - S''_{m,n}.$$

Hence
$$\lim_{m,n\to\infty} S_{m,n} = S' - S'',$$

and $\sum a_{m,n}$ converges.

If a double series is absolutely convergent it can be shown as in the case of an absolutely convergent simple series (§ 121) that the terms can be re-grouped and re-arranged in any way without affecting the convergence or the sum of the series. Moreover the terms in any row or any column are the terms of an absolutely convergent series. Then

$$\sum_{n=0}^{\infty} a_{m,n} = \sigma_m - t_m,$$

where σ_m is the sum of the positive terms of the mth row and t_m is the sum of the absolute values of the negative terms. Moreover the series $\sum \sigma_m$ and $\sum t_m$ are convergent since the sum of a convergent double series of positive or zero terms can be obtained by first adding the terms in the separate rows and then taking the sum of the series whose terms are these sums. It follows that the series

$$\sum (\sigma_m - t_m)$$

converges. We have just seen that

$$S = S' - S'',$$

where $S' = \sum \sigma_m$ and $S'' = \sum t_m$. Hence $\sum(\sigma_m - t_m) = S$. We would get the same result by first adding the elements in the separate columns and then taking the sum of the series formed by these sums.

That the result just established for absolutely convergent double series does not hold for double series in general can be seen from the following series:

$$\begin{array}{|l}
1 + 1 + 1 + 1 + \cdots \\
+1 - 1 - 1 - 1 + \cdots \\
+1 - 1 + 0 + 0 + \cdots \\
+1 - 1 + 0 + 0 + \cdots \\
\cdots\cdots\cdots\cdots\cdots\cdots \\
\cdots\cdots\cdots\cdots\cdots\cdots
\end{array}$$

in which all the terms are zero, except those in the first two rows and the first two columns. Now $S_{m,n} = 2$ if m and n both exceed 1. Hence the series converges to 2. But the series formed by the terms in either of the first two rows or the first two columns does not converge.

141. Multiplication of simple series. Let $\sum a_n$ and $\sum b_n$ be two absolutely convergent simple series with the sums S and S' respectively. Then the double series $\sum\limits_{m,n=0}^{\infty} a_m b_n$ is absolutely convergent, since $\sum\limits_{\mu=0}^{m}\sum\limits_{\nu=0}^{n} |a_\mu b_\nu| = \sum\limits_{\mu=0}^{m} |a_\mu| \cdot \sum\limits_{\nu=0}^{n} |b_\nu|$. Hence the terms of $\sum a_m b_n$ can be arranged and grouped in any way without affecting the sum. In particular, we can group the terms in the respective diagonals. This gives us

$$a_0 b_0 + (a_0 b_1 + a_1 b_0) + (a_0 b_2 + a_1 b_1 + a_2 b_0) + \cdots \\ + (a_0 b_n + a_1 b_{n-1} + \cdots + a_{n-1} b_1 + a_n b_0) + \cdots$$

and the sum of this series is SS' since $S_m S_n' = T_{m,n}$, where $T_{m,n}$ is the sum of the terms in the first $m + 1$ rows and the first $n + 1$ columns of $\sum a_m b_n$. We have therefore the theorem:

THEOREM 19. *If the series $S = \sum a_n$ and $S' = \sum b_n$ are absolutely convergent, the series $\sum c_n$, where $c_n = a_n b_0 + a_{n-1} b_1 + \cdots + a_0 b_n$, converges to SS'.*

The series $\sum c_n$ is called the *Cauchy product* of the series $\sum a_n$ and $\sum b_n$.

MISCELLANEOUS EXERCISES

1. Prove that if the sequence (a_n) is monotonic the sequence whose general term is $\dfrac{a_1 + a_2 + \cdots + a_n}{n}$ is also monotonic.

2. Prove that if $a_n = \dfrac{1}{n+1} + \dfrac{1}{n+2} + \cdots + \dfrac{1}{n+p}$, where p is any positive integer, the sequence (a_n) is convergent.

3. Prove that $na_n \underset{n \to \infty}{\longrightarrow} 0$ if the series $\sum a_n$ converges and $a_{n-1} > a_n > 0$.

4. Show that the series $\sum_{n=1}^{\infty} \dfrac{1}{n^3 + n^4 x^2}$ converges uniformly in every interval.

5. Show that Lambert's series $\sum_{1}^{\infty} \dfrac{x^n}{1 - x^n}$ converges if $|x| < 1$.

6. Show that $\sum \dfrac{\log n}{n^x}$ converges uniformly for $x > 1 + k$, $k > 0$.

7. Show that (a_n) is a null sequence if $a_n = na^n$, where $0 < a < 1$.

8. Show that $\sum \dfrac{a_n}{n}$ converges if $\sum a_n^2$ does. Is the converse of this true? Illustrate your answer by an example.

9. Show that the series $\sum f_n(x)$ converges absolutely and uniformly in the closed interval $(0, a)$ in case $f_0(x)$ is continuous in this interval and $f_n(x) = \int_0^x f_{n-1}(x)\,dx$ $(n = 1, 2, \cdots)$.

10. Show that the series $\sum \dfrac{\sin n\theta}{n}$ converges for every value of θ. Make use of the fact that $\sin \theta + \sin 2\theta + \cdots + \sin n\theta$ is bounded for all values of n and θ (see §153), and apply Abel's lemma.

11. Show that the series $\sum \dfrac{\cos n\theta}{n}$ converges for all values of θ except $\theta = 2k\pi$.

CHAPTER X

POWER SERIES

142. Definition. A series of the form $\sum_{n=0}^{\infty} a_n x^n$ in which the coefficients are independent of x is called a *power series in x*. We shall assume that the coefficients are all real and that x takes on only real values.

Every power series in x converges for $x = 0$. Some diverge for every other value of x; as, for example, the series $\sum n! x^n$. This is one extreme. At the other extreme are those series that converge for every finite value of x. The series $\sum \dfrac{x^n}{n!}$ is such a series. But in general a power series in x converges for some values of x different from zero and diverges for others; as, for example, the series $\sum x^n$, which converges for $-1 < x < 1$ and diverges for all other values of x.

143. Our first problem in connection with power series is to determine those values of x for which a given series converges and those values for which it diverges. To this end we now establish the following theorems:

Theorem 1. *If the terms of the power series $\sum a_n x^n$ are bounded for $x = x_0$—that is, if the sequence $(a_n x_0^n)$ is a bounded sequence—the series is absolutely convergent for every x that is less than x_0 in absolute value.*

By hypothesis there is a number M such that

$$|a_n x_0^n| < M$$

for every n. Now

$$|a_n x_1^n| = |a_n x_0^n| \cdot \left|\frac{x_1}{x_0}\right|^n < M \left|\frac{x_1}{x_0}\right|^n.$$

For any x_1 such that $|x_1| < |x_0|$ the series $\sum M \left|\dfrac{x_1}{x_0}\right|^n$

converges since $\left|\dfrac{x_1}{x_0}\right| < 1$, and therefore $\sum a_n x_1^n$ converges absolutely.

The conditions of the theorem are satisfied if the series $\sum a_n x_0^n$ converges. We can therefore say that *if $\sum a_n x^n$ converges for $x = x_0$ it converges absolutely for every x that is less than x_0 in absolute value.*

THEOREM 2. *If $\sum a_n x^n$ diverges for $x = x_0$ it diverges for every x that is greater than x_0 in absolute value.*

For if the series converged for $x = x_1$ where $|x_1| > |x_0|$ we know from the preceding theorem that it would converge for $x = x_0$.

THEOREM 3. *If $\sum a_n x^n$ converges for some values of x different from zero and diverges for others, there is a positive number R such that the series converges absolutely for every value of x less than R in absolute value, and diverges for every value of x greater than R in absolute value.*

We know from Theorem 2 that the set of values for which the series converges is bounded. Let R be the upper limit of this set (see Chapter 1, Theorem 5). Then $R > 0$ and the series converges absolutely for every value of x less than R in absolute value, and diverges for every value of x greater than R in absolute value. This is equivalent to saying that the origin is at the center of the interval within which the series converges.

In case the series converges only for $x = 0$ we shall say that $R = 0$; and in case the series converges for every finite value of x we shall say that R is infinite. In all cases we shall say that R is the *radius of convergence* of the series. Then when $-R < x < R$ the series converges absolutely, and when $x < -R$ or $x > R$ it diverges. It may converge or diverge when $x = \pm R$. Consider, for example, the three series $\sum x^n$, $\sum \dfrac{x^n}{n}$, and $\sum \dfrac{x^n}{n^2}$. It is easy to verify that in each case $R = 1$. But the first one diverges for $x = \pm 1$. The second one diverges for $x = 1$ and converges for $x = -1$, while the third converges absolutely for $x = \pm 1$.

144. Connection between the coefficients and the radius of convergence. If $x \neq 0$ the two sequences

$$a_1, \sqrt{a_2}, \cdots, \sqrt[n]{a_n}, \cdots$$

and

$$a_1 x, \sqrt{a_2} x, \cdots, \sqrt[n]{a_n} x, \cdots$$

are either both bounded or both unbounded. If they are both unbounded the general term of the series $\sum a_n x^n$ does not approach zero as n increases indefinitely and the series diverges for every value of x except zero. That is, $R = 0$. If they are both bounded let ω be the greatest limit of the first one. Then ωx is the greatest limit of the second one, and the series converges when $\omega x < 1$ and diverges when $\omega x > 1$. In other words, the series converges when $x < \frac{1}{\omega}$, and diverges when $x > \frac{1}{\omega}$. Hence $R = \frac{1}{\omega}$.

145. Continuity of a power series. If R is the radius of convergence of the power series

$$S(x) = \sum a_n x^n,$$

let R' be any positive number less than R. The series converges absolutely for $x = R'$. Hence for any positive ϵ there is an N such that when $n > N$

$$A_n R'^n + \cdots + A_{n+p} R'^{n+p} < \epsilon,$$

where $|a_i| = A_i$ and p is any positive integer, or zero. If $|x| = X \leq R'$, then

$$A_n X^n + \cdots + A_{n+p} X^{n+p}$$
$$\leq A_n R'^n + \cdots + A_{n+p} R'^{n+p} < \epsilon.$$

That is, the series is uniformly convergent in the interval $(-R', R')$, and $S(x)$ is continuous in any interval that is contained within the interval $(-R, R)$.

It follows that the power series $S(x) = \sum_{0}^{\infty} a_n x^n$ can be integrated term by term between limits that lie within the interval of convergence. That is,

$$\int_\alpha^\beta S(x)dx = \sum_0^\infty a_n \int_\alpha^\beta x^n dx,$$

where $-R < \alpha < \beta < R$.

This discussion throws no light on the behavior of $S(x)$ at the ends of the interval of convergence, unless the series converges absolutely when $x = R$. Then the series is uniformly convergent, and $S(x)$ is continuous, in the closed interval $(-R, R)$, with the reservation that at $x = -R$ only right-hand continuity is meant and only left-hand continuity at $x = R$. We cannot, for example, speak of the limit of $S(-R-h)$ as $h \to +0$, since $S(x)$ is undefined for values of $x < -R$. If the series converges when $x = R$, but not absolutely, a further examination is necessary. In this case let $0 < x \leq R$. We can write

$$\sum a_n x^n = \sum a_n R^n \left(\frac{x}{R}\right)^n.$$

The series $\sum a_n R^n$ converges by hypothesis and the sequence $\left(\frac{x}{R}\right)^n$ does not increase with n. By Abel's Test therefore the series converges uniformly at least in the interval $(-R', R)$, where $0 < R' < R$. Hence $S(x)$ is continuous at $x = R$. The interval of uniform convergence may, or may not, include $-R$. That depends upon the convergence or non-convergence of the series when $x = -R$.

146. The derivative of a power series. We consider the series $\sum a_n x^n$ with a radius of convergence equal to R and form a new series each of whose terms is the derivative of the corresponding term of the given series. The important fact in this connection is that the radius of convergence of this new series $\sum n a_n x^{n-1}$ is also R. We can see this as follows: The two series $\sum n a_n x^{n-1}$ and $\sum n a_n x^n$ have the same radius of convergence. The radius of the latter is the same as that of $\sum a_n x^n$ if the greatest limit of the sequence $(\sqrt[n]{a_n})$ is the same as that of $(\sqrt[n]{n a_n})$. It will be clear that these two greatest limits are the same if $\sqrt[n]{n} \to 1$ as $n \to \infty$, for in this case the limit of any infinite partial sequence of either sequence is a limit of an infinite partial sequence of the other one.

To complete the proof we have to establish the limit of $\sqrt[n]{n}$ as n becomes infinite. Let

$$\sqrt[n]{n} = 1 + y.$$

Then $y > 0$ if $n > 1$. Now

$$n = (1+y)^n = 1 + ny + \frac{n(n-1)}{2} y^2 + \cdots$$

and

$$n > 1 + \frac{n(n-1)}{2} y^2,$$

or

$$y < \sqrt{\frac{2}{n}}.$$

Hence $y \to 0$ as $n \to \infty$, and $\sqrt[n]{n} \to 1$.

Although the two series $\sum a_n x^n$ and $\sum n a_n x^{n-1}$ have the same radius of convergence R, they do not necessarily have the same convergence property at $x = \pm R$. For example, $\sum \frac{x^n}{n^2}$ converges at $x = \pm R = \pm 1$ while $\sum \frac{x^{n-1}}{n}$ diverges at $x = 1$ and converges at $x = -1$.

Since the two series $\sum a_n x^n$ and $\sum n a_n x^{n-1}$ have the same interval of convergence and since a power series converges uniformly in any interval that is within its interval of convergence, it follows that

$$S'(x) = \sum n a_n x^{n-1} \quad (-R < x < R).$$

EXERCISES

Find the interval of convergence of each of the following power series, and determine whether or not this interval is closed:

1. $\sum (-1)^{n-1} \frac{x^n}{n}$. 2. $\sum (nx)^n$. 3. $\sum (-1)^n \frac{x^{2n}}{(2n)!}$. 4. $\sum \frac{x^{2n}}{(2n)!}$.

5. $\sum (-1)^n \frac{x^{2n+1}}{(2n+1)!}$. 6. $\sum (-1)^n \frac{x^{2n+1}}{2n+1}$. 7. $\sum n! x^n$.

8. $\sum \dfrac{x^n}{2^n n^2}$. 9. $\sum \dfrac{[(n-1)!]^2 x^n}{(2n+2)!}$.

10. Find the interval of convergence of the hypergeometric series
$$1 + \sum_{n=1}^{\infty} \dfrac{\alpha(\alpha+1)\cdots(\alpha+n-1)\beta(\beta+1)\cdots(\beta+n-1)}{n!\gamma(\gamma+1)\cdots(\gamma+n-1)} x^n.$$

11. What function does the hypergeometric series represent if $\alpha = -m$, $\beta = \gamma$, $x = -y$?

12. Show that $xF(1, 1, 2; -x) = \log(1+x)$ if $F(\alpha, \beta, \gamma; x)$ represents the hypergeometric series given in Exercise 10.

13. Show that $F\left(\dfrac{1}{2}, 1, \dfrac{3}{2}; x^2\right) = \dfrac{1}{2x} \log \dfrac{1+x}{1-x}$.

147. Dominant functions. The function $\varphi(x) = \sum b_n x^n$ is said to dominate the function $f(x) = \sum a_n x^n$ in the interval $(-\alpha, \alpha)$ if it converges in this interval and if $b_n \geqq |a_n|$ for all values of n. The interval of convergence of the second series includes the interval $(-\alpha, \alpha)$. The utility of dominant functions rests on the fact that if $P(a_0, a_1, \cdots, a_n)$ is a polynomial in $a_0, a_1, \cdots, a_n$ with positive coefficients, then
$$|P(a_0, a_1, \cdots, a_n)| \leqq P(b_0, b_1, \cdots, b_n).$$

A given power series with a radius of convergence greater than zero has many dominant functions. A simple one can be obtained as follows:

Let r be a positive number less than R, the radius of convergence of the series $\sum a_n x^n$. Since this series converges for $x = r$, there is a number M such that $|a_n r^n| < M$, or $|a_n| < \dfrac{M}{r^n}$, for all values of n. Hence the series
$$M + \dfrac{Mx}{r} + \cdots + \dfrac{Mx^n}{r^n} + \cdots,$$
which converges when $|x| < r$, dominates the given series within the interval $(-r, r)$. Now r is any positive number less than R, and if we decrease r we can take a smaller value for M. But M cannot be less than $|a_0|$, although it can be equal to this value if the latter is different from zero. To see that we can take $M = |a_0| > 0$ consider the dominant

function given above, and select ρ subject to the condition $0 < \rho < r\frac{|a_0|}{M}$. For $n \geqq 1$

$$|a_n \rho^n| = |a_n r^n|\left(\frac{\rho}{r}\right)^n < M\frac{\rho}{r}\left(\frac{\rho}{r}\right)^{n-1} < |a_0|,$$

or

$$\frac{|a_0|}{\rho^n} > |a_n|.$$

Hence

$$\frac{|a_0|}{1-\frac{x}{\rho}} = |a_0| + |a_0|\frac{x}{\rho} + \cdots + |a_0|\left(\frac{x}{\rho}\right)^n + \cdots$$

dominates $f(x)$ within the interval $(-\rho, \rho)$. Obviously any number greater than $|a_0|$ can be taken in place of M to form a dominant function. If a_n $(n > 0)$ is the first coefficient different from zero and M_1 is any positive number we can select $k(> 1)$ such that $\frac{M}{k^n} < M_1$. Then $|a_n|r_1^n < \frac{M}{k^n} < M_1$, where $r_1 = \frac{r}{k}$. For any $m > n$ we have $|a_m|r_1^m < \frac{M}{k^m} < \frac{M}{k^n} < M_1$. Hence when $a_0 = 0$ we can take for M any positive number.

148. Substitution of one series in another. We have now to consider the problem: Having given z as a power series in y, and y as a power series in x, to express z as a power series in x and to determine a lower limit for the radius of convergence of this last series.

Suppose that the two series

$$z = f(y) = a_0 + a_1 y + \cdots + a_n y^n + \cdots \qquad (1)$$

and

$$y = \varphi(x) = b_0 + b_1 x + \cdots + b_n x^n + \cdots \qquad (2)$$

have R and r for their respective radii of convergence. If in (1) we substitute for y and its powers the series for y in terms of x and its respective powers, we obtain the double series

$$\left.\begin{array}{l} a_0 + a_1 b_0 + a_2 b_0^2 + \cdots + a_n b_0^n + \cdots \\ + a_1 b_1 x + 2 a_2 b_0 b_1 x + \cdots + n a_n b_0^{n-1} b_1 x + \cdots \\ + a_1 b_2 x^2 + a_2(b_1^2 + 2 b_0 b_2) x^2 + \cdots \\ \cdots\cdots\cdots\cdots\cdots\cdots\cdots\cdots\cdots\cdots\cdots\cdots\cdots \\ \cdots\cdots\cdots\cdots\cdots\cdots\cdots\cdots\cdots\cdots\cdots\cdots\cdots \end{array}\right\} \quad (A)$$

The term in the $(m + 1)$th row and the $(n + 1)$th column of the double series (A) is the product of a_n by the sum of the terms in the expansion of y^n that involve x^m. Hence if we add this series by columns we get the value of z, provided that x is taken sufficiently near to zero. But if the series is absolutely convergent we get the same sum if we add by rows (§ 140), and this sum will express z as a power series in x. This is what we want. Our problem then is to determine a positive number such that the double series (A) converges absolutely when x is less than this number in absolute value.

For the absolute convergence of the double series it is necessary that the simple series in the first row converge absolutely. This condition may be satisfied when $|b_0| = R$. It is certainly satisfied when $|b_0| < R$. We shall assume that the inequality holds since it turns out that this is sufficient to insure the absolute convergence of the double series for certain values of x different from zero. To see this we consider that the series (2) is dominated by the function $\dfrac{m}{1 - \dfrac{x}{\rho}}$, where m is any positive number greater than $|b_0|$ and ρ is a suitably chosen positive number less than r. We can select m and ρ in this way and at the same time satisfy the condition $m < R$. Let R' be a positive number between m and R. Then (1) is dominated by the function

$$\frac{M}{1 - \dfrac{y}{R'}} = M + M \frac{y}{R'} + \cdots + M \left(\frac{y}{R'}\right)^n + \cdots,$$

if M is a suitably chosen positive number.

If in the right member of this equation we replace y by the series $\sum_{n=0}^{\infty} m \left(\dfrac{x}{\rho}\right)^n$ we get the double series

$$\left.\begin{array}{l} M + M \dfrac{m}{R'} + \cdots + M \left(\dfrac{m}{R'}\right)^n + \cdots \\[4pt] + M \dfrac{m}{R'} \cdot \dfrac{x}{\rho} + \cdots + nM \left(\dfrac{m}{R'}\right)^n \dfrac{x}{\rho} + \cdots \\[4pt] + \cdots\cdots\cdots\cdots\cdots\cdots\cdots\cdots\cdots\cdots\cdots \\ \cdots\cdots\cdots\cdots\cdots\cdots\cdots\cdots\cdots\cdots\cdots \end{array}\right\} \quad (B)$$

Now each coefficient in the double series (A) is a polynomial with positive coefficients in the coefficients of series (1) and (2); and the coefficients in the double series (B) are the same polynomials in the coefficients of series that dominate (1) and (2). Hence (A) converges absolutely if (B) does. But in order that (B) converge absolutely the terms in any column must form an absolutely convergent series. Looking at the second column we see that then we must have $|x| < \rho$, and when this condition is satisfied the sum of the absolute values of the terms in the $(n+1)$th column is

$$M \left[\dfrac{m}{R'\left(1 - \dfrac{|x|}{\rho}\right)}\right]^n.$$

In order that (B) converge absolutely it is necessary that the series

$$\sum M \left[\dfrac{m}{R'\left(1 - \dfrac{x}{\rho}\right)}\right]^n$$

converge, and for this it is necessary and sufficient that

$$m < R'\left(1 - \dfrac{|x|}{\rho}\right),$$

or

$$|x| < \rho\left(1 - \dfrac{m}{R'}\right). \qquad (C)$$

When this condition is satisfied the former one, $|x| < \rho$, is also satisfied.

Now (C) is not only necessary for the absolute convergence of (B). It is also sufficient. For when it is satisfied the sum of the absolute values of the terms in the $(n + 1)$th column converges to, say, C_n. Moreover $\sum C_n$ converges to a limit, say, L. Then for an arbitrary positive ϵ there is an N such that when $n > N$ we have

$$|C_0 + C_1 + \cdots + C_n - L| < \frac{\epsilon}{2}.$$

For a fixed $n > N$ we can take l sufficiently great, say, $l > N_1$ to validate the inequalities

$$|C_{l,i} - C_i| < \frac{\epsilon}{2(n+1)} \quad (i = 0, 1, 2, \cdots, n),$$

where $C_{l,i}$ is the sum of the absolute values of the first $l + 1$ terms in the $(i + 1)$th column. For these values of l and n the sum of the absolute values of the terms in the first $l + 1$ rows and the first $n + 1$ columns differs from L by an amount less than ϵ in absolute value. Hence (C) is a sufficient condition for the absolute convergence of (A).

For values of x satisfying this condition the series (2) is convergent and

$$|\varphi(x)| < \frac{m}{1 - \frac{|x|}{\rho}}.$$

Moreover

$$\frac{|x|}{\rho} < 1 - \frac{m}{R'}.$$

Hence $|\varphi(x)| < R'$.

Since R' can be taken as near to R as we wish, condition (C) is equivalent to

$$|x| < \rho\left(1 - \frac{m}{R}\right). \tag{C'}$$

In this formula it is not sufficient to take $\rho < r$ and $|b_0| < m < R$. For example, if

$$z = \log(1 + y) = y - \frac{y^2}{2} + \cdots + (-1)^{n-1}\frac{y^n}{n} + \cdots$$

and
$$y = \frac{1}{2} + \sin x = \frac{1}{2} + x - \frac{x^3}{3!} + \cdots$$
$$+ (-1)^n \frac{x^{2n+1}}{(2n+1)!} + \cdots$$

we have $R = 1$ and $b_0 = \frac{1}{2}$. Suppose then that we take $m = \frac{3}{4}$. Since r is infinite, $\rho = 8 < r$. The inequality (C') becomes
$$|x| < 8(1 - \tfrac{3}{4}) = 2.$$

Now $\frac{\pi}{2} < 2$ and $y = \frac{3}{2}$ when $x = \frac{\pi}{2}$. But the series for z is divergent for this value of y. This seeming contradiction is due to the fact that the values used for m and ρ are not related in the way required by § 147. They must satisfy the condition $|b_n \rho^n| < m$ for all values of n in addition to the other conditions. Here $|b_n| = \frac{1}{n!}$ and $\frac{8^n}{n!} > \frac{3}{4}$ for some values of n. It follows however from the article referred to that for $m = \frac{3}{4}$ it is possible to select a ρ that will satisfy both of the required conditions.

149. Division by a power series. We conclude from the preceding article that the function
$$f(x) = \frac{1}{1 + a_1 x + \cdots + a_n x^n + \cdots},$$
can be expressed as a power series in x whose **radius of convergence is greater than zero**, provided that the **radius of convergence of the denominator is greater than zero**. For if we put
$$y = a_1 x + \cdots + a_n x^n + \cdots,$$
we have
$$f(x) = \frac{1}{1+y} = 1 - y + y^2 + \cdots + (-1)^n y^n + \cdots$$
$$= 1 - a_1 x + (a_1^2 - a_2)x^2 + \cdots.$$

This last series converges when
$$|x| < \rho(1-m),$$
where $0 < m < 1$ and ρ is a suitably chosen positive number.

We can make use of this result to show that the quotient of two power series whose radii of convergence are greater than zero can be expressed as a power series whose radius of convergence is greater than zero, provided that the constant term of the denominator is not zero. For if

$$\frac{f(x)}{g(x)} = \frac{a_0 + a_1 x + \cdots + a_n x^n + \cdots}{b_0 + b_1 x + \cdots + b_n x^n + \cdots},$$

$b_0 \neq 0$, we can write

$$\frac{f(x)}{g(x)} = (a_0 + a_1 x + \cdots + a_n x^n + \cdots)$$
$$\cdot \frac{1}{b_0 + b_1 x + \cdots + b_n x^n + \cdots}$$
$$= (a_0 + a_1 x + \cdots + a_n x^n + \cdots)$$
$$\cdot (c_0 + c_1 x + \cdots + c_n x^n + \cdots)$$
$$= d_0 + d_1 x + \cdots + d_n x^n + \cdots;$$

or
$$a_0 + a_1 x + \cdots + a_n x^n + \cdots$$
$$= (b_0 + b_1 x + \cdots)(d_0 + d_1 x + \cdots).$$

The values of the coefficients d_i can easily be computed from this relation in terms of the a's and b's.

If $b_0 = 0$, let $g(x) = x^k g_1(x)$, where $g_1(0) \neq 0$. Then

$$\frac{f(x)}{g_1(x)} = d_0 + d_1 x + \cdots + d_n x^n + \cdots$$

and

$$\frac{f(x)}{g(x)} = \frac{d_0}{x^k} + \frac{d_1}{x^{k-1}} + \cdots + \frac{d_{k-1}}{x} + d_k + d_{k+1} x$$
$$+ d_{k+2} x^2 + \cdots.$$

The right member of this equation is not a power series in x.

EXERCISES

Evaluate the following integrals approximately by first expanding the integrands in power series in x to three terms and then integrating term by term:

1. $\int_0^x e^{-x^2} dx$. 2. $\int_0^x \sin x^2 dx$. 3. $\int_0^x \sqrt{(1-x^2)(1-\tfrac{1}{4}x^2)} dx$.

4. $\int_0^x \dfrac{dx}{\sqrt{(1-x^2)(1-\tfrac{1}{4}x^2)}}$. 5. $\int_0^x \sqrt{1+x^4} dx$. 6. $\int_0^x \dfrac{dx}{\sqrt{1-\tfrac{1}{2}\sin^2 x}}$.

7. Get the expansion of $\sec x$ from the expansion of $\cos x$ in powers of x.

8. From the expansions of $\sin x$ and $\cos x$ in powers of x get the expansion of $\tan x$.

9. Expand $e^x \sin x$ in powers of x directly. Then expand the factors separately and multiply the resulting series. Compare the two results.

10. In the expansion of e^t in powers of t substitute for t the expansion of $\sin x$ in powers of x, and arrange the result in powers of x.

11. Substitute for t in Exercise 10 the expansion in powers of x of $\tan x$.

12. Expand the integrand of $\int_0^x \dfrac{dx}{\sqrt{1-x^2}}$ in powers of x. Then integrate the resulting series term by term and compare the result with the expansion of $\arcsin x$.

13. Do the same for $\int_0^x \dfrac{dx}{1+x^2}$ and $\arctan x$.

150. Double power series.

A series of the form

$$\sum_{m=0}^{\infty} \sum_{n=0}^{\infty} a_{m,n} x^m y^n,$$

where the $a_{m,n}$ are independent of x and y, is called a *double power series* in x and y. It is sometimes written in the form $\sum_{m,n} a_{m,n} x^m y^n$.

Theorem 4. *If the terms of the series $\sum a_{m,n} x_0^m y_0^n$ are bounded, the series converges absolutely for any x and y such that $|x| < |x_0|$ and $|y| < |y_0|$. The convergence is uniform throughout the region $|x| \leq X < |x_0|, |y| \leq Y < |y_0|$.*

By hypothesis there is an M such that $|a_{m,n}x_0^m y_0^n| < M$ for all positive integral and zero values of m and n. Then

$$\sum |a_{m,n}x^m y^n| = \sum |a_{m,n}x_0^m y_0^n| \cdot \left|\frac{x}{x_0}\right|^m \cdot \left|\frac{y}{y_0}\right|^n$$
$$< M \sum \left|\frac{x}{x_0}\right|^m \cdot \left|\frac{y}{y_0}\right|^n.$$

If $|x| < |x_0|$ and $|y| < |y_0|$, the two series $\sum \left|\frac{x}{x_0}\right|^m$ and $\sum \left|\frac{y}{y_0}\right|^n$ converge, and therefore the series $\sum \left|\frac{x}{x_0}\right|^m \left|\frac{y}{y_0}\right|^n$ also converges (§ 141). Hence for these values of x and y the given series converges absolutely. Moreover if $|x| \leq X < |x_0|$ and $|y| \leq Y < |y_0|$ there is for an arbitrary positive ϵ a number N such that when $m > N$ and $n > N$ we have,

$$\sum_{m=N+1}^{N+p} \sum_{n=N+1}^{N+p} |a_{m,n}| X^m Y^n < \epsilon$$

for any positive integral value of p. But

$$\left|\sum_{m=N+1}^{N+p} \sum_{n=N+1}^{N+p} a_{m,n} x^m y^n\right| \leq \sum \sum |a_{m,n}| \cdot |x|^m \cdot |y|^n$$
$$\leq \sum \sum |a_{m,n}| X^m Y^n < \epsilon.$$

Since this inequality holds for every x and y subject to the given conditions, the series converges uniformly in the closed rectangle bounded by the lines $x = \pm X$, $y = \pm Y$. This completes the proof of the theorem.

THEOREM 5. *If the terms of the series $\sum a_{m,n}x_0^m y_0^n$ are bounded, the series $\sum a_{m,n}x^m y^n$ can be integrated term by term over any region contained within the rectangle bounded by the lines $x = \pm X$, $y = \pm Y$, where $0 < X < |x_0|$, and $0 < Y < |y_0|$.*

We leave the proof of this theorem to the reader.

THEOREM 6. *If the terms of the series $\sum a_{m,n}x_0^m y_0^n$ are bounded, the series $\sum a_{m,n}x^m y^n$ can be differentiated partially term by term with respect to either x or y within the rectangle* *bounded by the lines* $x = \pm X$, $y = \pm Y$, *where* $X < |x_0|$ *and* $Y < |y_0|$.

At a point within the rectangle the series $\sum a_{m,n} x^m y^n$ converges absolutely. Hence

$$f(x, y) = \sum_{m,n} a_{m,n} x^m y^n = \sum_m \left(\sum_n a_{m,n} y^n\right) x^m,$$

and

$$\frac{\partial f(x,y)}{\partial x} = \sum_m \left(\sum_n a_{m,n} y^n\right) m x^{m-1} \quad (\S\ 146).$$

We are here looking upon this expression for $\dfrac{\partial f(x,y)}{\partial x}$ as a simple power series in x, the coefficient of x^{m-1} being $\sum_n a_{m,n} y^n \cdot m$. But this simple power series is equal to the double power series $\sum m a_{m,n} x^{m-1} y^n$ provided the latter converges absolutely. It is easy to see that this condition is satisfied. For this purpose it is only necessary to convince ourselves that the terms of the series $\sum m A_{m,n} X^{m-1} Y^n$ are bounded ($A_{m,n} = |a_{m,n}|$). We know that

$$\varphi(x) = \sum_m \left(\sum_n A_{m,n} Y^n\right) x^m$$

is a power series in x whose radius of convergence is greater than X. Hence

$$\sum_m \left(\sum_n A_{m,n} Y^n\right) m X^{m-1}$$

converges (§ 146), and its terms are therefore bounded. But

$$\sum_n A_{m,n} Y^n m X^{m-1} \geqq |m a_{m,n} X^{m-1} Y^n|$$

for every value of m and n, and these latter terms are therefore bounded. Hence the series

$$\sum m a_{m,n} x^{m-1} y^n$$

converges absolutely for every set of values of x and y for which $|x| < X$ and $|y| < Y$, and we have

$$\frac{\partial f(x,y)}{\partial x} = \sum_{m,n} m a_{m,n} x^{m-1} y^n.$$

at every point inside the rectangle bounded by the lines $x = \pm |x_0|$, $y = \pm |y_0|$. For any such set of values of x and y there is an X and a Y such that $|x| < X < |x_0|$ and $|y| < Y < |y_0|$. There is a similar formula for $\dfrac{\partial f(x, y)}{\partial y}$ under the same conditions.

It would be natural to conclude as in § 143 that if $\sum\limits_{m, n} a_{m, n} x^m y^n$ converges for $x = x_0$, $y = y_0$, it converges absolutely when $|x| < |x_0|$ and $|y| < |y_0|$. But this is not the case, as we shall show. In the article referred to the argument was based on the fact that if the series $\sum\limits_{n} a_n x^n$ converges for $x = x_0$ the terms $a_n x_0^n$ are bounded. The double series $\sum a_{m, n} x^m y^n$ however may converge for $x = x_0$ and $y = y_0$ and the terms $a_{m, n} x_0^m y_0^n$ form an unbounded set. For example, the series in which $a_{m, 0} = 2^m$, $a_{m, 1} = -2^m$, $a_{0, n} = 2^n$, $a_{1, n} = -2^n$, $a_{m, n} = 0$ when $n > 1, m > 1$, and $a_{m, n}$ is arbitrary when neither m nor n exceeds 1 converges for $x = 1$ and $y = 1$. But for these values of the variables the terms of the series form an unbounded set. Moreover the series diverges when $x = \tfrac{1}{2} = y$. Since however the terms are bounded for these values of the variables, the series converges absolutely when $|x| < \tfrac{1}{2}$ and $|y| < \tfrac{1}{2}$.

MISCELLANEOUS EXERCISES

1. Expand $\dfrac{e^x - e^{-x}}{2}$ and $\dfrac{e^x + e^{-x}}{2}$ in powers of x and determine the intervals of convergence. The functions are known as the *hyperbolic sine* and *hyperbolic cosine* of x respectively and are represented by the respective symbols sinh x and cosh x.

2. By definition tanh $x = \dfrac{\sinh x}{\cosh x} = \dfrac{e^x - e^{-x}}{e^x + e^{-x}}$. Expand this in powers of x.

Establish the following relations by expanding into power series where necessary and forming the Cauchy product:

3. $e^x \cdot e^y = e^{x+y}$. 4. $2 \cos^2 x = 1 + \cos 2x$. 5. $\sin^2 x + \cos^2 x = 1$.

6. Show that only odd powers of x appear in the expansion of an odd function in powers of x; and only even powers in the expansion of an even function.

7. Find the interval of convergence of the series
$$J_m(x) = x^m \sum (-1)^n \frac{x^{2n}}{2^{m+2n}(n!)[(m+n)!]}$$
and show that
$$J_{m+1}(x) = \frac{2m}{x} J_m(x) - J_{m-1}(x).$$

8. Show that $F(\alpha, \beta, \gamma; x)$ satisfies the differential equation
$$x(x-1)\frac{d^2y}{dx^2} + [(\alpha+\beta+1)x - \gamma]\frac{dy}{dx} + \alpha\beta y = 0.$$
For the definition of $F(\alpha, \beta, \gamma; x)$ see Exercise 10, § 146.

CHAPTER XI

TRIGONOMETRIC SERIES AND SERIES OF ORTHOGONAL FUNCTIONS

151. Schwarz's inequality. Let $x_1, x_2, \cdots, x_n$ and $y_1, y_2, \cdots, y_n$ be two sets of n real numbers each. From the fact that the quadratic form

$$\sum_{i=1}^{n} (\lambda x_i + \mu y_i)^2$$

is greater than, or equal to, zero for all real values of λ and μ, we have immediately

$$\left(\sum_{i=1}^{n} x_i y_i\right)^2 \leq \sum x_i^2 \sum y_i^2. \qquad (1)$$

This important formula is known as *Schwarz's inequality*.

In the same order of ideas the inequality

$$\int_a^b [\lambda f(x) + \mu g(x)]^2 dx \geq 0,$$

or

$$\lambda^2 \int_a^b \overline{f(x)}^2 dx + 2\lambda\mu \int_a^b f(x) \cdot g(x) dx + \mu^2 \int_a^b \overline{g(x)}^2 dx \geq 0,$$

where $f(x)$ and $g(x)$ are integrable functions and λ and μ are independent of x, leads to the formula

$$\left[\int_a^b f(x) \cdot g(x) dx\right]^2 \leq \int_a^b \overline{f(x)}^2 dx \int_a^b \overline{g(x)}^2 dx. \qquad (2)$$

152. Definition. A series of the form

$$S_m(x) = \tfrac{1}{2}a_0 + \sum_{n=1}^{m} (a_n \cos nx + b_n \sin nx),$$

where a_n and b_n are real constants, is called a *trigonometric series*. If m is finite the series is said to have $m + 1$ terms.

Every such series is a continuous function of x with the period 2π. But not every such function of x can be represented as a finite trigonometric series. Such a representation can in general be only approximate. In deciding which of all such representations with a given number of terms we shall consider the best one, we have to give effect to the difference $f(x) - S_m(x)$ for every value of x in an interval of length 2π, say the interval $(0, 2\pi)$. This consideration suggests that we take the integral $\int_0^{2\pi}[f(x) - S_m(x)]dx$ as a measure of this approximation. But this integral might be small in absolute value, or even zero, in cases where the difference $f(x) - S_m(x)$ is fairly large throughout a part of the interval, if it is nearly as large in absolute value and of opposite sign in other parts of the interval. In order to avoid this difficulty we agree to say that that trigonometric series $S_m(x)$ gives the best possible approximate representation of $f(x)$ in the interval $(0, 2\pi)$ of all trigonometric series with the same number of terms if it makes the integral

$$I_m = \int_0^{2\pi}[f(x) - S_m(x)]^2 dx \tag{3}$$

a minimum.

We have first to consider whether such a best approximate representation exists. Can the coefficients a_n and b_n be so determined as to make I_m a minimum? A necessary condition for this is that all the first partial derivatives $\dfrac{\partial I_m}{\partial a_n}$ and $\dfrac{\partial I_m}{\partial b_n}$ $(n = 0, 1, 2, \cdots, m)$ shall equal zero. Now

$$\frac{\partial I_m}{\partial a_n} = -2\int_0^{2\pi} f(x)\cdot\cos nx\, dx + 2\int_0^{2\pi} S_m(x)\cdot\cos nx\, dx$$

and

$$\frac{\partial I_m}{\partial b_n} = -2\int_0^{2\pi} f(x)\sin nx\, dx + 2\int_0^{2\pi} S_m(x)\sin nx\, dx$$

for $n \neq 0$, and

$$\frac{\partial I_m}{\partial a_0} = -\int_0^{2\pi} f(x)dx + \int_0^{2\pi} S_m(x)dx.$$

Moreover

$$\begin{aligned}
\int_0^{2\pi} \cos nx\, dx &= \begin{matrix} 0 & (n \neq 0), \\ 2\pi & (n = 0) \end{matrix}, \quad \int_0^{2\pi} \sin nx\, dx = 0 \\
\int_0^{2\pi} \cos^2 nx\, dx &= \int_0^{2\pi} \sin^2 nx\, dx = \pi \quad (n \neq 0) \\
\int_0^{2\pi} \sin mx \cos nx\, dx &= 0 \\
\int_0^{2\pi} \sin mx \sin nx\, dx &= 0 \quad (m \neq n) \\
\int_0^{2\pi} \cos mx \cos nx\, dx &= 0 \quad (m \neq n)
\end{aligned} \quad (4)$$

Hence

$$\begin{aligned}
\frac{\partial I_m}{\partial a_0} &= -\int_0^{2\pi} f(x)\, dx + \pi a_0 \\
\frac{\partial I_m}{\partial a_n} &= -2\int_0^{2\pi} f(x) \cos nx\, dx + 2\pi a_n \quad (n \neq 0) \\
\frac{\partial I_m}{\partial b_n} &= -2\int_0^{2\pi} f(x) \sin nx\, dx + 2\pi b_n
\end{aligned} \quad (5)$$

In order that these partial derivatives shall equal zero we must have

$$\begin{aligned}
a_n &= \frac{1}{\pi}\int_0^{2\pi} f(x) \cos nx\, dx, \\
b_n &= \frac{1}{\pi}\int_0^{2\pi} f(x) \sin nx\, dx.
\end{aligned} \quad (6)$$

It is clear from formulae (5) that the second partial derivative of I_m with respect to a_0 is π, and with respect to any one of the other coefficients is 2π, while all the mixed second partial derivatives are zero. Hence if we denote by a_0, a_n, b_n the values of the coefficients given by (6) and by $\bar{a}_0, \bar{a}_n, \bar{b}_n$ any other values, we have

$$I_m(\bar{a}, \bar{b}) = I_m(a, b) + \pi\left[\frac{(\bar{a}_0 - a_0)^2}{2} + \sum_{n=1}^{m}\{(\bar{a}_n - a_n)^2 + (\bar{b}_n - b_n)^2\}\right].$$

It follows that $I_m(a, b)$ is the minimum value of I_m the existence of which has been in question.

The coefficients given by (6) are called the *Fourier constants* of the function. If we substitute them in $S_m(x)$ we obtain the formula

$$\int_0^{2\pi} f(x) S_m(x) dx = \pi \left[\frac{a_0^2}{2} + \sum_{n=1}^{m} (a_n^2 + b_n^2) \right]$$
$$= \int_0^{2\pi} [S_m(x)]^2 dx.$$

From this it follows that

$$I_m(a, b) = \int_0^{2\pi} \overline{f(x)}^2 dx - \pi \left[\frac{a_0^2}{2} + \sum_{n=1}^{m} (a_n^2 + b_n^2) \right]. \quad (7)$$

It is clear from (7) that the series $\frac{a_0^2}{2} + \sum_{n=1}^{\infty} (a_n^2 + b_n^2)$ converges since $I_m(a, b) \geq 0$, and that therefore the series Σa_n^2 and Σb_n^2 converge.

Since the integral in the right member of (7) is independent of m, and since the series in this right member has no negative terms, we have

$$I_{m+1} \leq I_m.$$

That is, we get better and better approximations by increasing the value of m, unless all the coefficients from a certain point on are zero, in which case $I_{m+1} = I_m$. This raises the question as to whether the infinite series

$$\frac{a_0}{2} + \sum_{n=1}^{\infty} (a_n \cos nx + b_n \sin nx)$$

converges to $f(x)$. Whatever the answer to this question may be, the series is called the *Fourier series* of the function $f(x)$.

We have been assuming that the function $f(x)$ is continuous in the interval $(0, 2\pi)$, and of period 2π. But only the values of the function in this interval have been considered. The restriction that $f(x)$ be periodic has therefore been unnecessary up to this point. We shall use it however in § 153. And we did not need to assume continuity—it was only necessary to assume that $f(x)$ is integrable in the interval $(0, 2\pi)$, inasmuch as the integrability of $\overline{f(x)}^2$, $f(x) \cos nx$, and $f(x) \sin nx$ follows from this assumption. It is no limitation to confine our attention to an interval of length 2π. For if the interval is (a, b) the substitution $x' = \dfrac{2\pi(x-a)}{b-a}$ replaces it by the interval $(0, 2\pi)$.

153. If in formulae (6) we replace the variable of integration by α and then substitute the values of these constants in the expression for $S_m(x)$, we get

$$S_m(x) = \frac{1}{\pi} \int_0^{2\pi} f(\alpha) \left[\frac{1}{2} + \cos(\alpha - x) + \cdots + \cos m(\alpha - x) \right] d\alpha.$$

But

$$\sin \frac{\theta}{2} \left(\frac{1}{2} + \cos \theta + \cdots + \cos m\theta \right)$$

$$= \frac{1}{2} \sin \frac{\theta}{2} + \sum_{n=1}^{m} \frac{1}{2} \left[\sin \left(n + \frac{1}{2} \right) \theta - \sin \left(n - \frac{1}{2} \right) \theta \right]$$

$$= \frac{\sin \left(m + \frac{1}{2} \right) \theta}{2},$$

and therefore

$$\frac{1}{2} + \cos \theta + \cdots + \cos m\theta = \frac{\sin \left(m + \frac{1}{2} \right) \theta}{2 \sin \frac{\theta}{2}}.$$

Then

$$S_m(x) = \frac{1}{\pi} \int_0^{2\pi} f(\alpha) \frac{\sin \left(m + \frac{1}{2} \right)(\alpha - x)}{2 \sin \frac{1}{2}(\alpha - x)} d\alpha.$$

If we put $\alpha = x + t$,

$$S_m(x) = \frac{1}{\pi} \int_{-x}^{2\pi-x} f(x+t) \frac{\sin\left(m+\frac{1}{2}\right)t}{2\sin\frac{1}{2}t} dt$$

$$= \frac{1}{\pi} \int_0^{2\pi} f(x+t) \frac{\sin\left(m+\frac{1}{2}\right)t}{2\sin\frac{1}{2}t} dt,$$

since the integrand is of period 2π. Now

$$\frac{\sin\left(m+\frac{1}{2}\right)t}{2\sin\frac{1}{2}t} = \frac{\sin\frac{t}{2}\sin\left(m+\frac{1}{2}\right)t}{2\sin^2\frac{1}{2}t}$$

$$= \frac{\cos mt - \cos(m+1)t}{4\sin^2\frac{1}{2}t}.$$

Hence

$$S_m(x) = \frac{1}{2\pi} \int_0^{2\pi} f(x+t) \frac{\cos mt - \cos(m+1)t}{2\sin^2\frac{1}{2}t} dt.$$

If we put $\Sigma_m(x) = \dfrac{S_0(x) + S_1(x) + \cdots + S_m(x)}{m+1}$, we have

$$\Sigma_m(x) = \frac{1}{2\pi(m+1)} \int_0^{2\pi} f(x+t) \frac{1 - \cos(m+1)t}{2\sin^2\frac{1}{2}t} dt$$

$$= \frac{1}{2\pi(m+1)} \int_0^{2\pi} f(x+t) \left[\frac{\sin\frac{m+1}{2}t}{\sin\frac{1}{2}t}\right]^2 dt. \quad (8)$$

Our first problem now is to determine the behavior of $\Sigma_m(x)$ as m increases without limit. If in formula (8) we take $f(x) = 1$, we get

$$1 = \frac{1}{2\pi(m+1)} \int_0^{2\pi} \left[\frac{\sin \frac{m+1}{2} t}{\sin \frac{1}{2} t} \right]^2 dt, \qquad (9)$$

since in this case $a_0 = 2$ and $a_n = b_n = 0$ when $n > 0$; and therefore $S_m(x) = \Sigma_m(x) = 1$ for all values of m. Hence, for the $f(t)$ of formula (8),

$$\Sigma_m(x) - f(x)$$

$$= \frac{1}{2\pi(m+1)} \int_0^{2\pi} [f(x+t) - f(x)] \left[\frac{\sin \frac{m+1}{2} t}{\sin \frac{1}{2} t} \right]^2 dt. \qquad (10)$$

We shall for the present assume that $f(x)$ is continuous in the closed interval $(0, 2\pi)$ and therefore uniformly continuous. There is then a positive δ associated with every positive ϵ such that for every x in the interval and every t that is less than δ in absolute value we have

$$|f(x+t) - f(x)| < \epsilon.$$

For $x > 0$, $x + t$ becomes greater than 2π as t varies from 0 to 2π. It is necessary therefore that $f(x)$ be defined for values of x outside of the interval $(0, 2\pi)$. This is provided for by the assumption that $f(x)$ has the period 2π, or that $f(x + 2\pi) = f(x)$. In order that it be continuous it is necessary that it satisfy the condition $f(0) = f(2\pi)$. Now

$$\int_0^{2\pi} [f(x+t) - f(x)] \left[\frac{\sin \frac{m+1}{2} t}{\sin \frac{1}{2} t} \right]^2 dt$$

$$= \int_0^{\delta} + \int_{\delta}^{2\pi-\delta} + \int_{2\pi-\delta}^{2\pi}. \qquad (11)$$

If we denote the common integrand in (11) by $\varphi(t)$, and if

§ 153]   TRIGONOMETRIC SERIES   271

$|f(x)| < M$ in the interval $(0, 2\pi)$, we have

$$\left|\frac{1}{2\pi(m+1)}\int_0^\delta \varphi(t)dt\right| \leq \frac{\epsilon}{2\pi(m+1)}\int_0^\delta \left[\frac{\sin\frac{m+1}{2}t}{\sin\frac{1}{2}t}\right]^2 dt$$

$$< \frac{\epsilon}{2\pi(m+1)}\int_0^{2\pi}\left[\frac{\sin\frac{m+1}{2}t}{\sin\frac{1}{2}t}\right]^2 dt = \epsilon$$

by (9). Similarly we get

$$\left|\frac{1}{2\pi(m+1)}\int_{2\pi-\delta}^{2\pi}\varphi(t)dt\right|$$

$$\leq \frac{\epsilon}{2\pi(m+1)}\int_{2\pi-\delta}^{2\pi}\left[\frac{\sin\frac{m+1}{2}t}{\sin\frac{1}{2}t}\right]^2 dt$$

$$< \frac{\epsilon}{2\pi(m+1)}\int_0^{2\pi}\left[\frac{\sin\frac{m+1}{2}t}{\sin\frac{1}{2}t}\right]^2 dt = \epsilon.$$

Also

$$\left|\frac{1}{2\pi(m+1)}\int_\delta^{2\pi-\delta}\varphi(t)dt\right|$$

$$< \frac{M}{\pi(m+1)}\int_\delta^{2\pi-\delta}\left[\frac{\sin\frac{m+1}{2}t}{\sin\frac{1}{2}t}\right]^2 dt$$

$$< \frac{M}{\pi(m+1)\sin^2\frac{1}{2}\delta}\int_\delta^{2\pi-\delta} dt < \frac{2M}{(m+1)\sin^2\frac{1}{2}\delta}.$$

In other words,

$$|\Sigma_m(x) - f(x)| < 2\epsilon + \frac{2M}{(m+1)\sin^2\frac{1}{2}\delta},$$

But for all sufficiently large values of m we shall have
$$\frac{2M}{(m+1)\sin^2\frac{1}{2}\delta} < \epsilon,$$
and therefore
$$|\Sigma_m(x) - f(x)| < 3\epsilon$$
for every x in $(0, 2\pi)$.

We have now proved that if $f(x)$ is continuous in the closed interval $(0, 2\pi)$ with $f(x) = f(x + 2\pi)$ then $\Sigma_m(x) \to f(x)$ as $m \to \infty$. Moreover this approach is uniform in this interval. It follows that
$$\int_0^{2\pi}[f(x) - \Sigma_m(x)]^2 dx \to 0$$
as $m \to \infty$. But we have seen (§ 152) that for a given m
$$\int_0^{2\pi}[f(x) - S_m(x)]^2 dx \leq \int_0^{2\pi}[f(x) - \Sigma_m(x)]^2 dx,$$
since $\Sigma_m(x)$ is a trigonometric series with $m + 1$ terms. Hence by virtue of (7)
$$\frac{a_0^2}{2} + \sum_{n=1}^{\infty}(a_n^2 + b_n^2) = \frac{1}{\pi}\int_0^{2\pi}\overline{f(x)}^2 dx. \tag{12}$$

154. Equation (12) has been established on the assumption that $f(x)$ is continuous for $0 \leq x \leq 2\pi$, with $f(x) = f(x + 2\pi)$. Suppose now that we have a function $g(x)$ which has at most a finite number of discontinuities $\alpha_1, \alpha_2, \cdots, \alpha_k$ in this interval and is of period 2π. These discontinuities are to be such that the limits $g(\alpha_i - 0)$ and $g(\alpha_i + 0)$ exist for $i = 1, 2, \cdots, k$; that is, they are to be of the *first kind*. We shall take $\frac{g(\alpha_i + 0) + g(\alpha_i - 0)}{2}$ as the value of the function at α_i; and $\frac{g(+0) + g(2\pi - 0)}{2}$ as the value of $g(0)$ and of $g(2\pi)$, if such discontinuities occur at the ends of the interval. If we connect the points on the curve $y = g(x)$ whose abscissae are $\alpha_i - \delta$ and $\alpha_i + \delta$, where $\delta > 0$, by a straight line and replace the curve in each interval by the corresponding line, we obtain the graph of a continuous function $f(x)$ with period 2π, it being understood that if such discontinuities occur at the ends of the interval we draw straight lines from $[0, g(0)]$ to $[\delta, g(\delta)]$ and from $[2\pi - \delta, g(2\pi - \delta)]$ to $[2\pi, g(2\pi)]$. If a_n and b_n are the Fourier constants of $f(x)$ we know from § 153 that

$$\int_0^{2\pi} \left[f(x) - \frac{a_0}{2} - \sum_{\nu=1}^{n} (a_\nu \cos \nu x + b_\nu \sin \nu x) \right]^2 dx$$

can be made as small as we please by a suitable choice of n. Now

$$M' = \int_0^{2\pi} \left[g(x) - \frac{a_0}{2} - \sum_{\nu=1}^{n} (a_\nu \cos \nu x + b_\nu \sin \nu x) \right]^2 dx$$

$$= \int_0^{2\pi} \left[g(x) - f(x) + f(x) - \frac{a_0}{2} \right.$$
$$\left. - \sum_{\nu=1}^{n} (a_\nu \cos \nu x + b_\nu \sin \nu x) \right]^2 dx$$

$$= \int_0^{2\pi} [g(x) - f(x)]^2 dx$$
$$+ \int_0^{2\pi} \left[f(x) - \frac{a_0}{2} - \sum_{\nu=1}^{n} (a_\nu \cos \nu x + b_\nu \sin \nu x) \right]^2 dx$$
$$+ 2 \int_0^{2\pi} [g(x) - f(x)]$$
$$\cdot \left[f(x) - \frac{a_0}{2} - \sum_{\nu=1}^{n} (a_\nu \cos \nu x + b_\nu \sin \nu x) \right] dx.$$

By virtue of (151) the absolute value of the last integral in the right member of this equation does not exceed the square root of the product of the other two integrals. By taking δ sufficiently near to zero we can make $\int_0^{2\pi} [g(x) - f(x)]^2 dx$ less than any pre-assigned positive number, and we have just seen that

$$\int_0^{2\pi} \left[f(x) - \frac{a_0}{2} - \sum_{\nu=1}^{n} (a_\nu \cos \nu x + b_\nu \sin \nu x) \right]^2 dx$$

can be made arbitrarily near to zero by taking n sufficiently large. Hence we can make M' as near to zero as we wish. But if c_r and d_r are the Fourier constants of $g(x)$ we know from § 152 that

$$\int_0^{2\pi} \left[g(x) - \frac{c_0}{2} - \sum_{r=1}^{n} (c_r \cos \nu x + d_r \sin \nu x) \right]^2 dx \leq M'.$$

We conclude from this that

$$\frac{c_0^2}{2} + \sum_{r=1}^{n} (c_r^2 + d_r^2) \to \frac{1}{\pi} \int_0^{2\pi} \overline{g(x)}^2 dx \quad \text{as} \quad n \to \infty.$$

155. We are now in a position to prove that if $f(x)$ is continuous in $(0, 2\pi)$ with $f(x) = f(x + 2\pi)$, and if it has a derivative which is continuous except for a finite number of discontinuities of the first kind, it is the sum of its Fourier series.

We denote the Fourier constants of $f(x)$ by a_n and b_n, and those of $f'(x)$ by a_n' and b_n'. Integration by parts gives us for $n > 0$

$$a_n = \frac{1}{\pi} \int_0^{2\pi} f(x) \cos nx\, dx$$

$$= -\frac{1}{n\pi} \int_0^{2\pi} f'(x) \sin nx\, dx = -\frac{b_n'}{n},$$

$$b_n = \frac{1}{\pi} \int_0^{2\pi} f(x) \sin nx\, dx = \frac{1}{n\pi} \int_0^{2\pi} f'(x) \cos nx\, dx = \frac{a_n'}{n}.$$

Now the series $\sum a_r'^2$ and $\sum b_r'^2$ converge (§ 152), and moreover, by Schwarz's inequality,

$$\left(\sum_{r=m}^{n} |a_r \cos \nu x| \right)^2 \leq \left(\sum_{r=m}^{n} \frac{|b_r'|}{\nu} \right)^2 \leq \sum_{r=1}^{n} b_r'^2 \sum_{r=m}^{n} \frac{1}{\nu^2},$$

$$\left(\sum_{r=m}^{n} |b_r \sin \nu x| \right)^2 \leq \left(\sum_{r=m}^{n} \frac{|a_r'|}{\nu} \right)^2 \leq \sum_{r=1}^{n} a_r'^2 \sum_{r=m}^{n} \frac{1}{\nu^2}.$$

Since $\sum \frac{1}{\nu^2}$ converges it follows that the series

$$\frac{a_0}{2} + \sum (a_r \cos \nu x + b_r \sin \nu x)$$

converges absolutely and uniformly in the interval $(0, 2\pi)$. But we do not know to what it converges. If we call the sum $\varphi(x)$ the two functions $f(x)$ and $\varphi(x)$ have the same Fourier constants. (Why?) Moreover the series

$$\left[\frac{a_0}{2} + \sum_{\nu=1}^{\infty} (a_\nu \cos \nu x + b_\nu \sin \nu x)\right] f(x)$$

also converges uniformly in this interval. Hence

$$\int_0^{2\pi} f(x)\varphi(x)dx = \pi\left[\frac{a_0^2}{2} + \sum (a_\nu^2 + b_\nu^2)\right].$$

Now $f(x)$ and $\varphi(x)$ are both continuous and therefore by § 153

$$\int_0^{2\pi} \overline{f(x)}^2 dx = \int_0^{2\pi} \overline{\varphi(x)}^2 dx = \pi\left[\frac{a_0^2}{2} + \sum(a_\nu^2 + b_\nu^2)\right].$$

From this it follows that

$$\int_0^{2\pi} [f(x) - \varphi(x)]^2 dx = 0.$$

Since this integrand is continuous, we conclude that

$$f(x) = \varphi(x)$$

throughout the interval. This completes the proof of the theorem:

THEOREM 1. *The function $f(x)$ is the sum of its Fourier series, and the convergence is absolute and uniform in the interval $(0, 2\pi)$ if it is continuous in this interval, with $f(x) = f(x + 2\pi)$, and has a derivative which is continuous except for a finite number of discontinuities of the first kind.*

We have formulated this theorem with a view to its use in the proof of the next theorem. Otherwise we need not have restricted $f'(x)$ as much as we have.

Where have we made use of the condition $f(x) = f(x + 2\pi)$?

Where have we made use of the condition that $f(x)$ be continuous? How do we know that $\varphi(x)$ is continuous?

156. The restriction that $f(x)$ be continuous can be removed in the following way: We assume merely that it is continuous except for a finite number of discontinuities of the first kind, and has a derivative with the same property; and we consider first the special function $h(x)$ defined as follows:

$$h(x) = \frac{\pi - x}{2} \quad \text{for} \quad 0 < x < 2\pi,$$
$$h(0) = 0,$$
$$h(x) = h(x + 2\pi).$$

The Fourier series for $h(x)$ is

$$\sum_{\nu=1}^{\infty} \frac{\sin \nu x}{\nu}.$$

We have first to show that this series converges to $h(x)$. We cannot use the argument of § 155 since $h(x)$ is not continuous at the origin. Now

$$S_n(x) = \sum_{\nu=1}^{n} \frac{\sin \nu x}{\nu} = \sum_{\nu=1}^{n} \frac{2 \sin \frac{x}{2} \sin \nu x}{2\nu \sin \frac{x}{2}}$$

$$= \frac{\cos \frac{1}{2} x + \sum_{\nu=2}^{n} \left(\frac{1}{\nu} - \frac{1}{\nu - 1} \right) \cos \frac{2\nu - 1}{2} x - \frac{1}{n} \cos \frac{2n + 1}{2} x}{2 \sin \frac{x}{2}},$$

as may be seen by applying the formula $2 \sin \alpha \cdot \sin \beta = \cos(\alpha - \beta) - \cos(\alpha + \beta)$.

$$R_n(x) = \sum_{\nu=n+2}^{\infty} \frac{\left(\frac{1}{\nu} - \frac{1}{\nu - 1} \right) \cos \frac{2\nu - 1}{2} x}{2 \sin \frac{1}{2} x}$$

$$- \lim_{\nu \to \infty} \frac{\cos \frac{2\nu + 1}{2} x}{2\nu \sin \frac{1}{2} x} + \frac{\cos \frac{2n + 1}{2} x}{2(n + 1) \sin \frac{1}{2} x}.$$

If we restrict x to the interval $(\delta, 2\pi - \delta)$, where $0 < \delta < \pi$, we have

$$|R_n(x)| < \frac{1}{2\sin\frac{1}{2}\delta}\left[\sum_{\nu=n+2}^{\infty}\left(\frac{1}{\nu-1}-\frac{1}{\nu}\right)+\frac{1}{n+1}\right]$$

$$= \frac{1}{(n+1)\sin\frac{1}{2}\delta}.$$

Hence the series

$$\sum_{\nu=1}^{\infty}\frac{\sin \nu x}{\nu}$$

is uniformly convergent in this interval.

Consider now the function

$$H(x) = \frac{\pi x}{2} - \frac{x^2}{4}$$

for $0 \leq x \leq 2\pi$, with the understanding that $H(x + 2\pi) = H(x)$. Its Fourier series is

$$\frac{\pi^2}{6} - \sum_{\nu=1}^{\infty}\frac{\cos \nu x}{\nu^2}.$$

By the theorem of § 155 we know that this series converges absolutely and uniformly to $H(x)$. That is,

$$H(x) = \frac{\pi^2}{6} - \sum_{\nu=1}^{\infty}\frac{\cos \nu x}{\nu^2}.$$

If we differentiate this series term by term, we obtain the series $\sum \frac{\sin \nu x}{\nu}$, which we have seen to be uniformly convergent in the interval $(\delta, 2\pi - \delta)$. Moreover in this interval

$$H'(x) = h(x).$$

Hence

$$h(x) = \sum_{\nu=1}^{\infty}\frac{\sin \nu x}{\nu}.$$

If now $f(x)$ has a discontinuity of the first kind at $x = \xi$ $(0 \leq \xi \leq 2\pi)$, and we put $s(\xi) = f(\xi + 0) - f(\xi - 0)$,

then
$$F(x) = f(x) - \frac{s(\xi)}{\pi} h(x - \xi)$$
is continuous at $x = \xi$, as well as at all the points of continuity of $f(x)$. For
$$F(\xi + 0) = f(\xi + 0) - \frac{s(\xi)}{2} = \frac{f(\xi + 0) + f(\xi - 0)}{2}$$
and
$$F(\xi - 0) = f(\xi - 0) + \frac{s(\xi)}{2} = \frac{f(\xi + 0) + f(\xi - 0)}{2}.$$

Moreover the discontinuities of $F'(x)$ are at the same points and of the same kind as the discontinuities of $f'(x)$.

If $f(x)$ has only a finite number of discontinuities $\xi_1, \xi_2, \cdots, \xi_k$ in the interval $(0, 2\pi)$ and these are all of the first kind, then
$$F(x) = f(x) - \sum_{i=1}^{k} \frac{s(\xi_i)}{\pi} h(x - \xi_i)$$

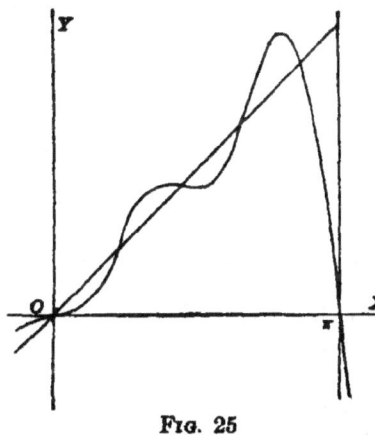

Fig. 25

is continuous throughout the interval, and its derivative has only a finite number of discontinuities which are all of the first kind. The Fourier series of $F(x)$ converges therefore to $F(x)$ absolutely and uniformly in the interval $(0, 2\pi)$. In any interval that does not include any of the discontinuities of $f(x)$ the function $\sum_{i=1}^{k} \frac{s(\xi_i)}{\pi} h(x - \xi_i)$ is represented uniformly by its Fourier series. Moreover
$$f(x) = F(x) + \sum_{i=1}^{k} \frac{s(\xi_i)}{\pi} h(x - \xi_i)$$
is the sum of the Fourier series of the $k + 1$ functions in the right member. This sum is therefore the Fourier series of $f(x)$. It converges uniformly in any interval that does not

include any of the discontinuities of $f(x)$. We have therefore proved the theorem:

Theorem 2. *If $f(x)$ is sectionally smooth in the interval $(0, 2\pi)$, it is the sum of its Fourier series* [1] *and the convergence is uniform in any partial interval that does not contain any of the discontinuities of $f(x)$.*

The figure shows the approximation curve

$$y = 2\left(\sin x - \frac{1}{2}\sin 2x + \frac{1}{3}\sin 3x - \frac{1}{4}\sin 4x\right)$$

for the function defined in the interval $(0, \pi)$ as follows: $f(x) = x$ for $0 \leq x < \pi$ and $f(\pi) = 0$.

157. The following theorem concerns an important property of Fourier series:

Theorem 3. *If the Fourier series of a function $f(x)$ that is sectionally continuous in the interval $(0, 2\pi)$ is integrated term by term between the limits ξ and x that lie in this interval, the resulting series converges to $\int_\xi^x f(x)dx$ uniformly with respect to x.*

The proof is simple. In the first place, we know that the function $F(x) = \int_0^x \left[f(x) - \frac{1}{2}a_0\right]dx$ is continuous in the given interval and is sectionally smooth (see § 67). Moreover $F(0) = 0$ and $F(2\pi) = \int_0^{2\pi} f(x)dx - \pi a_0 = 0$. Hence if A_n and B_n are the Fourier constants of $F(x)$ the series

$$\frac{1}{2}A_0 + \Sigma(A_n \cos nx + B_n \sin nx)$$

converges uniformly to $F(x)$. But for $n > 0$

$$A_n = \frac{1}{\pi}\int_0^{2\pi} F(x) \cos nx\, dx = -\frac{1}{n}b_n,$$

$$B_n = \frac{1}{\pi}\int F(x) \sin nx\, dx = \frac{1}{n}a_n.$$

---

[1] At a point where a sectionally smooth function is differentiable its derivative is continuous. (See Exercise 17, p. 38.)

Then
$$F(x) - F(\xi) = \sum_1^\infty [A_n(\cos nx - \cos n\xi) + B_n(\sin nx - \sin n\xi)]$$
$$= \sum \left[-\frac{b_n}{n}(\cos nx - \cos n\xi) + \frac{a_n}{n}(\sin nx - \sin n\xi) \right]$$
$$= \sum \int_\xi^x (a_n \cos nx + b_n \sin nx) dx.$$

On the other hand,
$$F(x) - F(\xi) = \int_0^x \left[f(x) - \frac{1}{2} a_0 \right] dx - \int_0^\xi \left[f(x) - \frac{1}{2} a_0 \right] dx$$
$$= \int_\xi^x \left[f(x) - \frac{1}{2} a_0 \right] dx.$$

Hence
$$\int_\xi^x f(x) dx = \int_\xi^x \frac{1}{2} a_0 dx + \sum_1^\infty \int_\xi^x (a_n \cos nx + b_n \sin nx) dx.$$

If $f(x)$ has a point of discontinuity in the interval $(0, 2\pi)$ its Fourier series does not converge uniformly in this interval. And yet we have just seen that we can integrate it term by term in the ordinary sense. This is an example of the fact that uniform convergence, although a sufficient condition for term-by-term integration, is not a necessary one. It may even be that the Fourier series for $f(x)$ diverges. In this case the proof just given would still be valid. We see then that there are divergent series that are integrable term by term—that is, the series of the integrals of the terms of the given series converges to the integral of a certain function. Moreover, in this case the convergence is uniform. This is a surprising fact. But it should be no more surprising than the fact that term-by-term differentiation of uniformly convergent series sometimes gives rise to divergent series (§ 123). We have here two aspects of the same phenomenon.

158. If $f(x)$ is of period 2π we can replace the interval $(0, 2\pi)$ by the interval $(-\pi, \pi)$. If now $f(x)$ is an even function, $f(x) \sin nx$ is odd and $b_n = \dfrac{1}{\pi}\int_{-\pi}^{\pi} f(x) \sin nx\, dx = 0$. If $f(x)$ is odd, $a_n = \dfrac{1}{\pi}\int_{-\pi}^{\pi} f(x) \cos nx\, dx = 0$. We see then that the Fourier series of an even function contains no terms involving $\sin nx$, and the Fourier series of an odd function contains no terms involving $\cos nx$. But any function satisfying the conditions of Theorem 2 can be represented by a cosine series or by a sine series in the interval $(0, \pi)$ by extending the definition of the function to the interval $(-\pi, 0)$ in such a way as to make it even or odd. It can also be represented by a series containing both sines and cosines by suitably defining it in the interval $(-\pi, 0)$.

159. An application in the theory of heat. We consider the infinite solid bounded by the planes $x = 0$, $x = \pi$, and $y = 0$, which lies on the positive side of the last plane. If the first two of these planes are kept at the temperature 0 and every point $(x, 0, z)$, where $0 < x < \pi$, at the temperature $u = f(x)$, the temperature at the different points of the solid will vary for a time and then remain constant. This constant temperature is called the *steady temperature* of the solid. It varies from point to point. We assume that $f(x)$ satisfies the conditions of Theorem 2. It is evident that the steady temperature at the two points (x, y, z_1) and (x, y, z_2) will be the same, and that we can confine our attention to the temperature in a plane parallel to the (x, y)-plane.[1]

The steady temperature is governed by the partial differential equation

$$\frac{\partial^2 u}{\partial x^2} + \frac{\partial^2 u}{\partial y^2} = 0 \qquad (13)$$

subject to the boundary conditions:

[1] This is essentially the first problem considered by Fourier in his *Théorie de la chaleur*. Cf. Carslaw, *The Conduction of Heat*, p. 89.

(a) $u = 0$ when $x = 0$ and when $x = \pi$.
(b) $u = f(x)$ when $y = 0$ and $0 < x < \pi$.
(c) $u \to 0$ uniformly with respect to x as $y \to \infty$ when $0 \leq x \leq \pi$.

Condition (c) is a physical consequence of the other two.

It is easy to verify that

$$u = e^{-ny} \sin nx$$

is a solution of (13) that satisfies the conditions (a) and (c) in case n is a positive integer. The same is true of any finite sum of the form

$$u = \sum_{\nu=1}^{n} b_\nu e^{-\nu y} \sin \nu x, \tag{14}$$

where the b_ν are constants. But a solution of this form does not in general satisfy condition (b).

If now we extend the interval of definition of $f(x)$ to the interval $(-\pi, 0)$ by imposing the condition $f(-x) = -f(x)$, the Fourier constants of $f(x)$ are

$$a_n = \frac{1}{\pi} \int_{-\pi}^{\pi} f(x) \cos nx\, dx = 0$$

and

$$b_n = \frac{1}{\pi} \int_{-\pi}^{\pi} f(x) \sin nx\, dx = \frac{2}{\pi} \int_{0}^{\pi} f(x) \sin nx\, dx.$$

By Theorem 2

$$f(x) = \sum_{\nu=1}^{\infty} b_\nu \sin \nu x.$$

For $y \geq y_0 > 0$ the series $\sum_{\nu=1}^{\infty} e^{-\nu y}$ and $\sum_{\nu=1}^{\infty} \nu e^{-\nu y}$ are convergent series of positive terms. Moreover there is a number M such that $|f(x)| < M$ for all values of x in the interval $(0, \pi)$. Hence

$$|b_n| < \frac{2}{\pi} \int_{0}^{\pi} M\, dx = 2M,$$

and the series $\sum_{\nu=1}^{\infty} b_\nu e^{-\nu y} \sin \nu x$ converges for every value of x.

If we differentiate it term by term partially with respect to x we obtain the series

$$\sum \nu b_\nu e^{-\nu y} \cos \nu x \tag{15}$$

and if we differentiate partially with respect to y we obtain the series

$$-\sum \nu b_\nu e^{-\nu y} \sin \nu x. \tag{16}$$

Now (15) is uniformly convergent with respect to x in any interval provided $y \geqq y_0$; and (16) is uniformly convergent with respect to y in the interval (y_0, l), where l is any number greater than y_0, for any x. It follows from this that

$$\frac{\partial u}{\partial x} = \sum \nu b_\nu e^{-\nu y} \cos \nu x \tag{17}$$

and

$$\frac{\partial u}{\partial y} = -\sum \nu b_\nu e^{-\nu y} \sin \nu x. \tag{18}$$

Moreover term-by-term differentiation of the right member of (17) with respect to x and of the right member of (18) with respect to y is permissible, since

$$-\sum \nu^2 b_\nu e^{-\nu y} \sin \nu x$$

is uniformly convergent with respect to x and

$$\sum \nu^2 b_\nu e^{-\nu y} \sin \nu x$$

is uniformly convergent with respect to y. Hence

$$\frac{\partial^2 u}{\partial x^2} = -\sum \nu^2 b_\nu e^{-\nu y} \sin \nu x$$

and

$$\frac{\partial^2 u}{\partial y^2} = \sum \nu^2 b_\nu e^{-\nu y} \sin \nu x;$$

and therefore

$$\frac{\partial^2 u}{\partial x^2} + \frac{\partial^2 u}{\partial y^2} = 0.$$

In other words,
$$u = \sum_{\nu=1}^{\infty} b_\nu e^{-\nu y} \sin \nu x$$

is a solution of (13) in the interior of the infinite rectangle bounded by the lines $x = 0$, $x = \pi$, and $y = 0$, and lying in the first quadrant. This solution satisfies boundary conditions (a) and (b). As to condition (c) we observe that

$$\left| \sum_{\nu=1}^{\infty} b_\nu e^{-\nu y} \sin \nu x \right| < 2M \sum e^{-\nu y} = \frac{2M}{e^y - 1}.$$

But $\dfrac{2M}{e^y - 1} \to 0$ as $y \to \infty$. Therefore u satisfies condition (c). That it is the only solution which satisfies all of these conditions follows from the theory of linear partial differential equations of the second order. Therefore the value of $u(x, y)$ at any point (x, y) in the interior of the rectangle is the steady temperature at that point.

160. An exponential form for Fourier series. In view of the fact that

$$\sin \nu x = \frac{e^{i\nu x} - e^{-i\nu x}}{2i} \quad \text{and} \quad \cos \nu x = \frac{e^{i\nu x} + e^{-i\nu x}}{2}$$

we can write

$$a_\nu \cos \nu x + b_\nu \sin \nu x = a_\nu \frac{e^{i\nu x} + e^{-i\nu x}}{2} + b_\nu \frac{e^{i\nu x} - e^{-i\nu x}}{2i}$$

$$= \frac{1}{2}(a_\nu - ib_\nu)e^{i\nu x} + \frac{1}{2}(a_\nu + ib_\nu)e^{-i\nu x}.$$

Hence
$$f(x) = \sum_{\nu = -\infty}^{\infty} \alpha_\nu e^{i\nu x},$$

where

$$\alpha_\nu = \frac{1}{2}(a_\nu - ib_\nu) \text{ for } \nu > 0, \quad \alpha_\nu = \frac{1}{2}(a_\nu + ib_\nu) \text{ for } \nu < 0,$$

and

$$\alpha_0 = \frac{1}{2}a_0. \qquad (\nu = 0, \pm 1, \pm 2, \cdots).$$

EXERCISES

Get the first three non-vanishing terms in the Fourier expansion of each of the following functions:

1. $f(x) = -1$ for $-\pi < x < 0$; $f(x) = 1$ for $0 < x < \pi$; $f(0) = 0$.
2. $f(x) = |x|$. 3. $f(x) = \tfrac{1}{2}x$ for $0 \leqq x < \pi$; $f(x) = \tfrac{1}{2}x - \pi$ for $\pi < x \leqq 2\pi$.
4. $f(x) = \dfrac{\pi^2}{2} - \pi x$ for $0 \leqq x \leqq \pi$; $f(x) = \pi x - \dfrac{3\pi^2}{2}$ for $\pi \leqq x \leqq 2\pi$.
5. $f(x) = 1$ for $0 \leqq x \leqq \pi$ (sine series).
6. $f(x) = x$ for $0 \leqq x \leqq \pi$ (sine series).
7. $f(x) = x$ for $0 \leqq x \leqq \pi$ (cosine series).
8. $f(x) = 0$ for $-\pi < x \leqq 0$; $f(x) = 1$ for $0 < x < \pi$.
9. $f(x) = x \sin x$.
10. $f(x) = \cos x$ for $0 \leqq x \leqq \pi$ (sine series).
11. $f(x) = \sin x$ for $0 \leqq x \leqq \pi$ (cosine series).
12. $f(x) = e^x$ for $0 \leqq x \leqq \pi$ (sine series).
13. $f(x) = e^x$ for $0 \leqq x \leqq \pi$ (cosine series).
14. Integrate the series in Exercise 10 term by term from 0 to x.
15. Integrate the series in Exercise 11 term by term from $\dfrac{\pi}{2}$ to x.
16. Can the series in Exercise 11 be differentiated term by term?
17. Can the series in Exercise 2 be differentiated term by term?
18. Expand $f(x) = \sin tx$ (t not an integer).
19. From the result in Exercise 18 get a formula for $\dfrac{\pi}{\cos \pi t}$.
20. Get a formula for $\pi \cot \pi t$ by expanding $\cos tx$.
21. Get the Fourier series for x^2 by integrating the series for x term by term from 0 to x. $\left(\dfrac{\pi^2}{12} = \sum_1^\infty (-1)^{n-1} \dfrac{1}{n^2} \right)$.

161. Dirichlet integrals.
Integrals of the form

$$\int_{-a}^{a} \varphi(t) \frac{\sin kt}{\sin t} dt \quad \text{or} \quad \int_{-a}^{a} \varphi(t) \frac{\sin kt}{t} dt$$

are known as *Dirichlet integrals*.

We are here interested in a certain property of the second of these forms; namely, that

$$\lim_{k \to \infty} \frac{1}{\pi} \int_{-a}^{a} f(x + t) \frac{\sin kt}{t} dt = \frac{1}{2}[f(x + 0) + f(x - 0)],$$

where a is any positive number, provided that $f(x)$ and $f'(x)$ are each continuous in any finite interval, except for a finite number of discontinuities which are of the first kind.

Now

$$\int_{-a}^{a} f(x+t)\frac{\sin kt}{t}dt = \int_{-a}^{-\eta} + \int_{-\eta}^{\eta} + \int_{\eta}^{a},$$

where η is a small positive number. If we apply integration by parts to the indefinite integral

$$\int f(x+t)\frac{\sin kt}{t}dt,$$

we obtain the formula

$$\int f(x+t)\frac{\sin kt}{t}dt = -\frac{1}{k}\frac{f(x+t)}{t}\cos kt$$
$$+ \frac{1}{k}\int \cos kt \cdot \frac{d}{dt}\frac{f(x+t)}{t} \cdot dt.$$

We cannot substitute the limits η and a in this formula because of the possible discontinuities of $f(x+t)$. If these are at the points ξ_i ($i = 1, 2, \cdots, p$), we can apply the formula to each of the intervals $(\eta, \xi_1), (\xi_1, \xi_2), \cdots, (\xi_p, a)$ and then combine the results. This gives us

$$\int_{\eta}^{a} f(x+t)\frac{\sin kt}{t}dt = -\frac{1}{k}\left[\frac{f(x+t)}{t}\cos kt\Big|_{\eta}^{a} + S\right]$$
$$+ \frac{1}{k}\int_{\eta}^{a} \cos kt \frac{d}{dt}\frac{f(x+t)}{t}dt, \quad (19)$$

where S is the sum of the jumps of the function $\frac{f(x+t)}{t}$
$\cdot \cos kt$ within the interval (η, a). From (19) it follows that

$$\lim_{k\to\infty} \int_{\eta}^{a} f(x+t)\frac{\sin kt}{t}dt = 0. \qquad (20)$$

We see in a similar way that

$$\lim_{k\to\infty} \int_{-a}^{-\eta} f(x+t)\frac{\sin kt}{t}dt = 0. \qquad (21)$$

The integral

$$\int_{-\eta}^{\eta} f(x+t)\frac{\sin kt}{t}dt = \int_{-\eta}^{0} + \int_{0}^{\eta}.$$

Now

$$\int_0^\eta f(x+t)\frac{\sin kt}{t}dt - f(x+0)\int_0^\eta \frac{\sin kt}{t}dt$$
$$= \int_0^\eta \frac{f(x+t) - f(x+0)}{t}\sin kt\, dt.$$

There is a positive number M_1 such that $|f'(x+t)| < M_1$ in the interval $0 \leq t \leq \eta$. For these values of t we have by virtue of the first mean value theorem of Differential Calculus

$$\left|\frac{f(x+t) - f(x+0)}{t}\right| < M_1,$$

and therefore

$$\left|\int_0^\eta f(x+t)\frac{\sin kt}{t}dt - f(x+0)\int_0^\eta \frac{\sin kt}{t}dt\right| < M_1\eta.$$

Similarly

$$\int_{-\eta}^0 f(x+t)\frac{\sin kt}{t}dt - f(x-0)\int_{-\eta}^0 \frac{\sin kt}{t}dt$$
$$= \int_{-\eta}^0 \frac{f(x+t) - f(x-0)}{t}\sin kt\, dt,$$

and $\left|\dfrac{f(x+t) - f(x-0)}{t}\right| < M_2$, where $|f'(x+t)| < M_2$ in the interval $-\eta \leq t \leq 0$. Hence

$$\left|\int_{-\eta}^0 f(x+t)\frac{\sin kt}{t}dt - f(x-0)\int_{-\eta}^0 \frac{\sin kt}{t}dt\right| < M_2\eta$$

and

$$\left|\int_{-\eta}^\eta f(x+t)\frac{\sin kt}{t}dt\right.$$
$$\left. - [f(x+0) + f(x-0)]\int_0^\eta \frac{\sin kt}{t}dt\right| < (M_1 + M_2)\eta. \quad (22)$$

We have here taken advantage of the fact that

$$\int_0^\eta \frac{\sin kt}{t}dt = \int_{-\eta}^0 \frac{\sin kt}{t}dt.$$

If $kt = z$

$$\lim_{k \to \infty} \int_0^\eta \frac{\sin kt}{t} dt = \lim_{k \to \infty} \int_0^{k\eta} \frac{\sin z}{z} dz = \int_0^\infty \frac{\sin z}{z} dz = \frac{\pi}{2}, \quad (23)$$

as will be shown in § 214.

As $\eta \to 0$, we can keep M_1 and M_2 fixed. We see therefore from (20), (21), (22), and (23) that

$$\frac{1}{\pi} \int_{-a}^a f(x+t) \frac{\sin kt}{t} dt$$

$$\underset{k \to \infty}{\longrightarrow} \frac{1}{2}[f(x+0) + f(x-0)] = f(x). \quad (24)$$

162. The Fourier integral. It follows from (24) that

$$\pi f(x) = \lim_{k \to \infty} \int_{-a}^a f(x+t) \int_0^k \cos utdudt$$

$$= \lim_{k \to \infty} \int_0^k du \int_{-a}^a f(x+t) \cos utdt$$

$$= \int_0^\infty du \int_{-a}^a f(x+t) \cos utdt. \quad (25)$$

For any $A > a$

$$\int_{-A}^A f(x+t) \cos utdt = \int_{-A}^{-a} + \int_{-a}^a + \int_a^A ;$$

or

$$\int_{-A}^A - \int_{-a}^a = \int_{-A}^{-a} + \int_a^A .$$

Hence

$$\int_0^k du \int_{-A}^A f(x+t) \cos utdt - \int_0^k du \int_{-a}^a f(x+t) \cos utdt$$

$$= \int_0^k du \int_{-A}^{-a} + \int_0^k du \int_a^A$$

$$= \int_{-A}^{-a} f(x+t)dt \int_0^k \cos utdu$$

$$+ \int_a^A f(x+t)dt \int_0^k \cos utdu.$$

If the improper integral $\int_{-\infty}^{\infty} |f(x)|\, dx$ exists, then

$$\left| \int_0^k du \int_{-A}^{A} f(x+t) \cos ut\, dt - \int_0^k du \int_{-a}^{a} f(x+t) \cos ut\, dt \right|$$

$$\leq \left| \int_{-A}^{-a} f(x+t) \frac{\sin kt}{t}\, dt \right| + \left| \int_a^A f(x+t) \frac{\sin kt}{t}\, dt \right|$$

$$\leq \frac{1}{a} \left[\int_{-A}^{-a} |f(x+t)|\, dt + \int_a^A |f(x+t)|\, dt \right]$$

$$\leq \frac{1}{a} \int_{-\infty}^{\infty} |f(x+t)|\, dt = \frac{C}{a},$$

where $\int_{-\infty}^{\infty} |f(x+t)|\, dt = C$. If we keep k fixed and let A increase without limit, we have

$$\left| \int_0^k du \int_{-\infty}^{\infty} f(x+t) \cos ut\, dt \right.$$
$$\left. - \int_0^k du \int_{-a}^{a} f(x+t) \cos ut\, dt \right| \leq \frac{C}{a} < \epsilon \quad (26)$$

for a sufficiently large a, depending upon the value of the arbitrary positive number ϵ.

The number a having been selected subject to this condition, there is a positive number k_1 such that, for $k > k_1$, we have

$$\left| \int_0^k du \int_{-a}^{a} f(x+t) \cos ut\, dt - \pi f(x) \right| < \epsilon. \quad (27)$$

By combining (26) and (27) we find that for $k > k_1$

$$\left| \int_0^k du \int_{-\infty}^{\infty} f(x+t) \cos ut\, dt - \pi f(x) \right| < 2\epsilon.$$

Hence the left member of this inequality approaches zero as $k \to \infty$, and we have

$$\frac{1}{\pi} \int_0^{\infty} du \int_{-\infty}^{\infty} f(x+t) \cos ut\, dt = f(x). \quad (28)$$

The improper repeated integral in (28) whose existence has just been established is known as *Fourier's integral*, and the formula as *Fourier's integral formula*.

We can write (28) in the form

$$\frac{1}{\pi}\int_0^\infty du \int_{-\infty}^\infty f(t)\cos u(t-x)dt = f(x). \tag{29}$$

But $\cos u(t-x)$ is an even function of u. Hence

$$f(x) = \frac{1}{2\pi}\int_{-\infty}^\infty du \int_{-\infty}^\infty f(t)\cos u(t-x)dt. \tag{30}$$

On the other hand $\sin u(t-x)$ is an odd function of u. Hence

$$0 = \frac{1}{2\pi}\int_{-\infty}^\infty du \int_{-\infty}^\infty f(t)\sin u(t-x)dt \tag{31}$$

in case this infinite integral exists. By combining (30) and (31) we find that

$$\pi f(x) = \frac{1}{2}\int_{-\infty}^\infty du \int_{-\infty}^\infty f(t)e^{-iu(t-x)}dt. \tag{32}$$

This is equivalent to the formula

$$f(x) = \frac{1}{\sqrt{2\pi}}\int_{-\infty}^\infty g(u)e^{iux}du \tag{33}$$

in case

$$g(u) = \frac{1}{\sqrt{2\pi}}\int_{-\infty}^\infty f(t)e^{-iut}dt. \tag{34}$$

Equations (33) and (34) exhibit an important reciprocal relation between the functions $f(x)$ and $g(x)$.

163. Weierstrass' theorem. The result established in § 153 leads to a simple proof of an important theorem due to Weierstrass:

THEOREM 4. *If $f(x)$ is continuous in the interval (a, b) there corresponds to every positive number ϵ a polynomial $P(x)$ such that $|f(x) - P(x)| < \epsilon$ for every x in (a, b).*

We know from § 152 that we can assume that $0 < a < b < 2\pi$. Then we can extend $f(x)$ beyond (a, b) in any way provided that we make it continuous and have $f(x + 2\pi) = f(x)$. This requires that we make $f(0) = f(2\pi)$. It was proved in § 153 that we can satisfy the inequality

$$|f(x) - \Sigma_m(x)| < \frac{\epsilon}{2}$$

uniformly throughout $(0, 2\pi)$ by taking m sufficiently large, it being understood that $\Sigma_m(x)$ is the arithmetic mean of the Fourier sums $S_0, S_1, \cdots, S_m$. It follows from its definition that $\Sigma_m(x)$ is a polynomial in a limited number of sines and cosines of multiples of x. Each of these latter can be expanded into an everywhere convergent power series in x. Hence there is a polynomial $P(x)$ which satisfies the inequality

$$|\Sigma_m(x) - P(x)| < \frac{\epsilon}{2}$$

uniformly in (a, b). It follows immediately that

$$|f(x) - P(x)| < \epsilon$$

uniformly in (a, b).

164. Orthogonal functions. The functions $f_0(x)$, $f_1(x)$, $\cdots$, $f_n(x)$, $\cdots$, limited or unlimited in number, are said to be *orthogonal* with respect to the interval (a, b) if

$$\int_a^b f_i(x)f_j(x)dx = 0$$

when $i \neq j$. For example, the functions $\cos nx$ and $\sin nx$ with which we have been dealing in this chapter are orthogonal with respect to the interval $(0, 2\pi)$. This can be seen by identifying $\cos nx$ with $f_{2n}(x)$ and $\sin nx$ with $f_{2n+1}(x)$, where $n = 0, 1, 2, \cdots$, and referring to formulae (4).

We say that the functions $f_0(x)$, $f_1(x)$, $\cdots$ are *linearly independent* if no relation of the form

$$c_0 f_0 + c_1 f_1 + \cdots + c_r f_r = 0$$

exists for any r unless all the c's are zero. If there is a relation of this kind for which the c's are not all zero, the functions are said to be *linearly dependent*.

165. The reader ought to have no difficulty in showing that if the functions are linearly dependent they are not orthogonal with respect to any interval. But a given set of linearly independent functions may, or may not, be orthogonal with respect to an interval over which they are all integrable. If they are not, there are certain homogeneous linear functions of them that are. For if f_0 and f_1 are not orthogonal with respect to the interval (a, b), that is, if $\int_a^b f_0 f_1 dx \neq 0$, let $\varphi(x) = f_0(x) + c f_1(x)$, where c is a constant. Then $\int_a^b f_0 \varphi(x) dx = \int_a^b f_0^2 dx + c \int_a^b f_0(x) f_1(x) dx$.

This will equal zero if $c = -\dfrac{\int_a^b f_0^2 dx}{\int_a^b f_0 f_1 dx}$. For this choice of c, $f_0(x)$ and $\varphi(x)$ are orthogonal. We then replace $f_1(x)$ by $\varphi(x)$. If $f_2(x)$ is not orthogonal to both $f_0(x)$ and $f_1(x)$ we can in a similar way replace it by a homogeneous function of $f_0(x)$, $f_1(x)$, and $f_2(x)$ that is orthogonal to both $f_0(x)$ and $f_1(x)$. And we can clearly proceed in this way until we have orthogonalized the whole set, if it is finite, or have gone as far as we wish in the case of an infinite set.

Suppose for example that the original functions are the polynomials $1, x, x^2, \cdots, x^n, \cdots$ and that the interval is $(-1, 1)$. Now x is orthogonal to 1, while x^2 is orthogonal to x but not to 1. We therefore put $\varphi(x) = c_0 + c_2 x^2$, and then

$$\int_{-1}^{1} 1 \cdot \varphi(x) dx = c_0 \int_{-1}^{1} dx + c_2 \int_{-1}^{1} x^2 dx = 0$$

if $c_2 = -3 c_0$. In our original set of functions we replace x^2 by $1 - 3x^2$. We denote it by $f_2(x)$ and x by $f_1(x)$. Then the three functions 1, f_1, and f_2 are orthogonal, while f_1 contains only odd powers of x and f_2 only even powers.

Suppose now that we have obtained in this way the polynomials $1, f_1, f_2, \cdots, f_{n-1}$ which form an orthogonal set and are such that f_i is of degree i and contains only even powers of x when i is even and only odd powers when i is odd.

We wish so to determine the constants $c_0, c_1, \cdots, c_n$ that
$$f_n(x) = c_0 + c_1 f_1(x) + \cdots + c_{n-1} f_{n-1}(x) + c_n x^n$$
shall be orthogonal to each of the functions $1, f_1(x), f_2(x), \cdots, f_{n-1}(x)$. In order that $f_n(x)$ be orthogonal to 1 it is necessary that $\int_{-1}^{1} f_n(x) dx$ be zero. But
$$\int_{-1}^{1} f_n(x) dx = c_0 \int_{-1}^{1} dx + c_n \int_{-1}^{1} x^n dx,$$
since $f_1, f_2, \cdots, f_{n-1}$ are also orthogonal to 1. Hence
$$\left[c_0 x + c_n \frac{x^{n+1}}{n+1} \right]_{-1}^{1} = 0.$$

From this we see that $c_0 = 0$ or $c_0 + \frac{1}{n+1} c_n = 0$ according as n is odd or even. In order that $f_n(x)$ be orthogonal to $f_i(x)$ we must have $c_i \int_{-1}^{1} \overline{f_i(x)}^2 dx + c_n \int_{-1}^{1} x^n f_i(x) dx = 0$. Hence if i and n are not both even or both odd $c_i = 0$. If they are either both even or both odd
$$c_i = -c_n \frac{\int_{-1}^{1} x^n f_i(x) dx}{\int_{-1}^{1} \overline{f_i(x)}^2 dx},$$
and c_i is uniquely determined in terms of c_n. Moreover $f_n(x)$ contains only even powers of x if n is even and only odd powers if n is odd. Now the assumptions we have made for $1, f_1(x), f_2(x), \cdots, f_{n-1}(x)$ are satisfied by $1, f_1(x)$, and $f_2(x)$. Hence there is a set of polynomials with the following properties:

(a) It contains one, and only one, polynomial of each degree.

(b) The polynomials are orthogonal with respect to the interval $(-1, 1)$.

The polynomial of degree n of this set is uniquely determined up to a constant factor. For a certain choice of these factors these polynomials are:

$$P_0(x) = 1, \qquad P_1(x) = x, \qquad P_2(x) = \frac{3x^2 - 1}{2},$$

$$P_3(x) = \frac{5x^3 - 3x}{2},$$

and, in general,

$$P_n(x) = \frac{1 \cdot 3 \cdot 5 \cdots (2n-1)}{n!} \left[x^n - \frac{n(n-1)}{2(2n-1)} x^{n-2} \right.$$
$$\left. + \frac{n(n-1)(n-2)(n-3)}{2 \cdot 4 (2n-1)(2n-3)} x^{n-4} + \cdots \right],$$

where the last term in the square bracket is

$$(-1)^{n/2} \frac{n!}{2 \cdot 4 \cdots n(2n-1)(2n-3) \cdots (n+1)}$$

when n is even, and

$$(-1)^{(n-1)/2} \frac{n!}{2 \cdot 4 \cdots (n-1)(2n-1)(2n-3) \cdots (n+2)} x$$

when n is odd.

Now the polynomials

$$Q_0(x) = 1 \quad \text{and} \quad Q_n(x) = \frac{1}{2^n n!} \frac{d^n (x^2-1)^n}{dx^n} \quad (n = 1, 2, \cdots)$$

are orthogonal with respect to the interval $(-1, 1)$. For, if we put $(x^2 - 1)^n = u_n(x)$, then

$$\int_{-1}^{1} Q_n(x) \cdot x^m dx = \frac{1}{2^n n!} \int_{-1}^{1} u_n^{(n)}(x) x^m dx,$$

and successive integration by parts shows that this last integral is equal to zero when m is zero or a positive integer less than n. Hence since $Q_m(x)$ is a polynomial of degree m

$$\int_{-1}^{1} Q_n(x) Q_m(x) dx = 0 \quad (m < n).$$

It follows from this that $P_n(x)$ and $Q_n(x)$ differ at most by a constant factor. A direct comparison of the coefficients of the highest power of x in each shows that this constant is 1. Hence

$$P_n(x) = \frac{1}{2^n n!} \frac{d^n(x^2-1)^n}{dx^n}.$$

This formula is due to Rodigues. The polynomials are known as *Legendre's polynomials*. They are of great importance in mathematical physics. One application in the theory of potential will be given in § 167.

166. A recursion formula. We leave it to the reader to verify that any polynomial $f(x)$ of degree n can be written in the form

$$f(x) = c_0 P_0 + c_1 P_1 + \cdots + c_n P_n, \qquad (35)$$

where $c_0, c_1, \cdots, c_n$ are constants. Since $xP_n(x)$ is a polynomial of degree $n+1$, we have

$$xP_n(x) = c_0 P_0 + c_1 P_1 + \cdots + c_{n+1} P_{n+1}. \qquad (36)$$

If we multiply each side of (36) by $P_{n-2}(x)$ and then integrate each side of the resulting equation from -1 to 1, we get

$$\int_{-1}^{1} P_n(x) \cdot xP_{n-2}(x) dx = c_{n-2} \int_{-1}^{1} \overline{P_{n-2}(x)}^2 dx = 0.$$

But the last integral is not zero. Hence $c_{n-2} = 0$. We see in the same way that $c_i = 0$ ($i = 0, 1, 2, \cdots, n-3$). We have then

$$xP_n = c_{n-1} P_{n-1} + c_n P_n + c_{n+1} P_{n+1}. \qquad (37)$$

But xP_n does not contain x^n. Neither does P_{n-1} nor P_{n+1}. Hence $c_n = 0$. In order to determine c_{n-1} and c_{n+1}, we equate the coefficients of the highest power of x in the two sides of (37). This shows that $c_{n+1} = \dfrac{n+1}{2n+1}$. Leibniz's formula for the nth derivative of the product of two functions shows that $P_n(1) = 1$. Then for $x = 1$ (37) becomes

$$1 = c_{n-1} + \frac{n+1}{2n+1}$$

or

$$c_{n-1} = \frac{n}{2n+1}.$$

Hence

$$(n+1)P_{n+1} - (2n+1)xP_n + nP_{n-1} = 0. \qquad (38)$$

This recursion formula enables us to compute any number of Legendre polynomials when we know two consecutive ones.

167. An application to potential theory. The potential at the point whose polar coordinates are (t, θ) due to a unit of mass at the point $(1, 0)$ is $V = \dfrac{1}{\sqrt{1 - 2\cos\theta \cdot t + t^2}}$ (see § 110). If now $0 < t < 1$ we have

$$V = Q_0(x) + Q_1(x)t + \cdots + Q_n(x)t^n + \cdots, \qquad (39)$$

where we have written x for $\cos\theta$. It can be shown that the power series in t in the right member of (39) has the radius of convergence 1. For the values of t under consideration we can therefore differentiate term by term with respect to t. Hence

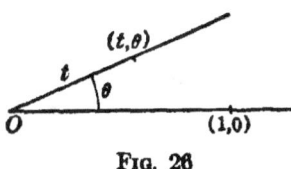

Fig. 26

$$\frac{x-t}{\sqrt{(1-2xt+t^2)^3}} = Q_1 + 2Q_2 t + \cdots + nQ_n t^{n-1} + \cdots$$

or

$$(x-t)(Q_0 + Q_1 t + \cdots + Q_n t^n + \cdots)$$
$$= (1 - 2xt + t^2)(Q_1 + 2Q_2 t + \cdots + nQ_n t^{n-1} + \cdots).$$

If we express each member of this equation as a power series in t and then equate the coefficients of t^n in the two members of the equation we find that

$$(n+1)Q_{n+1} - (2n+1)xQ_n + nQ_{n-1} = 0.$$

Moreover we see by direct verification that $Q_0(x) = 1 = P_0(x)$ and $Q_1(x) = x = P_1(x)$. Hence for every n we have $Q_n(x) = P_n(x)$. That is, the coefficients of the powers of t in (39) are the Legendre polynomials. They are for this reason often referred to as *Legendre coefficients*.

168. It is worth while noting that the theorems we have proved concerning the expansion of functions in Fourier series are special cases of theorems concerning the expansion of functions in series of orthogonal functions. We cannot go into the details of this theory here, but we cite one theorem concerning Legendre polynomials by way of illustration of this more general phase of the subject:[1]

If $f(x)$ is one-valued and continuous in the interval $(-1, 1)$ and has only a finite number of oscillations in this interval it can be expanded in a series of Legendre polynomials; that is,

$$f(x) = \sum q_n P_n(x), \quad \text{where} \quad q_n = \frac{2n+1}{2} \int_{-1}^{1} f(x) P_n(x) dx.$$

It will be observed that the law of formation of these coefficients is similar to the law of formation of the Fourier coefficients.

EXERCISES

Get three non-vanishing terms in the expansion of each of the following functions in terms of Legendre polynomials:

1. $f(x) = x + x^2$. 2. $f(x) = e^x$.
3. $f(x) = -x$ for $-1 \leq x \leq 0$; $f(x) = x$ for $0 \leq x \leq 1$.
4. Show by successive applications of the method of integration by parts that $\int_{-1}^{1} (x^2 - 1)^n dx = \frac{(-1)^n (n!)^2 2^{2n+1}}{(2n+1)!}$.
5. In the same way, starting with Rodrigues' form for $P_n(x)$, show that

$$\int_{-1}^{1} [P_n(x)]^2 dx = \frac{(-1)^n (2n)!}{2^{2n} (n!)^2} \int_{-1}^{1} (x^2 - 1)^n dx.$$

6. Show that

$$\int_{-1}^{1} [P_n(x)]^2 dx = \frac{2}{2n+1}.$$

---

[1] See, for example, Pierpont, *Theory of Functions of a Complex Variable*, p. 503.

7. Determine the coefficients $q_0, q_1, \cdots, q_m$ in such a way as to make the integral

$$\int_{-1}^{1} [f(x) - q_0 P_0(x) - q_1 P_1(x) \cdots q_m P_m(x)]^2 dx$$

a minimum.

8. Does the series $\sum_{\nu=0}^{\infty} \dfrac{q_\nu^2}{2\nu + 1}$ converge?

9. Assume that $f(x)$ can be expanded in a series of Legendre polynomials, $f(x) = \sum_{n=0}^{\infty} a_n P_n(x)$, that converges uniformly in the interval $(-1, 1)$. Determine the coefficients.

CHAPTER XII

IMPLICIT FUNCTIONS. FUNCTIONAL DETERMINANTS

169. Existence theorems. If x and y are connected by the relation $f(x, y) = 0$, it seems obvious that a value of x within a given range determines one or more definite values of y, or that within this range y is a function of x. That this conclusion is not always justified, however, can be seen from the following example: $xy = 0$. Here the value of y is not determined by the given relation when $x = 0$. In other words, y is not determined as a function of x within a range including $x = 0$. On the other hand, the value of y is determined by the value of x in any range if $x + y = 1$.

We wish to determine a set of conditions on the function $f(x, y)$ that will assure us that y is determined as a function of x by the relation $f(x, y) = 0$, and that will tell us something about the properties of this function. More generally we wish to determine the nature of the dependence of u on $x_1, x_2, \cdots, x_n$ that is determined by the relation

$$f(x_1, x_2, \cdots, x_n; u) = 0. \tag{1}$$

THEOREM 1. *Suppose that the function $f(x_1, x_2, \cdots, x_n; u)$ satisfies the following conditions:*

(1) It is continuous in the neighborhood of the point $M \equiv (a_1, a_2, \cdots, a_n; c)$.

(2) It is differentiable at this point.

(3) $f(a_1, a_2, \cdots, a_n; c) = 0$.

(4) $\dfrac{\partial f(x_1, x_2, \cdots, x_n; u)}{\partial u} \neq 0$ at M.

Then there is at least one function $u = \varphi(x_1, x_2, \cdots, x_n)$ which equals c at the point $(a_1, a_2, \cdots, a_n)$ and satisfies the equation

$$f(x_1, x_2, \cdots, x_n; u) = 0$$

identically in the neighborhood of this point. Moreover every function φ that satisfies the two conditions possesses all the first partial derivatives $\frac{\partial u}{\partial x_i}$ $(i = 1, 2, \cdots, n)$ at this point, and is continuous here.

In the proof we shall take $n = 2$ and put $x_1 = x$, $x_2 = y$, $a_1 = a$, $a_2 = b$. This will not affect the essential steps of the proof. Since $f(x, y; u) = 0$ at the point (a, b, c) and $\frac{\partial f(x, y, u)}{\partial u} \neq 0$ at this point $f(a, b, u)$ has one sign between $c - \delta$ and c, and the opposite sign between c and $c + \delta$, for sufficiently small values of δ. But $f(x, y, u)$ is by hypothesis continuous in the neighborhood of the point (a, b, c). There is therefore a positive number δ' such that

$$f(x, y; c - \delta) \quad \text{and} \quad f(x, y; c + \delta)$$

are of opposite signs when $|x - a| < \delta'$ and $|y - b| < \delta'$. If we give to x and y fixed values that satisfy these inequalities, there is a value of u between $c - \delta$ and $c + \delta$ such that $f(x, y, u) = 0$. This value of u depends upon our choice of x and y—that is, is a function $\varphi(x, y)$ of x and y which satisfies the given equation identically.

We have now to consider the value of $\varphi(a, b)$. We know that
$$f(a, b; c) = 0$$
and
$$f[a, b, \varphi(a, b)] = 0.$$

That is, $\varphi(a, b)$ and c are both roots of the equation $f(a, b, u) = 0$. But $\varphi(a, b)$ lies between $c - \delta$ and $c + \delta$. From the way δ was chosen we know that there are not two roots of this equation between $c - \delta$ and $c + \delta$. Hence $\varphi(a, b) = c$.

We have already seen that for a sufficiently small positive δ there is a δ' such that when $|x - a| < \delta'$ and $|y - b| < \delta'$ we have $\varphi(x, y)$ between $c - \delta$ and $c + \delta$. Hence $\varphi(x, y)$ is continuous at the point (a, b).

Since $f(x, y, u)$ is differentiable at the point (a, b, c),

$$\Delta f = f(x, y, u) - f(a, b, c) = [f_x(a, b, c) + \epsilon_1]\Delta x \\ + [f_y(a, b, c) + \epsilon_2]\Delta y + [f_u(a, b, c) + \epsilon_3]\Delta u = 0,$$

where Δu, ϵ_1, ϵ_2, and ϵ_3 approach zero with Δx and Δy. Then $\varphi(x, y)$ is differentiable at (a, b) since

$$\Delta u = -\frac{[f_x(a, b, c) + \epsilon_1]\Delta x + [f_y(a, b, c) + \epsilon_2]\Delta y}{f_u(a, b, c) + \epsilon_3}.$$

If we put $\Delta y = 0$ and let Δx approach zero we see that

$$\lim_{\Delta x \to 0} \frac{\Delta u}{\Delta x} = \frac{\partial u}{\partial x} = -\frac{f_x(a, b, c)}{f_u(a, b, c)}$$

and similarly

$$\lim_{\Delta y \to 0} \frac{\Delta u}{\Delta y} = \frac{\partial u}{\partial y} = -\frac{f_y(a, b, c)}{f_u(a, b, c)}.$$

We cannot conclude from the given conditions that the solution $u = \varphi(x, y)$ is unique. But if we add the additional condition

$$f_u(x, y, u) \neq 0$$

in the neighborhood of the point (a, b, c), it follows that there is only one continuous solution that reduces to c for $x = a$, $y = b$. For, if there were two, u and u_1, we should have

$$f(x, y, u) - f(x, y, u_1) = (u - u_1)f_u(x, y, U) = 0,$$

where U is between u and u_1. But $u - u_1 \neq 0$. Hence

$$f_u(x, y, U) = 0,$$

and this is contrary to the hypothesis.

Under this additional assumption we can apply the preceding discussion to the point (x, y, u) and conclude that $\varphi(x, y)$ is differentiable in the neighborhood of the point (a, b). When the conditions of the theorem are satisfied we say that equation (1) defines u as an *implicit* function of $x_1, x_2, \cdots, x_n$.

170. Theorem 2. *Suppose that the n functions f_1, f_2, $\cdots$, f_n of the $m + n$ variables x_1, x_2, $\cdots$, x_m; u_1, u_2, $\cdots$, u_n satisfy the following conditions:*

(1) They are continuous in the neighborhood of the point $(a_1, a_2, \cdots, a_m; c_1, c_2, \cdots, c_n)$.

(2) They are all equal to zero at this point.

(3) They are all differentiable at this point.

(4) The determinant

$$D = \begin{vmatrix} \dfrac{\partial f_1}{\partial u_1} & \dfrac{\partial f_1}{\partial u_2} & \cdots & \dfrac{\partial f_1}{\partial u_n} \\ \cdots & \cdots & & \cdots \\ \cdots & \cdots & & \cdots \\ \dfrac{\partial f_n}{\partial u_1} & \dfrac{\partial f_n}{\partial u_2} & \cdots & \dfrac{\partial f_n}{\partial u_n} \end{vmatrix}$$

is not zero at this point.

Then there is at least one system of functions u_1, u_2, $\cdots$, u_n of the m variables x_1, x_2, $\cdots$, x_m which are equal to c_1, c_2, $\cdots$, c_n respectively when $x_1 = a_1$, $x_2 = a_2$, $\cdots$, $x_m = a_m$, and which satisfy the equations $f_1 = 0$, $f_2 = 0_0$, $\cdots$, $f_n = 0$ identically in the neighborhood of the point $(a_1, a_2, \cdots, a_m)$.

If $n = 1$ this theorem is the same as Theorem 1 which has been proved. In order to prove it in general we assume that it has been proved for $n - 1$ functions of $n - 1$ dependent variables. Now

$$D = D_1 \frac{\partial f_1}{\partial u_1} + D_2 \frac{\partial f_2}{\partial u_1} + \cdots + D_n \frac{\partial f_n}{\partial u_1}, \tag{2}$$

where D_i is the co-factor of the element in the first column and ith row of D. Since D is not zero at M, not every D_i is zero at this point. We assume that D_1 is not zero here. Then by virtue of our assumption that the theorem is true for $n - 1$ functions in $n - 1$ unknowns there are $n - 1$ functions

$$u_2 = U_2(x_1, x_2, \cdots, x_m; u_1), \quad u_3 = U_3(x_1, x_2, \cdots, x_m; u_1),$$
$$\cdots, u_n = U_n(x_1, x_2, \cdots, x_m; u_1) \tag{3}$$

§ 170] IMPLICIT FUNCTIONS 303

of the $m + 1$ independent variables $x_1, x_2, \cdots, x_m; u_1$ which have the following properties:

(a) $U_i(a_1, a_2, \cdots, a_m; c_1) = c_i$ $(i = 2, \cdots, n)$
(b) Satisfy identically the equations

$$\left.\begin{aligned} f_2(x_1, x_2, \cdots, x_m; u_1; U_2, U_3, \cdots, U_n) &= 0 \\ \cdots\cdots\cdots\cdots\cdots\cdots\cdots\cdots\cdots\cdots\cdots\cdots\cdots\cdots \\ \cdots\cdots\cdots\cdots\cdots\cdots\cdots\cdots\cdots\cdots\cdots\cdots\cdots\cdots \\ f_n(x_1, x_2, \cdots, x_m; u_1; U_2, U_3, \cdots, U_n) &= 0 \end{aligned}\right\} \quad (4)$$

(c) Are differentiable at M.

If we substitute these functions for $u_2, u_3, \cdots, u_n$ respectively in f_1 we obtain a function $\Phi(x_1, x_2, \cdots, x_m; u_1)$ of $x_1, x_2, \cdots, x_m; u_1$. We have now to show that the equation $\Phi = 0$ determines u_1 as a function of $x_1, x_2, \cdots, x_m$ with certain properties. To this end we can apply Theorem 1 if we know that $\dfrac{\partial \Phi}{\partial u_1} \neq 0$ at the point $(a_1, a_2, \cdots, a_m; c_1)$. Now

$$\frac{\partial \Phi}{\partial u_1} = \frac{\partial f_1}{\partial u_1} + \frac{\partial f_1}{\partial U_2}\frac{\partial U_2}{\partial u_1} + \cdots + \frac{\partial f_1}{\partial U_n}\frac{\partial U_n}{\partial u_1}. \quad (5)$$

Moreover, since equations (4) are identities in the x's and u_1,

$$\left.\begin{aligned} \frac{\partial f_2}{\partial u_1} + \frac{\partial f_2}{\partial U_2}\frac{\partial U_2}{\partial u_1} + \cdots + \frac{\partial f_2}{\partial U_n}\frac{\partial U_n}{\partial u_1} &= 0 \\ \cdots\cdots\cdots\cdots\cdots\cdots\cdots\cdots\cdots\cdots\cdots\cdots\cdots\cdots \\ \cdots\cdots\cdots\cdots\cdots\cdots\cdots\cdots\cdots\cdots\cdots\cdots\cdots\cdots \\ \frac{\partial f_n}{\partial u_1} + \frac{\partial f_n}{\partial U_2}\frac{\partial U_2}{\partial u_1} + \cdots + \frac{\partial f_n}{\partial U_n}\frac{\partial U_n}{\partial u_1} &= 0 \end{aligned}\right\} \quad (6)$$

Now $U_j = u_j$ $(j = 2, 3, \cdots, n)$ and therefore $\dfrac{\partial f_i}{\partial U_j} = \dfrac{\partial f_i}{\partial u_j}$.

Moreover $\sum_{i=1}^{n} \dfrac{\partial f_i}{\partial u_1} D_i = D$ and $\sum_{i=1}^{n} \dfrac{\partial f_i}{\partial u_j} D_i = 0$. Hence if we multiply equation (5) through by D_1 and equations (6) by $D_2, D_3, \cdots, D_n$ respectively, and add, we find that

$$\frac{\partial \Phi}{\partial u_1} D_1 = D, \quad \text{or} \quad \frac{\partial \Phi}{\partial u_1} = \frac{D}{D_1}.$$

Since D is not zero at M, $\dfrac{\partial \Phi}{\partial u_1} \neq 0$ at M. Hence u_1 is determined as a function of $x_1, x_2, \cdots, x_m$ that is differentiable at M and that satisfies the equation $\Phi = 0$ identically. If we substitute this value of u_1 for u_1 in equations (3) we obtain $u_2, u_3, \cdots, u_n$ as differentiable functions of $x_1, x_2, \cdots, x_m$.

If the partial derivatives that occur in D are continuous in the neighborhood of M, neither D nor D_1 vanishes in this neighborhood. Hence $\dfrac{\partial \Phi}{\partial u_1}$ does not vanish here and the solution of $\Phi = 0$ is unique. We are assuming that when D_1 does not vanish in the neighborhood of M the continuous solutions $U_2, U_3, \cdots, U_n$ are unique. Hence the continuous solutions of the original equations are unique.

171. Derivatives of implicit functions. The existence and differentiability of implicit functions having been established under certain conditions in Theorems 1 and 2, we consider now the problem of determining explicit representations of the partial derivatives of these functions. If $f(x_1, x_2, \cdots, x_n; u)$ satisfies the conditions of Theorem 1 and u is defined by the equation $f(x_1, x_2, \cdots, x_n; u) = 0$, we know that the derivatives $\dfrac{\partial u}{\partial x_i}$ $(i = 1, 2, \cdots, n)$ exist and are given by the formulae

$$\frac{\partial u}{\partial x_i} = -\frac{\dfrac{\partial f}{\partial x_i}}{\dfrac{\partial f}{\partial u}}.$$

We arrive at the same formulae if we look upon the equation

$$f(x_1, x_2, \cdots, x_n; u) = 0$$

as stating that the composite function in the left member is identically zero, and form its various partial derivatives in accordance with the rule for the differentiation of composite functions. In doing this we assume the existence of $\dfrac{\partial u}{\partial x_i}$.

Similarly, if the n equations

$$f_i(x_1, x_2, \cdots, x_m; u_1, u_2, \cdots, u_n) = 0 \quad (i = 1, 2, \cdots, n)$$

satisfy the conditions of Theorem 2 we have n composite functions all identically zero. If we differentiate them partially according to the rule for the differentiation of composite functions, we obtain the equations

$$\frac{\partial f_i}{\partial x_j} + \frac{\partial f_i}{\partial u_1}\frac{\partial u_1}{\partial x_j} + \cdots + \frac{\partial f_i}{\partial u_n}\frac{\partial u_n}{\partial x_j} = 0 \quad \left(\begin{matrix} i = 1, 2, \cdots, n \\ j = 1, 2, \cdots, m \end{matrix}\right),$$

on the assumption that the partial derivatives $\dfrac{\partial u_i}{\partial x_j}$ all exist. That they do in fact exist can be shown as in the case $n = 1$. Since the determinant

$$\begin{vmatrix} \dfrac{\partial f_1}{\partial u_1} & \cdots & \dfrac{\partial f_1}{\partial u_n} \\ \cdots & & \cdots \\ \dfrac{\partial f_n}{\partial u_1} & \cdots & \dfrac{\partial f_n}{\partial u_n} \end{vmatrix} \neq 0$$

we can solve these equations for $\dfrac{\partial u_i}{\partial x_j}$. The higher partial derivatives of the u's can be obtained by a further application of the rule for differentiating composite functions.

EXAMPLES.

(a) The equations

$$x - y + z = 1 \quad \text{and} \quad x^2 + y^2 + z^2 = 4$$

define y and z as functions of x. Find $\dfrac{dy}{dx}, \dfrac{dz}{dx}$, and $\dfrac{d^2y}{dx^2}$ without solving for y and z.

(b) The equation

$$\frac{x^2}{a^2} - \frac{y^2}{b^2} - \frac{z^2}{c^2} = 1$$

defines z as a function of x and y. Without solving for z find $\dfrac{\partial z}{\partial x}, \dfrac{\partial z}{\partial y}, \dfrac{\partial^2 z}{\partial x^2}, \dfrac{\partial^2 z}{\partial x \partial y}$, and $\dfrac{\partial^2 z}{\partial y^2}$.

(c) The equations
$$f(x, y, u, v) = 0 \quad \text{and} \quad \varphi(x, y, u, v) = 0$$
define u and v as functions of x and y; and also x and y as functions of u and v. Find the first partial derivatives of u and v with respect to x and y, and of x and y with respect to u and v. Show that [1]
$$\frac{\partial u}{\partial x}\frac{\partial x}{\partial u} + \frac{\partial v}{\partial x}\frac{\partial x}{\partial v} = 1,$$
$$\frac{\partial u}{\partial x}\frac{\partial y}{\partial u} + \frac{\partial v}{\partial x}\frac{\partial y}{\partial v} = 0.$$

(d) Given $x - (x^2 + y^2)u = 0$, $y - (x^2 + y^2)v = 0$. Verify the formulae of Example (c).

(e) If $\Phi(x, y, z) = 0$, where $\Phi(x, y, z)$ satisfies the conditions of § 170, we have $\dfrac{\partial z}{\partial x} = -\dfrac{\dfrac{\partial \Phi}{\partial x}}{\dfrac{\partial \Phi}{\partial z}}$ and $\dfrac{\partial z}{\partial y} = -\dfrac{\dfrac{\partial \Phi}{\partial y}}{\dfrac{\partial \Phi}{\partial z}}$, and the equation of the tangent plane of the surface $\Phi(x, y, z) = 0$ at the point (x_1, y_1, z_1) is
$$\left(\frac{\partial \Phi}{\partial x}\right)_1 (x - x_1) + \left(\frac{\partial \Phi}{\partial y}\right)_1 (y - y_1) + \left(\frac{\partial \Phi}{\partial z}\right)_1 (z - z_1) = 0$$
(see Eq. 12, § 35). If the three partial derivatives that occur in this equation are all zero at (x_1, y_1, z_1) this point is called a *singular point* of the surface.

172. Applications of the preceding theory. We add two important applications of the theory discussed in §§ 169 and 170.

(a) *Inversion.* If $x_1, x_2, \cdots, x_n$ are n differentiable functions of $u_1, u_2, \cdots, u_n$ whose Jacobian does not vanish

[1] De la Vallée-Poussin, *Cours d'analyse infinitésimale*, 5th ed., Vol. 1, pp. 146, 147.

identically, they determine $u_1, u_2, \cdots, u_n$ as differentiable functions of $x_1, x_2, \cdots, x_n$. That is, from the equations

$$x_1 = f_1(u_1, u_2, \cdots, u_n)$$
$$x_2 = f_2(u_1, u_2, \cdots, u_n)$$
$$\cdots\cdots\cdots\cdots\cdots\cdots\cdots$$
$$x_n = f_n(u_1, u_2, \cdots, u_n)$$

we derive the equations

$$u_1 = \varphi_1(x_1, x_2, \cdots, x_n)$$
$$u_2 = \varphi_2(x_1, x_2, \cdots, x_n)$$
$$\cdots\cdots\cdots\cdots\cdots\cdots\cdots$$
$$u_n = \varphi_n(x_1, x_2, \cdots, x_n).$$

(b) *Tangents to a skew curve.* Consider the curve of intersection of the surfaces

$$F(x, y, z) = 0,$$
$$\Phi(x, y, z) = 0. \qquad (7)$$

Its tangent at the point (x_0, y_0, z_0) lies in the tangent plane of each surface at this point. It is therefore the intersection of these planes. If we assume that the conditions of Theorem 2 are satisfied, and in particular that the Jacobian

$$\begin{vmatrix} \dfrac{\partial F}{\partial y} & \dfrac{\partial F}{\partial z} \\ \dfrac{\partial \Phi}{\partial y} & \dfrac{\partial \Phi}{\partial z} \end{vmatrix}$$

does not vanish at this point, we can verify this conclusion analytically. The equations of the curve define y and z as functions of x:

$$y = f(x), \qquad z = \varphi(x).$$

The equations of the tangent to the curve at the given point are

$$\frac{x - x_0}{1} = \frac{y - y_0}{f'(x_0)} = \frac{z - z_0}{\varphi'(x_0)}. \qquad (8)$$

(See § 35.) From equations (7) we have

$$\left.\begin{array}{l}\dfrac{\partial F}{\partial x}+\dfrac{\partial F}{\partial y}f'(x)+\dfrac{\partial F}{\partial z}\varphi'(x),\\[6pt]\dfrac{\partial \Phi}{\partial x}+\dfrac{\partial \Phi}{\partial y}f'(x)+\dfrac{\partial \Phi}{\partial z}\varphi'(x).\end{array}\right\} \quad (9)$$

If in these equations we put $x = x_0$, $y = y_0$, and $z = z_0$, and replace $f'(x)$ and $\varphi'(x)$ by their values as given by (8), we obtain the equations of the tangent line in the form

$$\left.\begin{array}{l}\left(\dfrac{\partial F}{\partial x}\right)_0 (x-x_0)+\left(\dfrac{\partial F}{\partial y}\right)_0 (y-y_0) \\[6pt] \qquad +\left(\dfrac{\partial F}{\partial z}\right)_0 (z-z_0)=0, \\[10pt] \left(\dfrac{\partial \Phi}{\partial x}\right)_0 (x-x_0)+\left(\dfrac{\partial \Phi}{\partial y}\right)_0 (y-y_0) \\[6pt] \qquad +\left(\dfrac{\partial \Phi}{\partial z}\right)_0 (z-z_0)=0.\end{array}\right\} \quad (10)$$

That is, the tangent line is the intersection of the tangent planes of the surfaces at the common point (x_0, y_0, z_0).

173. Functional determinants. Let

$$\begin{array}{l} u_1 = f_1(x_1, x_2, \cdots, x_n) \\ u_2 = f_2(x_1, x_2, \cdots, x_n) \\ \quad \cdots \\ u_n = f_n(x_1, x_2, \cdots, x_n) \end{array} \quad (11)$$

be n differentiable functions of the n independent variables $x_1, x_2, \cdots, x_n$. The determinant

$$D = \begin{vmatrix} \dfrac{\partial f_1}{\partial x_1} & \cdots & \dfrac{\partial f_1}{\partial x_n} \\ \cdots & & \cdots \\ \dfrac{\partial f_n}{\partial x_1} & \cdots & \dfrac{\partial f_n}{\partial x_n} \end{vmatrix}$$

is called the *functional determinant*, or *Jacobian*, of the u's with respect to the x's. It is represented by the symbol $\dfrac{D(u_1, u_2, \cdots, u_n)}{D(x_1, x_2, \cdots, x_n)}$, or, more simply, by the symbol

$$\left(\begin{matrix} u_1, u_2, \cdots, u_n \\ x_1, x_2, \cdots, x_n \end{matrix}\right).$$

Thus the determinant described in Condition 4 of Theorem 2 is a functional determinant.

Suppose that the x's are connected with the n variables $y_1, y_2, \cdots, y_n$ by the n equations

$$\begin{aligned} x_1 &= \varphi_1(y_1, y_2, \cdots, y_n) \\ &\cdots\cdots\cdots\cdots\cdots\cdots \\ &\cdots\cdots\cdots\cdots\cdots\cdots \\ x_n &= \varphi_n(y_1, y_2, \cdots, y_n) \end{aligned} \qquad (12)$$

and that the functions and their first partial derivatives are continuous at a given point. If in (11) we substitute for the x's their values as given by (12), we obtain an expression of the u's in terms of the y's. Now the element in the ith row and jth column of $\dfrac{D(u_1, \cdots, u_n)}{D(y_1, \cdots, y_n)}$ is

$$\frac{\partial u_i}{\partial y_j} = \frac{\partial u_i}{\partial x_1}\frac{\partial x_1}{\partial y_j} + \cdots + \frac{\partial u_i}{\partial x_n}\frac{\partial x_n}{\partial y_j}.$$

But this is the element in the ith row and jth column of the product of $\dfrac{D(u_1, \cdots, u_n)}{D(x_1, \cdots, x_n)}$ and $\dfrac{D(x_1, \cdots, x_n)}{D(y_1, \cdots, y_n)}$. We have therefore the formula

$$\frac{D(u_1, \cdots, u_n)}{D(y_1, \cdots, y_n)} = \frac{D(u_1, \cdots, u_n)}{D(x_1, \cdots, x_n)} \cdot \frac{D(x_1, \cdots, x_n)}{D(y_1, \cdots, y_n)}.$$

If $n = 1$ this formula is the same as the usual formula for the derivative of a composite function: $\dfrac{du}{dy} = \dfrac{du}{dx}\dfrac{dx}{dy}$. From this point of view the Jacobian of n functions of n variables is a natural generalization of the derivative of a single function of a single variable.

Equations (12) define what is called a *transformation*. By means of them a function, or set of functions, of the x's can be transformed into a function, or set of functions, of the y's. If $D \neq 0$ the transformation is said to be *reversible*, since the equations can be solved for the y's in terms of the x's. It follows from what has been proved that if a reversible transformation is applied to a set of n functions of n variables the Jacobian of the resulting set of functions vanishes, or not, according as the Jacobian of the original set of functions vanishes, or not.

If the equations (12) define a reversible transformation they can be solved for the y's in terms of the x's and the equations expressing the y's in terms of the x's define a transformation that is called the *inverse* of (12). If we apply a reversible transformation to a set of functions (11) and then the inverse transformation to the resulting set of functions, we get back to the original functions. It follows therefore that the Jacobian of a reversible transformation is the reciprocal of the Jacobian of its inverse.

EXERCISES

1. Show directly that if we have $f(x, y, u, v) = 0$ and $\varphi(x, y, u, v) = 0$, the Jacobian of u and v with respect to x and y is the reciprocal of the Jacobian of x and y with respect to u and v, it being assumed that neither of these Jacobians is zero.

2. The transformation $u = \dfrac{x}{x^2 + y^2}$, $v = \dfrac{y}{x^2 + y^2}$ is called an *inversion*. Show that it is a reversible transformation and find the equations of its inverse.

3. Show that an inversion transforms a straight line through the origin into a straight line through the origin, and a straight line not through the origin into a circle through the origin.

4. Into what does an inversion transform a circle?

5. Find the Jacobian of the u and v of Exercise 2 with respect to x and y, and the Jacobian of x and y with respect to u and v.

6. Show that the product of the two Jacobians of Exercise 5 is identically equal to 1.

7. Show that an inversion in the plane preserves angles. That is, that if the curves $f_1(x, y) = 0$ and $f_2(x, y) = 0$ intersect at the angle α,

the curves

$$f_1\left(\frac{u}{u^2+v^2}, \frac{v}{u^2+v^2}\right) = F_1(u, v) = 0$$

and

$$f_2\left(\frac{u}{u^2+v^2}, \frac{v}{u^2+v^2}\right) = F_2(u, v) = 0$$

also intersect at the angle α.

8. The equations

$$u = \frac{x}{x^2+y^2+z^2}, \quad v = \frac{y}{x^2+y^2+z^2}, \quad w = \frac{z}{x^2+y^2+z^2}$$

define an inversion in space. Find the inverse transformation.

9. Into what does an inversion in space transform a sphere?

10. Are the conditions of Theorem 1 satisfied at every point of the curve $x^2 - y^2 + 4 = 0$ if y is the dependent variable? Does the equation determine y as a one-valued function of x?

11. Are the conditions of Theorem 1 satisfied at every point of the curve of Exercise 10 if x is the dependent variable?

12. Are the conditions of Theorem 1 satisfied at every point of the curve $x^3 + y^3 - 3axy = 0$ if y is the dependent variable? If x is the dependent variable?

13. What must be the relation between a and b in order that the transformation $u = a\left(x + \frac{x}{x^2+y^2}\right)$, $v = b\left(y - \frac{y}{x^2+y^2}\right)$ shall transform the circles $x^2 + y^2 = r^2$ into a set of confocal ellipses?

174. An important functional determinant.

Let $f(x, y, z)$ be a function of the three variables x, y, and z, all of whose partial derivatives of the second order exist. The functional determinant of its first partial derivatives

$$h = \begin{vmatrix} \dfrac{\partial^2 f}{\partial x^2} & \dfrac{\partial^2 f}{\partial x \partial y} & \dfrac{\partial^2 f}{\partial x \partial z} \\ \dfrac{\partial^2 f}{\partial y \partial x} & \dfrac{\partial^2 f}{\partial y^2} & \dfrac{\partial^2 f}{\partial y \partial z} \\ \dfrac{\partial^2 f}{\partial z \partial x} & \dfrac{\partial^2 f}{\partial z \partial y} & \dfrac{\partial^2 f}{\partial z^2} \end{vmatrix}$$

is called the *Hessian* of $f(x, y, z)$. If we put

$$\left.\begin{array}{l} x = \alpha X + \beta Y + \gamma Z, \\ y = \alpha' X + \beta' Y + \gamma' Z, \\ z = \alpha'' X + \beta'' Y + \gamma'' Z, \end{array}\right\} \quad (13)$$

where the coefficients of X, Y, and Z are constants such that the determinant

$$\Delta = \begin{vmatrix} \alpha & \beta & \gamma \\ \alpha' & \beta' & \gamma' \\ \alpha'' & \beta'' & \gamma'' \end{vmatrix} \neq 0,$$

and in $f(x, y, z)$ replace x, y, and z by their equals as given by (13), we obtain a function of X, Y, and Z which we shall denote by $F(X, Y, Z)$. This function F has a Hessian H which is the same as $\Delta^2 h$ when its variables are replaced by their equals in terms of x, y, and z. Because of this fact h is said to be a *covariant of f of index* 2.

We have now to establish this covariant property of h. By § 173 we have

$$H = \frac{D\left(\dfrac{\partial F}{\partial X},\, \dfrac{\partial F}{\partial Y},\, \dfrac{\partial F}{\partial Z}\right)}{D(X, Y, Z)}$$

$$= \frac{D\left(\dfrac{\partial F}{\partial X},\, \dfrac{\partial F}{\partial Y},\, \dfrac{\partial F}{\partial Z}\right)}{D(x, y, z)} \cdot \frac{D(x, y, z)}{D(X, Y, Z)}.$$

Now by virtue of (13) we have, keeping in mind that $F(X, Y, Z) = f(x, y, z)$,

$$\frac{\partial F}{\partial X} = \frac{\partial f}{\partial x}\frac{\partial x}{\partial X} + \frac{\partial f}{\partial y}\frac{\partial y}{\partial X} + \frac{\partial f}{\partial z}\frac{\partial z}{\partial X} = \alpha\frac{\partial f}{\partial x} + \alpha'\frac{\partial f}{\partial y} + \alpha''\frac{\partial f}{\partial z},$$

$$\frac{\partial F}{\partial Y} = \frac{\partial f}{\partial x}\frac{\partial x}{\partial Y} + \frac{\partial f}{\partial y}\frac{\partial y}{\partial Y} + \frac{\partial f}{\partial z}\frac{\partial z}{\partial Y} = \beta\frac{\partial f}{\partial x} + \beta'\frac{\partial f}{\partial y} + \beta''\frac{\partial f}{\partial z},$$

$$\frac{\partial F}{\partial Z} = \frac{\partial f}{\partial x}\frac{\partial x}{\partial Z} + \frac{\partial f}{\partial y}\frac{\partial y}{\partial Z} + \frac{\partial f}{\partial z}\frac{\partial z}{\partial Z} = \gamma\frac{\partial f}{\partial x} + \gamma'\frac{\partial f}{\partial y} + \gamma''\frac{\partial f}{\partial z}.$$

Therefore

$$\frac{D\left(\dfrac{\partial F}{\partial X},\, \dfrac{\partial F}{\partial Y},\, \dfrac{\partial F}{\partial Z}\right)}{D\left(\dfrac{\partial f}{\partial x},\, \dfrac{\partial f}{\partial y},\, \dfrac{\partial f}{\partial z}\right)} = \begin{vmatrix} \alpha & \alpha' & \alpha'' \\ \beta & \beta' & \beta'' \\ \gamma & \gamma' & \gamma'' \end{vmatrix} = \Delta.$$

Moreover

$$\frac{D\left(\dfrac{\partial F}{\partial X},\dfrac{\partial F}{\partial Y},\dfrac{\partial F}{\partial Z}\right)}{D\left(\dfrac{\partial f}{\partial x},\dfrac{\partial f}{\partial y},\dfrac{\partial f}{\partial z}\right)} \cdot \frac{D\left(\dfrac{\partial f}{\partial x},\dfrac{\partial f}{\partial y},\dfrac{\partial f}{\partial z}\right)}{D(x,y,z)}$$

$$= \frac{D\left(\dfrac{\partial F}{\partial X},\dfrac{\partial F}{\partial Y},\dfrac{\partial F}{\partial Z}\right)}{D(x,y,z)}.$$

Hence

$$H = \frac{D\left(\dfrac{\partial F}{\partial X},\dfrac{\partial F}{\partial Y},\dfrac{\partial F}{\partial Z}\right)}{D\left(\dfrac{\partial f}{\partial x},\dfrac{\partial f}{\partial y},\dfrac{\partial f}{\partial z}\right)} \cdot \frac{D\left(\dfrac{\partial f}{\partial x},\dfrac{\partial f}{\partial y},\dfrac{\partial f}{\partial z}\right)}{D(x,y,z)} \cdot \frac{D(x,y,z)}{D(X,Y,Z)}$$

$$= \Delta h \Delta = \Delta^2 h.$$

This formula holds also for functions of n variables.

175. THEOREM 3. *Let $u_1, u_2, \cdots, u_n$ be n differentiable functions $f_1(x_1, \cdots, x_n), f_2(x_1, \cdots, x_n), \cdots, f_n(x_1, \cdots, x_n)$ of the n independent variables $x_1, x_2, \cdots, x_n$, and let the first partial derivatives of the u's with respect to the x's be continuous at a point M. A necessary and sufficient condition for the existence of any non-identical relation*

$$\varphi(u_1, u_2, \cdots, u_n) = 0$$

connecting them which involves only the u's and not the x's explicitly is that their functional determinant D shall vanish identically.

Suppose in the first place that D is different from zero at a point M. Then it is different from zero in the neighborhood of M, and we know from Theorem 2 that there are functions

$$\begin{aligned}
x_1 &= \psi_1(u_1, u_2, \cdots, u_n) \\
&\cdots\cdots\cdots\cdots\cdots\cdots \\
&\cdots\cdots\cdots\cdots\cdots\cdots \\
x_n &= \psi_n(u_1, u_2, \cdots, u_n)
\end{aligned} \quad (14)$$

that satisfy the equations

$$u_1 = f_1(x_1, x_2, \cdots, x_n)$$
$$\cdots\cdots\cdots\cdots\cdots\cdots$$
$$\cdots\cdots\cdots\cdots\cdots\cdots \quad (15)$$
$$u_n = f_n(x_1, x_2, \cdots, x_n)$$

identically in the neighborhood of M. This shows that we can assign an arbitrary set of values to the u's in the neighborhood of the point $(u_1^{(0)}, u_2^{(0)}, \cdots, u_n^{(0)})$ which corresponds to the point M and obtain from (14) a set of values of the x's in the neighborhood of the point M such that these values of the u's and x's satisfy (15). Hence a non-identical relation of the form $\varphi(u_1, u_2, \cdots, u_n) = 0$ would give rise to a non-identical relation connecting the independent variables $x_1, x_2, \cdots, x_n$. We conclude therefore that the condition of the theorem is a necessary one for the existence of a relation of the type described.

In order to prove the sufficiency of the condition, we assume that D vanishes identically. We consider first the situation in which there is a first minor D_1 of D that does not vanish identically. There will be no loss of generality in assuming that the minor of the element in the last row and last column is such a one.

Now if we put

$$\left.\begin{aligned} y_1 &= f_1(x_1, x_2, \cdots, x_n) \\ &\cdots\cdots\cdots\cdots\cdots\cdots \\ &\cdots\cdots\cdots\cdots\cdots\cdots \\ y_{n-1} &= f_{n-1}(x_1, x_2, \cdots, x_n) \\ y_n &= x_n \end{aligned}\right\} \quad (16)$$

the Jacobian of the y's with respect to the x's is

$$\begin{vmatrix} \dfrac{\partial f_1}{\partial x_1} & \cdots & \dfrac{\partial f_1}{\partial x_n} \\ \cdots & \cdots & \cdots \\ \cdots & \cdots & \cdots \\ \dfrac{\partial f_{n-1}}{\partial x_1} & \cdots & \dfrac{\partial f_{n-1}}{\partial x_n} \\ 0 & 0 \cdots & 1 \end{vmatrix}.$$

This in turn is equal to D_1 which by hypothesis is not zero in the neighborhood of a point M. We can therefore solve equations (16) for the x's in terms of the y's, and then eliminate the x's from (15). This gives us the equations

$$\begin{aligned} u_1 &= y_1 \\ &\cdots\cdots\cdots \\ &\cdots\cdots\cdots \\ u_{n-1} &= y_{n-1} \\ u_n &= \varphi_n(y_1, y_2, \cdots, y_n). \end{aligned} \qquad (17)$$

We know from § 173 that the Jacobian of the u's with respect to the y's vanishes identically since D does. But this Jacobian is

$$\begin{vmatrix} 1 & 0 & 0 & \cdots & 0 & 0 \\ 0 & 1 & 0 & \cdots & 0 & 0 \\ \cdots & \cdots & \cdots & \cdots & \cdots & \cdots \\ \cdots & \cdots & \cdots & \cdots & \cdots & \cdots \\ 0 & 0 & \cdots & & 1 & 0 \\ \dfrac{\partial \varphi_n}{\partial y_1} & \dfrac{\partial \varphi_n}{\partial y_2} & \cdots & & \dfrac{\partial \varphi_n}{\partial y_{n-1}} & \dfrac{\partial \varphi_n}{\partial y_n} \end{vmatrix} = \dfrac{\partial \varphi_n}{\partial y_n}.$$

Hence $\dfrac{\partial \varphi_n}{\partial y_n}$ vanishes identically. We conclude that φ_n does not contain y_n, and that therefore

$$u_n = \varphi_n(u_1, u_2, \cdots, u_{n-1}). \qquad (18)$$

This is a relation between the u's that does not contain any of the x's explicitly. Moreover this is the only such relation, since if there were another one, such as

$$\Phi(u_1, u_2, \cdots, u_n) = 0, \qquad (19)$$

we could eliminate u_n from the two and obtain a relation connecting $u_1, u_2, \cdots, u_{n-1}$. But there is no such relation since by hypothesis the first minor of the element in the last row and last column of D does not vanish identically. Of course there do exist relations like (19) that are dependent upon (18). In this case the result of the elimination of u_n is an identity.

If every first minor of D vanishes identically, but at least one second minor does not, we can assume that the one obtained by striking out the last two rows and last two columns of D is such a second minor. Denote it by D_2. If now we border D_2 by the $(n-1)$th column and the $(n-1)$th row of D we obtain a minor D_1 of D of order $n-1$. By hypothesis

$$D_1 = \frac{D(u_1, u_2, \cdots, u_{n-1})}{D(x_1, x_2, \cdots, x_{n-1})} \equiv 0$$

and has a first minor which is not identically zero; namely D_2. It follows from what has been proved that u_{n-1} is a function of $u_1, u_2, \cdots, u_{n-2}$ that does not involve $x_1, x_2, \cdots, x_{n-1}$ explicitly, but that may involve x_n:

$$u_{n-1} = \psi_1(u_1, u_2, \cdots, u_{n-2}; x_n). \tag{20}$$

We have now to show that (20) does not contain x_n explicitly. For this purpose we form the determinant

$$D_1' = \begin{vmatrix} \dfrac{\partial u_1}{\partial x_1} & \cdots & \dfrac{\partial u_1}{\partial x_{n-2}} & \dfrac{\partial u_1}{\partial x_n} \\ \cdots & \cdots & \cdots & \cdots \\ \cdots & \cdots & \cdots & \cdots \\ \dfrac{\partial u_{n-2}}{\partial x_1} & \cdots & \dfrac{\partial u_{n-2}}{\partial x_{n-2}} & \dfrac{\partial u_{n-2}}{\partial x_n} \\ \dfrac{\partial u_{n-1}}{\partial x_1} & \cdots & \dfrac{\partial u_{n-1}}{\partial x_{n-2}} & \dfrac{\partial u_{n-1}}{\partial x_n} \end{vmatrix}.$$

Since this is a first minor of D it is by hypothesis identically zero. From (20) we derive the following equations by differentiation:

$$\frac{\partial u_{n-1}}{\partial x_i} = \sum_{j=1}^{n-2} \frac{\partial \psi_1}{\partial u_j} \frac{\partial u_j}{\partial x_i}, \quad (i = 1, 2, \cdots, n-2)$$

$$\frac{\partial u_{n-1}}{\partial x_n} = \sum_{j=1}^{n-2} \frac{\partial \psi_1}{\partial u_j} \frac{\partial u_j}{\partial x_n} + \frac{\partial \psi_1}{\partial x_n}.$$

If then in D_1' we multiply the elements in the first $n-2$ rows by $\dfrac{\partial \psi_1}{\partial u_1}, \dfrac{\partial \psi_1}{\partial u_2}, \cdots,$ and $\dfrac{\partial \psi_1}{\partial u_{n-2}}$ respectively, and subtract

§ 175]   IMPLICIT FUNCTIONS   317

the sum of the products from the elements of the last row, we obtain the determinant

$$\begin{vmatrix} \dfrac{\partial u_1}{\partial x_1} & \cdots & \dfrac{\partial u_1}{\partial x_{n-2}} & \dfrac{\partial u_1}{\partial x_n} \\ \vdots & & \vdots & \vdots \\ \dfrac{\partial u_{n-2}}{\partial x_1} & \cdots & \dfrac{\partial u_{n-2}}{\partial x_{n-2}} & \dfrac{\partial u_{n-2}}{\partial x_n} \\ 0 & \cdots & 0 & \dfrac{\partial \psi_1}{\partial x_n} \end{vmatrix},$$

which is equal to D_1' and therefore identically zero. But also $D_1' = D_2 \dfrac{\partial \psi_1}{\partial x_n}$, and D_2 is not identically zero. Hence $\dfrac{\partial \psi_1}{\partial x_n}$ is identically equal to zero and ψ_1 does not contain x_n. We have therefore in (20) a relation connecting the u's that does not explicitly involve any of the x's:

$$u_{n-1} = \psi_1(u_1, u_2, \cdots, u_{n-2}). \tag{21}$$

If we border D_2 by the nth column and the nth row of D we can obtain a relation of the form

$$u_n = \psi_2(u_1, u_2, \cdots, u_{n-2}). \tag{22}$$

These two relations are obviously independent since one of them contains u_n and the other one does not. Moreover there is no third relation independent of these connecting the u's. For if

$$\Theta(u_1, u_2, \cdots, u_{n-2}; u_{n-1}, u_n) = 0$$

were such a relation, we could eliminate u_{n-1} and u_n from this by means of (21) and (22) and obtain a relation connecting $u_1, u_2, \cdots, u_{n-2}$. But there is no such relation since D_2 does not vanish identically.

We can proceed in a similar way if all the second minors of D vanish identically, but not every third minor. The final result can be stated as follows:

Theorem 4. *If we have given n differentiable functions $u_1, u_2, \cdots, u_n$ of the n independent variables $x_1, x_2, \cdots, x_n$, all of whose first partial derivatives are continuous at a point, and if every minor of the Jacobian of the u's of order greater than k vanishes identically, while at least one of order k does not, there are $n - k$ independent relations (and no more) that connect the u's and do not involve the x's explicitly.*

EXERCISES

1. Show that $\dfrac{D(u, v, w)}{D(x, y, z)} = 0$ in case $u = x + y - z$, $v = xy - yz - zx$, $w = x^2 + y^2 + z^2$. Exhibit a relation connecting u, v, and w that does not involve x, y, and z explicitly. How many independent relations of this kind are there?

2. Show without computing it that the Jacobian of u and v with respect to x and y is identically zero in case $u = \sin(x^2 + y^2)$, $v = \cos(x^2 + y^2)$.

3. Show that a necessary and sufficient condition that the two lines $u = ax + by + c = 0$ and $v = a'x + b'y + c' = 0$ be parallel is that the Jacobian of u and v with respect to x and y be identically zero. When this Jacobian is identically zero, what is the relation between u and v?

CHAPTER XIII

APPLICATIONS TO GEOMETRY

176. Parametric equations of curves and surfaces. If x and y are the coordinates of any point on the straight line through the point (x_0, y_0) with the inclination α, then

$$x = x_0 + s \cos \alpha,$$
$$y = y_0 + s \sin \alpha,$$

where s is the distance from the fixed point (x_0, y_0) to the variable point (x, y) and is to be taken as positive. We have here a straight line represented by two equations which give the coordinates of any point on the line in terms of a certain parameter.

Suppose that in general we have the equations

$$x = f(t),$$
$$y = \varphi(t), \tag{1}$$

where at least one of the right members, say $f(t)$, has a derivative which does not vanish in the neighborhood of $t = t_0$. Then (§ 169) we can solve the first of these equations for t in terms of x and substitute this value for t in the second equation. This gives us

$$t = F(x),$$
$$y = \varphi[F(x)]. \tag{2}$$

Now any values of x and y that are given by a value of t sufficiently near to t_0 satisfy equation (2); and, conversely, any values of x and y within a suitably restricted range that satisfy (2) are given by (1) for some value of t near to t_0. But the second of equations (2) represents a curve in the neighborhood of the point (x_0, y_0), where $x_0 = f(t_0)$ and $y_0 = \varphi(t_0)$, and therefore equations of the form (1) represent a curve. These equations are called *parametric equations* of the curve.

If we have given the equations

$$x = f(t), \qquad y = \varphi(t), \qquad z = \psi(t), \tag{3}$$

where at least one of the right members, say $f(t)$, has a derivative which is continuous and does not vanish for $t = t_0$, we can solve the first of these equations for t in terms of x and substitute this value for t in the other two equations. This gives us

$$\left.\begin{array}{l} t = F(x), \\ y = \varphi[F(x)], \\ z = \psi[F(x)]. \end{array}\right\} \tag{4}$$

Any values of x, y, and z that are given by values of t sufficiently near to t_0 satisfy equations (4); and, conversely, any values of x, y, and z within a suitably restricted range that satisfy (4) are given by (3) for some value of t near to t_0. But the last two of equations (4) represent a curve in the neighborhood of the point (x_0, y_0, z_0), where $x_0 = f(t_0)$, $y_0 = \varphi(t_0)$, and $z_0 = \psi(t_0)$, and therefore a set of equations of the form (3) represents a curve.

We can see in a similar way that the equations

$$x = f(u, v), \qquad y = \varphi(u, v), \qquad \text{and} \qquad z = \psi(u, v), \tag{5}$$

where at least one of the functional determinants

$$\frac{D(f, \varphi)}{D(u, v)}, \qquad \frac{D(\varphi, \psi)}{D(u, v)}, \qquad \text{and} \qquad \frac{D(\psi, f)}{D(u, v)}$$

does not vanish for $u = u_0$ and $v = v_0$, represent a surface. They are called *parametric equations* of the surface with the parameters u and v.

EXERCISES

Represent the following loci in Cartesian coordinates:
1. $x = a \sin t$, $y = a \cos t$. 2. $x = a \sin t$, $y = b \cos t$.
3. $x = a \sin^3 t$, $y = a \cos^3 t$. 4. $x = a(\theta - \sin \theta)$, $y = a(1 - \cos \theta)$.

The result in Exercise 4 will make it clear to the reader why the cycloid is nearly always represented by equations in the parametric form.

5. $x = t \cos u$, $y = t \sin u$, $z = t^2$.
6. $x = a \sin u \cos v$, $y = b \sin u \sin v$, $z = c \cos u$.
7. $x = \sin^3 u \sin^3 v$, $y = \sin^3 u \cos^3 v$, $z = \cos^3 u$.

Plane Curves

177. Envelopes of families of curves. The loci of the equation
$$f(x, y, \alpha) = 0 \tag{6}$$
corresponding to all real values of α, or at least to all values within a given range, form what is called a *one-parameter family of curves*. If there is a curve that is tangent to all the curves of the family, it is called the *envelope* of the family. For example, the curve $y^2 = 4x$ is the envelope of the family
$$y - \alpha x - \frac{1}{\alpha} = 0,$$ as the reader can readily verify. In fact, any plane curve is the envelope of the one-

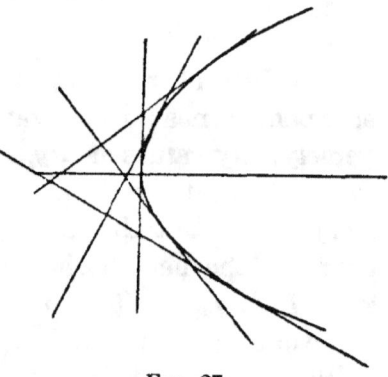

Fig. 27

parameter family of its tangent lines. But not every one-parameter family of curves has an envelope. For example, the family $x^2 + y^2 - \alpha^2 = 0$.

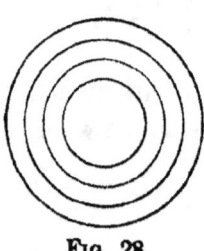

Fig. 28

If the curves $f(x, y, \alpha) = 0$ have an envelope, consider a particular value of α. It determines a particular curve of the family, and this curve has a particular point or points of contact with the envelope the coordinates of which are therefore functions of α; say, $x = \varphi(\alpha)$, $y = \psi(\alpha)$. We assume that $\varphi(\alpha)$ and $\psi(\alpha)$ are differentiable functions. They give us the parametric equations of the envelope.

In order to decide whether there is an envelope, or not, in a given case, and to determine it when it exists, we proceed as follows: Let (x, y) be a point on the envelope which we assume to exist. At this point the slope of the envelope is the same as the slope of the particular curve of the family that passes through this point. Denote this latter slope by $\frac{\delta y}{\delta x}$. Then

$$\frac{\partial f}{\partial x} \delta x + \frac{\partial f}{\partial y} \delta y = 0, \tag{7}$$

since α is a constant along a curve of the family. If in (6) we replace x and y by $\varphi(\alpha)$ and $\psi(\alpha)$ respectively, the left member becomes identically zero since every one of the given curves is tangent to the envelope. Hence

$$\frac{\partial f}{\partial x}\frac{d\varphi}{d\alpha} + \frac{\partial f}{\partial y}\frac{d\psi}{d\alpha} + \frac{\partial f}{\partial \alpha} = 0. \tag{8}$$

The slope of the envelope at (x, y) is $\dfrac{\frac{d\psi}{d\alpha}}{\frac{d\varphi}{d\alpha}}$, and therefore

$$\frac{\frac{d\psi}{d\alpha}}{\frac{d\varphi}{d\alpha}} = \frac{\delta y}{\delta x}. \tag{9}$$

It follows from (7), (8), and (9) that along the envelope

$$\frac{\partial f}{\partial \alpha} = 0. \tag{10}$$

If then we solve equations (6) and (10) for x and y in terms of α we shall obtain parametric equations of the envelope, in case there is an envelope. Or we could obtain the non-parametric equation of this curve by eliminating α by means of (6) and (10). Consider the example already referred to: $f(x, y, \alpha) = y - \alpha x - \frac{1}{\alpha}$. Here $f_\alpha = \frac{1}{\alpha^2} - x$, and the elimination of α gives us the equation $y^2 = 4x$. This is in fact the envelope, as we have seen. In the other case referred to, $f(x, y, \alpha) = x^2 + y^2 - \alpha^2$ and $f_\alpha = -2\alpha$. The

§ 178] APPLICATIONS TO GEOMETRY 323

equations $x^2 + y^2 - \alpha^2 = 0$ and $-2\alpha = 0$ do not determine x and y as functions of α. There is no envelope. There is still another possibility. Consider, for example, the curves $f(x, y, \alpha) = x^3 - (y - \alpha)^2 = 0$. Here $f_\alpha = 2(y - \alpha)$ and the elimination of α from the equations $f = 0$ and $f_\alpha = 0$ gives us $x = 0$. This is not the envelope of the given family. It is rather the locus of the singular points of members of the family.

That loci of singular points should appear when we combine equations (6) and (10), in case there are such points, is obvious from the following considerations: Along such a locus equations (6) and (8) hold. But at a singular point $\frac{\partial f}{\partial x} = \frac{\partial f}{\partial y} = 0$ (§ 37). Hence (10) also holds.

In practice then in order to find the envelope of a given one-parameter family of curves (6) we solve equations (6) and (10) for x and y in terms of α. We know that under certain conditions this will give us parametric equations of the envelope (see Th. 1, § 179). We might also eliminate α by means of equations (6) and (10). If the equation $R(x, y) = 0$ is the result of this elimination, it is satisfied by the coordinates of any point of the envelope. If there are singular points, their coordinates also satisfy the equation. The loci of all such points must be separated out unless these loci also belong to the envelope.

The same curve may be the envelope of many families of curves. For example, for every value of a the equation

$$\frac{(x - \alpha)^2}{a^2} + y^2 = 1$$

represents a one-parameter of curves with the lines $y = \pm 1$ for envelope.

178. Let

$$f(x, y, \alpha) = 0 \quad \text{and} \quad f(x, y, \alpha + h) = 0$$

be two curves of the given family. The intersections of these two curves are the same as the intersections of the curves

$$f(x, y, \alpha) = 0 \quad \text{and} \quad \frac{f(x, y, \alpha + h) - f(x, y, \alpha)}{h} = 0.$$

If we let $h \to 0$ the limiting positions of these intersection points are the points of intersection of the curves (6) and (10). In other words, the locus of the equation $R(x, y) = 0$ contains the locus of the limiting points of intersection of the two curves corresponding to α and $\alpha + h$ as $h \to 0$, if there is such a locus.

179. Sufficient conditions. There is nothing in the discussion of § 177 to assure us that the one-parameter family of curves $f(x, y, \alpha) = 0$ has an envelope. As a matter of fact, we have seen by an example that in some cases no envelope exists. Under certain circumstances, however, which we shall describe, we may be sure that there is one.

THEOREM 1. *The equations*

$$f(x, y, \alpha) = 0, \quad f_\alpha(x, y, \alpha) = 0 \tag{11}$$

define a curve which is tangent to each curve of the family $f(x, y, \alpha) = 0$ for values of α near to α_0 in case $f(x, y, \alpha)$ satisfies the following conditions:

(a) $f(x_0, y_0, \alpha_0) = 0, f_\alpha(x_0, y_0, \alpha_0) = 0$;

(b) $f(x, y, \alpha)$, $f_x(x, y, \alpha)$, $f_y(x, y, \alpha)$, $f_{x,\alpha}(x, y, \alpha)$, $f_{y,\alpha}(x, y, \alpha)$, and $f_{\alpha,\alpha}(x, y, \alpha)$ are continuous in the neighborhood of the point (x_0, y_0, α_0);

(c) $f_{\alpha,\alpha}(x_0, y_0, \alpha_0) \neq 0$,

$$\begin{vmatrix} f_x(x_0, y_0, \alpha_0) & f_y(x_0, y_0, \alpha_0) \\ f_{x,\alpha}(x_0, y_0, \alpha_0) & f_{y,\alpha}(x_0, y_0, \alpha_0) \end{vmatrix} \neq 0.$$

By virtue of the second part of Condition (c) we know from Theorem 2 of Chapter XII that there are two functions $\varphi(\alpha)$ and $\psi(\alpha)$ which are continuous, together with their first derivatives, in the neighborhood of α_0 and which reduce equations (11) to identities when we put $x = \varphi(\alpha)$ and $y = \psi(\alpha)$. Moreover $x_0 = \varphi(\alpha_0)$ and $y_0 = \psi(\alpha_0)$.

Since
$$f[\varphi(\alpha), \psi(\alpha), \alpha] \equiv 0$$
and
$$f_\alpha[\varphi(\alpha), \psi(\alpha), \alpha] \equiv 0,$$
we have
$$\left.\begin{array}{r}f_x\varphi' + f_y\psi' + f_\alpha = 0, \\ f_{\alpha, x}\varphi' + f_{\alpha, y}\psi' + f_{\alpha, \alpha} = 0.\end{array}\right\} \quad (12)$$

If $\varphi'(\alpha_0)$ and $\psi'(\alpha_0)$ were both zero, it would follow from the second of equations (12) that $f_{\alpha, \alpha}(x_0, y_0, \alpha_0) = 0$. But this contradicts the first part of Condition (c). We have seen that in the neighborhood of α_0 the function $f_\alpha(x, y, \alpha)$ vanishes. In this neighborhood therefore

$$f_x\varphi' + f_y\psi' = 0. \quad (13)$$

Hence the curve
$$x = \varphi(\alpha), \qquad y = \psi(\alpha) \quad (14)$$

has a continuously turning tangent in this neighborhood,[1] since its slope is

$$\frac{\psi'(\alpha)}{\varphi'(\alpha)} = -\frac{f_x}{f_y}.$$

On the other hand, since by virtue of the second part of Condition (c) f_x and f_y do not vanish simultaneously in the neighborhood under consideration, the first of the curves (11) has a continuously turning tangent whose slope is

$$\frac{dy}{dx} = -\frac{f_x}{f_y} = \frac{\psi'}{\varphi'}.$$

Hence this curve and the curve (14) are tangent for any value of α in a suitable neighborhood of α_0.

It was pointed out in § 177 that the family

$$x^3 - (y - \alpha)^2 = 0$$

has no envelope. We conclude therefore that the conditions of the theorem are not satisfied. This can be readily

---

[1] That is, a tangent whose inclination is a continuous function of the abscissa of the point of contact.

verified. We have

$$f(x, y, \alpha) = x^3 - (y - \alpha)^2 = 0,$$
$$f_\alpha(x, y, \alpha) = 2(y - \alpha) = 0,$$
$$f_x = 3x^2, \qquad f_y = -2(y - \alpha).$$

Hence $y = \alpha_0$ and $x = 0$ when $\alpha = \alpha_0$ and the determinant in the second part of Condition (c) vanishes for every value of α.

It should be pointed out that the conditions of this theorem have not been proved necessary for the existence of an envelope. And in fact they are not necessary. This can be seen from a consideration of the curves $x(x - \alpha)^4 - y^2 = 0$. The line $y = 0$ is tangent to every curve of the family, the point of contact for the curve corresponding to $\alpha = \alpha_0$ being $(\alpha_0, 0)$. This line is the envelope of the family. On the other hand $f_x(x_0, y_0, \alpha_0)$ and $f_y(x_0, y_0, \alpha_0)$ both equal zero and the second part of Condition (c) of the theorem is not satisfied. The theorem is nevertheless useful since its conditions are satisfied by many families of curves that are apt to come to the reader's attention.

180. An application in optics. If a bundle of parallel rays of light in the same plane strike the polished concave surface of a narrow circular band, every ray will be reflected from the surface in such a way that the angle between the incident and reflected rays will be bisected by the normal. The curve of light which appears is the envelope of these reflected rays. In order to determine the equation of this envelope we suppose that the original rays are parallel to the x-axis and that the center of the circle is at the origin, the radius of the circle being 1. If an incident ray make the angle θ with the normal it will be reflected along a path whose equation is

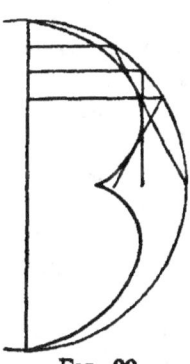

Fig. 29

$$y \cos 2\theta - x \sin 2\theta + \sin \theta = 0. \qquad (15)$$

We solve equation (15) and the equation

$$f_\theta(x, y, \theta) = -2y \sin 2\theta - 2x \cos 2\theta + \cos \theta = 0$$

for x and y in terms of θ. In the light of Theorem I we can be sure that the solution of (15) and the equation $f_\theta(x, y, \theta) = 0$ for x and y in terms of θ will give us parametric equations of the envelope. We leave it to the reader to verify that these equations are [1]

$$x = \frac{3}{4} \cos \theta - \frac{1}{4} \cos 3\theta,$$

$$y = \frac{3}{4} \sin \theta - \frac{1}{4} \sin 3\theta.$$

These are parametric equations of an epicycloid of two cusps.

EXERCISES

Find the envelope of each of the following one-parameter families of curves:

1. $x \cos \alpha + y \sin \alpha = 1$.   2. $y = mx + \sqrt{1 + m^2}$.
3. $\alpha^2 x^2 + 2\alpha y + 1 = 0$.   4. $(x - m)^2 + y^2 = 1$.
5. $\dfrac{x^2}{a^2} + \dfrac{y^2}{(1-a)^2} = 1$.
6. $y = -k^2 \sec^2 \alpha \cdot x^2 + l \tan \alpha \cdot x$, where the parameter is α.
7. A straight line moves in such a way that the axes intercept on it a constant length. Find its envelope.
8. Find the envelope of a straight line which moves in such a way that the product of its intercepts on the axes is constant.
9. Find the envelope of the family of circles that are tangent to the y-axis and have their centers on the curve $y^2 = x$.
10. Find the envelope of the family of ellipses whose axes lie on the coordinate axes and have the sum of their squares a constant.
11. A circle rolls along a straight line without sliding. Find the envelope of a fixed diameter of the circle. What are the highest and lowest points of the envelope if the base is horizontal?
12. The generating circle of a cycloid carries a fixed tangent which coincides with the x-axis when the circle is in its initial position. Find the envelope of this tangent, and show that it is below the x-axis when $0 < \theta < \dfrac{\pi}{2}$.

---

[1] See Serret-Scheffers, *Lehrbuch der Differential- und Integralrechnung*, Vol. 1, p. 403; or Osgood, *Introduction to the Calculus*, p. 274.

13. Find the envelope of the normals to the curve $y = x^2$.
14. Find the envelope of the normals to the curve $xy = 1$.
15. Find the envelope of the family $4xy = (a - x - y)^2$.
16. Find the envelope of the family of cubical parabolas $y = (x - \alpha)^3$. In what respect does this family differ from the other families of curves in the preceding exercises?
17. Show that the envelope described in § 180 is tangent to the circle $x^2 + y^2 = 1$ at the point $(0, 1)$.
18. Show that this envelope is farthest from the y-axis at the point for which $\theta = \dfrac{\pi}{4}$.

181. Curvature. When a point moves along the curve

$$y = f(x)$$

we say that its direction at any point is that of the tangent to the curve at this point taken in the sense of increasing or decreasing abscissa according as the abscissa of the point is increasing or decreasing. (At a point where the tangent is vertical we consider the change in the ordinate.) If A and B are two points on the curve with no point of inflection between them, we define the *curvature of the curve from A to B* to be the difference between the inclinations of the tangents to the curve at A and B. We define the *average curvature over the arc AB* to be the quotient of the curvature divided by the length of the arc AB. If α and $\alpha + \Delta\alpha$ are the inclinations of the tangents to the curve at A and B respectively and s and $s + \Delta s$ are the lengths of the arcs to these two points measured from an arbitrary point on the curve, the average curvature over the arc AB is $\dfrac{\Delta\alpha}{\Delta s}$. The curvature at the point A is defined as the limit of $\dfrac{\Delta\alpha}{\Delta s}$ as Δs approaches zero,

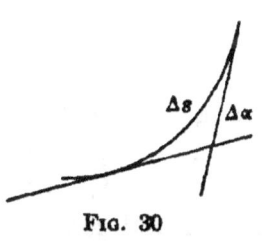

Fig. 30

in case this limit exists. This is the derivative of α with respect to s. In the case of a circle $\Delta\alpha$ is the same as the angle between the radii to the points A and B, and $\Delta s = R\Delta\alpha$, where R is the radius of the circle. Hence

$\frac{\Delta \alpha}{\Delta s}$ has the constant value $\frac{1}{R}$, and this is the value of $\frac{d\alpha}{ds}$ — that is, the value of the curvature, which is the same for all points of the circle. It follows from this that the curvature of a curve at a point is the reciprocal of the radius of a circle that has the same curvature. We therefore define the *radius of curvature of a curve at a point* to be the reciprocal of its curvature.

It is easy to derive an explicit expression for the curvature of the curve. If α is the inclination of the tangent,

$$\alpha = \arctan y',$$
$$\frac{d\alpha}{ds} = \frac{y''}{1 + y'^2} \frac{dx}{ds}.$$

But
$$\frac{dx}{ds} = \frac{1}{\sqrt{1 + y'^2}} \quad (\S\ 187).$$

Hence
$$\frac{d\alpha}{ds} = \frac{y''}{(1 + y'^2)^{3/2}}, \tag{16}$$

and the radius of curvature R is given by the formula

$$R = \frac{(1 + y'^2)^{3/2}}{y''}. \tag{17}$$

182. The osculating circle. The tangent to the curve

$$y = f(x)$$

at the point A is the line through A with an inclination α such that the inclination of the chord through A and A' approaches α as A' approaches A along the curve. Similarly there is associated with a non-singular, non-inflectional point A of a plane curve a circle of radius R and center C such that the radius of the circle through A and two other points A' and A'' approaches R and the center approaches C as A' and A'' approach A along the curve. This circle is called the *osculating circle* of the curve at the point in question. The existence of such a circle can be seen as follows:

Let
$$(x - \alpha')^2 + (y - \beta')^2 - R'^2 = 0 \qquad (18)$$
be the equation of the circle through A, A', and A''. If we put $f(x)$ for y in the left member of (18), this left member becomes a function of x. We shall represent it by the symbol $F(x)$. If x, x_1, and x_2 are the abscissae of A, A', and A'', respectively, and x_1 is between x and x_2, we have
$$F(x) = 0, \qquad F(x_1) = 0, \qquad F(x_2) = 0. \qquad (19)$$
If $F(x)$ has derivatives of the first two orders, we know from Rolle's theorem that $F'(x)$ vanishes for a value ξ of x between x and x_1; and also for a value ξ' between x_1 and x_2. That is,
$$F'(\xi) = 0, \qquad F'(\xi') = 0.$$
Then by the same theorem $F''(x)$ vanishes for a value ξ_1 of x between ξ and ξ'. In other words,
$$\left. \begin{array}{l} (x - \alpha')^2 + [f(x) - \beta']^2 - R'^2 = 0, \\ (\xi - \alpha') + [f(\xi) - \beta']f'(\xi) = 0, \\ 1 + \overline{f'(\xi_1)}^2 + [f(\xi_1) - \beta']f''(\xi_1) = 0. \end{array} \right\} \qquad (20)$$

Now as x_1 and x_2 approach x, ξ and ξ_1 approach the same limit. It follows from the last of these equations that β' approaches a limit β, where
$$y - \beta = -\frac{1 + y'^2}{y''}. \qquad (21)$$

Then it follows from the second of these equations that α' approaches a limit α, where
$$x - \alpha = \frac{y'(1 + y'^2)}{y''}; \qquad (22)$$

and from the first equation that R' approaches a limit R, where
$$(x - \alpha)^2 + (y - \beta)^2 = R^2. \qquad (23)$$

This shows the existence of the osculating circle and tells us its center and radius. We see from (21), (22), and (23) that

$$R = \pm \frac{(1 + y'^2)^{3/2}}{y''}.$$

Hence the radius of the osculating circle is equal to the radius of curvature. For this reason the osculating circle is called the *circle of curvature*, and its center the *center of curvature*, of the given curve at the point in question.

183. The evolute. As the point (x, y) moves along the curve the corresponding center of curvature traces out a locus that is called the *evolute* of the given curve.

THEOREM 2. *The evolute of a curve is the envelope of the normals to the curve.*

The equation of the normal is

$$X - x + (Y - y)y' = 0, \qquad (24)$$

where x and y are the coordinates of the point of contact and X and Y are the running coordinates. We take the parameter in this equation to be x, it being understood that $y = f(x)$ is the equation of the curve. We know from Theorem 1 that in general the loci (24) have an envelope. To find it we first equate to zero the derivative of the left member of (24) with respect to the parameter:

$$-1 + (Y - y)y'' - y'^2 = 0. \qquad (25)$$

This equation shows that Y is equal to the ordinate of the center of curvature as given by [1] (21), and then we see from (22) that X is the abscissa of this center. The parametric equations for X and Y therefore show that the envelope in question is the locus of the centers of curvature of the given curve; that is, the evolute of the given curve. The normals of a curve are accordingly tangent to its evolute.

184. An example. The equations of the cycloid are

$$x = r(\theta - \sin \theta),$$
$$y = r(1 - \cos \theta),$$

---

[1] In comparing equations (25) and (21) we must keep in mind that Y takes the place of β.

where r is the radius of the generating circle. Then
$$y' = \frac{\sin \theta}{1 - \cos \theta} \quad \text{and} \quad y'' = \frac{-1}{r(1 - \cos \theta)^2}.$$
Hence
$$R = \frac{\left[1 + \frac{\sin^2 \theta}{(1 - \cos \theta)^2}\right]^{3/2}}{\frac{1}{r(1 - \cos \theta)^2}} = 4r \sin \frac{\theta}{2}.$$

We can find the evolute by the method we have described, but in this particular case it is better to proceed as follows:
$$PM = 2r \sin \frac{\theta}{2}.$$

Hence $R = 2PM$. That is, the center of curvature P' lies on the circle symmetrical to the generating circle with respect to the point M, since $MP' = PM$.

arc $NP' = \pi r - MP' = \pi r - $ arc $MP = \pi r - OM$
$$= OD - OM = MD.$$

We conclude from this that the evolute of a cycloid is an equal cycloid.

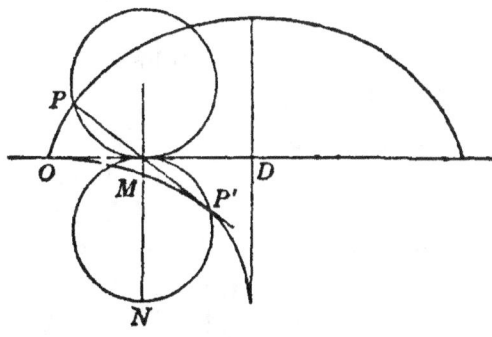

Fig. 31

EXERCISES

1. Find the radius of curvature of the parabola $y = x^2$ at the point (2, 4).

2. Find the radius of curvature of the parabola $y^2 = 4ax$ at either end of the latus rectum.

3. At what point is the curvature of the parabola a maximum? Does it have a minimum at any point?

4. Find the radius of curvature of the ellipse $\dfrac{x^2}{a^2} + \dfrac{y^2}{b^2} = 1$ at the points (x, y), $(0, b)$, and $(a, 0)$.

5. Is the center of curvature of an ellipse at a vertex ever at the opposite vertex?

6. At what points is the curvature of the ellipse a maximum and at what points is it a minimum?

Find the evolute of each of the following curves:

7. $\dfrac{x^2}{a^2} + \dfrac{y^2}{b^2} = 1.$ **8.** $x^{2/3} + y^{2/3} = a^{2/3}.$ **9.** $xy = a^2.$

10. $y^2 = 4ax.$ **11.** The catenary $y = \tfrac{1}{2}(e^x + e^{-x}).$

12. $x = \cos\theta + \theta\sin\theta,\; y = \sin\theta - \theta\cos\theta.$

13. Find the radius of curvature of the spiral of Archimedes, $r = a\theta$.

14. Find the radius of curvature of the cardioid $r = 2a(1 + \cos\theta)$.

15. Find the equation of the circle of curvature of the curve $x^3 + y^3 + 2y^2 - 4x + 3y = 0$ at the origin.

185. Intrinsic equation of a curve.

Every equation of a curve that we have considered so far has depended in some way upon the relation of the curve to the particular system of coordinates used. If we had changed the system, the equation of the curve would have been changed. It is possible however to describe a curve by an equation that expresses only properties of the curve that are independent of any system of coordinates. These properties are *intrinsic* properties of the curve, and such an equation is called an *intrinsic*, or *natural*, equation of the curve. It represents in fact a whole system of congruent curves.

Let k represent the curvature of a curve at a point whose abscissa with respect to a given system of rectangular coordinates is x, and let α be the inclination of the tangent to the curve at this point. We have already seen that k and α are the following functions of x:

$$k = \frac{y''}{(1 + y'^2)^{3/2}}, \qquad \alpha = \arctan y',$$

the equation of the curve with respect to this system of coordinates being $y = f(x)$. By means of these equations we can obtain $\dfrac{dk}{d\alpha}$ as a function of x, provided y has a third derivative with respect to x. Then we shall have

$$k = \varphi(x), \qquad \frac{dk}{d\alpha} = \psi(x). \tag{26}$$

If we can solve the first of these equations for x in terms of k, we can write the second one in the form

$$\frac{dk}{d\alpha} = \theta(k). \tag{27}$$

We have now to show that this equation (27) represents all the curves of a given congruent set determined by the function $\theta(k)$. If k is a constant, the curves are clearly either circles or straight lines. We exclude this case from our consideration. Now

$$\frac{dx}{ds} = \cos\alpha \quad\text{and}\quad \frac{dy}{ds} = \sin\alpha.$$

Moreover

$$\frac{ds}{d\alpha} = \frac{1}{k}.$$

Hence

$$x = \int \frac{1}{k} \cos\alpha\, d\alpha, \qquad y = \int \frac{1}{k} \sin\alpha\, d\alpha. \tag{28}$$

From (27) we have

$$\int \frac{dk}{\theta(k)} = \alpha - c \qquad (c \text{ a constant}). \tag{29}$$

It will be convenient to replace $\alpha - c$ in this equation by the parameter t. Then (29) determines k as a function of t. Equations (28) can be written in the form

$$x = \int_{t_0}^{t} \frac{1}{k} \cos(t+c)\, dt + c_1, \quad y = \int_{t_0}^{t} \frac{1}{k} \sin(t+c)\, dt + c_2, \tag{30}$$

where t_0 is a definite number. All the curves represented by (27) have parametric equations of the form (30). And conversely, every curve represented by equations (30) has the curvature k and a tangent with the inclination α. For

$$\frac{dx}{d\alpha} = \frac{dx}{dt} = \frac{1}{k}\cos(t+c) = \frac{1}{k}\cos\alpha$$
$$= \frac{1}{k}(\cos t \cos c - \sin t \sin c),$$

and

$$\frac{dy}{d\alpha} = \frac{dy}{dt} = \frac{1}{k}\sin(t+c) = \frac{1}{k}\sin\alpha$$
$$= \frac{1}{k}(\sin t \cos c + \cos t \sin c).$$

Hence the inclination is arc $\tan \dfrac{dy}{dx} = \alpha$ and the curvature is

$$\frac{d\alpha}{ds} = \frac{d\alpha}{dx} \cdot \frac{dx}{ds} = \frac{k}{\cos\alpha} \cdot \cos\alpha = k.$$

All of the curves represented by (30) can be obtained from the particular curve

$$\bar{x} = \int_{t_0}^{t} \frac{1}{k}\cos t\,dt, \qquad \bar{y} = \int \frac{1}{k}\sin t\,dt \qquad (31)$$

by means of the transformation

$$x = \bar{x}\cos c - \bar{y}\sin c + c_1,$$
$$y = \bar{x}\sin c + \bar{y}\cos c + c_2.$$

Hence all these curves are congruent [1] to the particular curve (31), and they therefore form a set of congruent curves.

The appearance of the arbitrary constant in (29) corresponds to the geometrical fact that c is arbitrary since the system of coordinates we are using is an arbitrary one. The important fact is that (27) does not contain an arbitrary constant and is therefore independent of such a system.

---

[1] In the sense that they can all be obtained from (31) by means of a rotation around the origin, or such a rotation followed by a translation.

SKEW CURVES

186. Length of an arc. Consider an arc, plane or twisted, given by the equations

$$y = \varphi(x) \quad \text{and} \quad z = \psi(x)$$

and connecting any two points A and B with abscissae a and b respectively, $a < b$. We take $n + 1$ points (inclusive of A and B) on the arc and connect consecutive points by straight lines. Then we form the sum

$$S = \sum_{i=1}^{n} c_i,$$

where c_i is the length of the ith chord. If S approaches a unique limit as we increase n and select the points in such a way as to make the maximum length of the chords approach zero, we call this limit the *length of the arc from A to B*.

In order to see whether there is a unique limit under these circumstances we assume that $\varphi(x)$ and $\psi(x)$, together with their first derivatives, are continuous in the interval (a, b). If the abscissae of the points of division are $x_1 = a$, $x_2, \cdots, x_n, x_{n+1} = b$, then

$$c_i = \sqrt{(x_{i+1} - x_i)^2 + (y_{i+1} - y_i)^2 + (z_{i+1} - z_i)^2},$$

where $y_j = \varphi(x_j)$ and $z_j = \psi(x_j)$. Now, by the mean value theorem,

$$y_{i+1} - y_i = \varphi(x_{i+1}) - \varphi(x_i) = (x_{i+1} - x_i)\varphi'(\xi_i),$$

where ξ_i lies between x_i and x_{i+1}; and

$$z_{i+1} - z_i = \psi(x_{i+1}) - \psi(x_i) = (x_{i+1} - x_i)\psi'(\eta_i),$$

where η_i also lies between x_i and x_{i+1}. Then

$$c_i = (x_{i+1} - x_i)\sqrt{1 + \varphi'^2(\xi_i) + \psi'^2(\eta_i)}$$

and

$$S = \sum_{i=1}^{n} (x_{i+1} - x_i)\sqrt{1 + \varphi'^2(\xi_i) + \psi'^2(\eta_i)}.$$

When we come to the investigation of the question as to the existence of a limit for S as the maximum length of the chords approaches zero, we are met by the difficulty that under the radical sign the arguments of φ'^2 and ψ'^2 are not the same. We get around this difficulty in the following way: Let M_i be the upper limit of $\varphi'^2(x)$ in the interval (x_i, x_{i+1}), and m_i the lower limit. Then

$$\varphi'^2(\xi_i) = \varphi'^2(\eta_i) + \theta_i(M_i - m_i),$$

where $-1 \leqq \theta_i \leqq 1$, and

$$S = \sum(x_{i+1} - x_i)\sqrt{1 + \varphi'^2(\eta_i) + \psi'^2(\eta_i) + \theta_i(M_i - m_i)}.$$

We have now the same argument in φ'^2 and ψ'^2; but, on the other hand, the disturbing term $\theta_i(M_i - m_i)$ has appeared. In order to deal with this new difficulty we consider that

$$\sqrt{1 + \varphi'^2(\eta_i) + \psi'^2(\eta_i) + \theta_i(M_i - m_i)} \\ - \sqrt{1 + \varphi'^2(\eta_i) + \psi'^2(\eta_i)} \\ = \frac{\theta_i(M_i - m_i)}{D},$$

where

$$D = \sqrt{1 + \varphi'^2(\eta_i) + \psi'^2(\eta_i) + \theta_i(M_i - m_i)} \\ + \sqrt{1 + \varphi'^2(\eta_i) + \psi'^2(\eta_i)}.$$

The denominator of this fraction is equal to, or greater than, one; and therefore the fraction is equal to, or less than, $\theta_i(M_i - m_i)$. Hence

$$\sqrt{1 + \varphi'^2(\eta_i) + \psi'^2(\eta_i) + \theta_i(M_i - m_i)} \\ = \sqrt{1 + \varphi'^2(\eta_i) + \psi'^2(\eta_i)} + \theta_i'(M_i - m_i),$$

where $|\theta_i'| \leqq 1$. Then

$$S = \sum(x_{i+1} - x_i)\sqrt{1 + \varphi'^2(\eta_i) + \psi'^2(\eta_i)} \\ + \sum(x_{i+1} - x_i)\theta_i'(M_i - m_i).$$

But since $\varphi'^2(x)$ is continuous in (a, b) the intervals (x_i, x_{i+1})

can all be taken sufficiently small to make $M_i - m_i$ less than an arbitrary positive ϵ for every interval, and therefore

$$\sum(x_{i+1} - x_i)\theta_i'(M_i - m_i) < \epsilon\sum(x_{i+1} - x_i) = \epsilon(b - a).$$

From this it follows that

$$s = \lim S = \int_a^b \sqrt{1 + \varphi'^2(x) + \psi'^2(x)}\, dx.$$

The discussion shows that the limit in question exists and is equal to the definite integral $\int_a^b \sqrt{1 + \varphi'^2(x) + \psi'^2(x)}\, dx$ irrespective of the manner in which the maximum length of the chords approaches zero provided that the conditions laid down are satisfied. More generally, if the function $\sqrt{1 + \varphi'^2(x) + \psi'^2(x)}$ is integrable in the interval (a, b) we say that the arc is *rectifiable* and that its length is $\int_a^b \sqrt{1 + \varphi'^2(x) + \psi'^2(x)}\, dx.$

187. The differential of arc. The length of arc from A to the point X whose abscissa is $x (a < x < b)$ is

$$s = \int_a^x \sqrt{1 + \varphi'^2(x) + \psi'^2(x)} \cdot dx.$$

Then

$$\frac{ds}{dx} = \sqrt{1 + \varphi'^2(x) + \psi'^2(x)}$$

and

$$ds = \sqrt{\overline{dx}^2 + \overline{\varphi'(x)dx}^2 + \overline{\psi'(x)dx}^2} = \sqrt{\overline{dx}^2 + \overline{dy}^2 + \overline{dz}^2}.$$

The length c of the chord from A to X is

$$c = \sqrt{\overline{\Delta x}^2 + \overline{\Delta y}^2 + \overline{\Delta z}^2},$$

where $x - x_1 = \Delta x$, $y - y_1 = \Delta y$, and $z - z_1 = \Delta z$. The reader should note carefully the difference in meanings of the two radicals $\sqrt{\overline{\Delta x}^2 + \overline{\Delta y}^2 + \overline{\Delta z}^2}$ and $\sqrt{\overline{dx}^2 + \overline{dy}^2 + \overline{dz}^2}$. If

Δs is the length of the corresponding arc, we have

$$\frac{\Delta s}{c} = \frac{\Delta s}{\sqrt{\overline{\Delta x}^2 + \overline{\Delta y}^2 + \overline{\Delta z}^2}} = \frac{\frac{\Delta s}{\Delta x}}{\sqrt{1 + \left(\frac{\Delta y}{\Delta x}\right)^2 + \left(\frac{\Delta z}{\Delta x}\right)^2}}$$

and

$$\lim_{\Delta s \to 0} \frac{\Delta s}{c} = \frac{\frac{ds}{dx}}{\sqrt{1 + \left(\frac{dy}{dx}\right)^2 + \left(\frac{dz}{dx}\right)^2}} = 1.$$

EXERCISES

1. The supports of a suspension bridge are 1200 feet apart and the suspending cable sags in the arc of a parabola 60 feet at its middle point. Find the length of the cable.

2. Find the length of the arch of the cycloid $x = a(\theta - \sin\theta)$, $y = a(1 - \cos\theta)$.

3. Find the length of the helix $x = a\cos t$, $y = a\sin t$, $z = t$ between the planes $z = 0$ and $z = 2$.

4. Find the length of the inner loop of the curve $r = a(\theta^2 - 1)$.

5. Find the length of the semicubical parabola $y^2 = x^3$ from the origin to the point $(1, 1)$.

6. Find the length of the cardioid $r = 3(1 - \cos\theta)$.

7. Find the length of the curve $x = \int_0^\alpha \sqrt{2\alpha^2 + 1}\, d\alpha$, $y = \int_1^\alpha \sqrt{4 - \alpha^2}\, d\alpha$ from the point $\alpha = 1$ to the point $\alpha = 2$.

8. Show that if s is the length of the arc of the curve $x^{\frac{1}{3}} + y^{\frac{1}{3}} = a^{\frac{1}{3}}$ measured from the y-axis to a point whose abscissa is x, then s^3 varies as x^2.

9. Find the length of the arc of the curve $4(x + y) = 3x^2 + 1$ that lies in the fourth quadrant.

188. The osculating plane. Let

$$x = f(t), \qquad y = \varphi(t), \qquad z = \psi(t) \qquad (32)$$

be parametric equations of a given twisted curve. The equation

$$A_1(X - x) + B_1(Y - y) + C_1(Z - z) = 0, \qquad (33)$$

where X, Y, and Z are the running coordinates, is the equation of a plane through the point (x, y, z). We assume that this point is on the curve (32). Let t be the value of the parameter that corresponds to this point and $t + h$ the value that corresponds to another point on the curve. Since the direction cosines of the tangent to the curve at the point (x, y, z) are proportional to $f'(t)$, $\varphi'(t)$, and $\psi'(t)$ (§ 21), it is necessary and sufficient that the coefficients A_1, B_1, and C_1 satisfy the relation

$$A_1 f'(t) + B_1 \varphi'(t) + C_1 \psi'(t) = 0 \qquad (34)$$

in order that the plane (33) contain the tangent. If the plane also contains the point to which $t + h$ corresponds, then

$$A_1[f(t + h) - f(t)] + B_1[\varphi(t + h) - \varphi(t)] \\ + C_1[\psi(t + h) - \psi(t)] = 0. \qquad (35)$$

If we apply Taylor's expansion to the three functions $f(t + h)$, $\varphi(t + h)$, and $\psi(t + h)$, we can write equation (35) in the form

$$A_1 \left\{ h f'(t) + \frac{h^2}{2} [f''(t) + \epsilon_1] \right\} \\ + B_1 \left\{ h \varphi'(t) + \frac{h^2}{2} [\varphi''(t) + \epsilon_2] \right\} \\ + C_1 \left\{ h \psi'(t) + \frac{h^2}{2} [\psi''(t) + \epsilon_3] \right\} = 0,$$

where ϵ_1, ϵ_2, and ϵ_3 approach zero as h does. A combination of equation (34) with this one shows that

$$A_1[f''(t) + \epsilon_1] + B_1[\varphi''(t) + \epsilon_2] \\ + C_1[\psi''(t) + \epsilon_3] = 0. \qquad (36)$$

Equations (34) and (36) determine A_1, B_1, and C_1 up to a constant factor which we can take to be 1. This gives us

$$A_1 = \varphi'(\psi'' + \epsilon_3) - \psi'(\varphi'' + \epsilon_2),$$
$$B_1 = \psi'(f'' + \epsilon_1) - f'(\psi'' + \epsilon_3),$$
$$C_1 = f'(\varphi'' + \epsilon_2) - \varphi'(f'' + \epsilon_1).$$

As $h \to 0$
$$A_1 \to A = \varphi'\psi'' - \psi'\varphi'',$$
$$B_1 \to B = \psi'f'' - f'\psi'',$$
$$C_1 \to C = f'\varphi'' - \varphi'f''.$$
(37)

The plane
$$A(X - x) + B(Y - y) + C(Z - z) = 0,$$
where A, B, and C are given by (37), is called the *osculating plane* of the given curve at the point (x, y, z). It is the limiting position of a plane through the tangent line and a point on the curve as this point approaches the point of contact. The equation can also be written in the form
$$\begin{vmatrix} X - x & Y - y & Z - z \\ f'(t) & \varphi'(t) & \psi'(t) \\ f''(t) & \varphi''(t) & \psi''(t) \end{vmatrix} = 0. \tag{38}$$

For example, the osculating plane of the curve
$$x = t, \quad y = t^2, \quad z = t^3,$$
at the point corresponding to t is
$$3t^2 X - 3tY + Z = t^3.$$

EXERCISES

1. Find the equation of the osculating plane of the helix $x = \cos t$, $y = \sin t$, $z = t$ at the point $t = t_1$.

2. Show that the normal to the cylinder on which the helix of Exercise 1 lies at a point on the helix lies in the osculating plane of the helix.

3. Find the equation of the osculating plane of the conical helix $x = t \cos t$, $y = t \sin t$, $z = kt$ at the point $t = t_1$.

4. Find the equation of the cone on which the helix of Exercise 3 lies.

5. Find the osculating plane of the curve $y = x$, $z = 1 - x^2$ at the point $(1, 1, 0)$.

6. Find the equation of the osculating plane of the curve $xyz = 1$, $y = x$ at the point $(1, 1, 1)$.

7. Find the equation of the osculating plane of the curve $x = 2 \cos t$, $y = \sin t$, $z = 3t$ at the point $t = 2$.

8. The curve of Exercise 7 lies on a cylinder whose elements are parallel to the z-axis. Does the normal to this surface at a point on the curve lie in the osculating plane of the curve at this point?

9. Find the osculating plane of the curve $x = a \cos t$, $y = a \sin t$, $z = a \cos t$ at the point $t = t_1$.

SURFACES

189. Envelopes of one-parameter families of surfaces. Consider the one-parameter family of surfaces

$$F(x, y, z, \alpha) = 0, \tag{39}$$

where α can have any real value, or at least any value within a given range. It may be that there is a surface E that is tangent to each surface of the family along a curve. If there is such a surface it is called the *envelope* of the given family of surfaces, and the curve of tangency on each surface is called a *characteristic curve*.

The question we now consider is whether there is a curve on each surface of the family such that the locus of these curves is a surface which is tangent to each of the given surfaces along that one of these curves that lies on this surface. Suppose that E is such a locus, and consider a point P of contact of it with a surface of the family. This latter surface is determined by P and α is determined by the surface. Hence α is a function of the coordinates x, y, and z of P. It is constant along the characteristic through P. We assume that this function possesses first partial derivatives with respect to x, y, and z. Then equation (39) represents the envelope E provided that we replace the constant α by this function.

Let

$$x = f(u, v), \qquad y = \varphi(u, v), \qquad z = \psi(u, v)$$

be parametric equations of the surface E. Since α is a function of x, y, and z, it is a function of u and v. If we introduce these functions for x, y, z, and α in (39), we shall obtain an identity in u and v. Hence

$$\frac{\partial F}{\partial x}\frac{\partial x}{\partial u} + \frac{\partial F}{\partial y}\frac{\partial y}{\partial u} + \frac{\partial F}{\partial z}\frac{\partial z}{\partial u} + \frac{\partial F}{\partial \alpha}\frac{\partial \alpha}{\partial u} = 0,$$

$$\frac{\partial F}{\partial x}\frac{\partial x}{\partial v} + \frac{\partial F}{\partial y}\frac{\partial y}{\partial v} + \frac{\partial F}{\partial z}\frac{\partial z}{\partial v} + \frac{\partial F}{\partial \alpha}\frac{\partial \alpha}{\partial v} = 0.$$

§189] APPLICATIONS TO GEOMETRY

From this it follows that

$$\frac{\partial F}{\partial x}dx + \frac{\partial F}{\partial y}dy + \frac{\partial F}{\partial z}dz + \frac{\partial F}{\partial \alpha}d\alpha = 0, \qquad (40)$$

since

$$dx = \frac{\partial x}{\partial u}du + \frac{\partial x}{\partial v}dv,$$

$$dy = \frac{\partial y}{\partial u}du + \frac{\partial y}{\partial v}dv,$$

$$dz = \frac{\partial z}{\partial u}du + \frac{\partial z}{\partial v}dv,$$

$$d\alpha = \frac{\partial \alpha}{\partial u}du + \frac{\partial \alpha}{\partial v}dv.$$

Now the condition that E be tangent at P to the surface of the family that passes through P is

$$\frac{\partial F}{\partial x}dx + \frac{\partial F}{\partial y}dy + \frac{\partial F}{\partial z}dz = 0,$$

since $\frac{\partial F}{\partial x}$, $\frac{\partial F}{\partial y}$, and $\frac{\partial F}{\partial z}$ are proportional to the direction cosines of the normal to this surface of the family, and dx, dy, and dz are proportional to the direction cosines of a line through P in the tangent plane to E. Hence on E we have

$$\frac{\partial F}{\partial \alpha} = 0. \qquad (41)$$

If then there is an envelope, it is represented by the equation

$$R(x, y, z) = 0, \qquad (42)$$

which is the result of eliminating α from equations (39) and (41). But the loci of singular points of the surfaces of the family would also be given by this equation, since at such a singular point

$$\frac{\partial F}{\partial x} = 0, \qquad \frac{\partial F}{\partial y} = 0, \qquad \frac{\partial F}{\partial z} = 0.$$

If in equations (39) and (41) α has a constant value, the two equations represent the characteristic on the surface of the family that corresponds to this value of α.

190. Developable surfaces. Consider a one-parameter family of planes that has an envelope. This envelope is called a *developable surface*. Let the equation of the planes be

$$u(\alpha)x + v(\alpha)y + w(\alpha)z + \theta(\alpha) = 0. \tag{43}$$

This equation and the equation

$$u'(\alpha)x + v'(\alpha)y + w'(\alpha)z + \theta'(\alpha) = 0 \tag{44}$$

determine for an arbitrary value of α a characteristic in case $uv' - u'v$, $vw' - v'w$, and $wu' - w'u$ are not all zero for this value of α. Equations (43) and (44) together with

$$u''(\alpha)x + v''(\alpha)y + w''(\alpha)z + \theta''(\alpha) = 0 \tag{45}$$

determine x, y, and z in terms of α for those values of α for which

$$\begin{vmatrix} u & v & w \\ u' & v' & w' \\ u'' & v'' & w'' \end{vmatrix} \neq 0.$$

Under these conditions these equations determine a curve Γ. This curve is the envelope of the characteristics determined by (43) and (44). To see this we differentiate (43) and (44), taking account of (44) and (45). This gives us

$$\left. \begin{array}{l} u\,dx + v\,dy + w\,dz = 0, \\ u'\,dx + v'\,dy + w'\,dz = 0. \end{array} \right\} \tag{46}$$

The direction cosines of the tangent to Γ at a given point are proportional to dx, dy, and dz as given by (46). That is, these direction cosines are proportional to $vw' - v'w$, $wu' - w'u$, and $uv' - u'v$. But these are also proportional to the direction cosines of the characteristic through the point in question. Hence Γ *is the envelope of the family of characteristics*. It is called the *edge of regression* of the envelope.

191. The osculating planes of Γ.

The equation of the osculating plane of Γ at a given point can be written in the form (see Eq. 38)

$$\begin{vmatrix} X-x & Y-y & Z-z \\ dx & dy & dz \\ d^2x & d^2y & d^2z \end{vmatrix} = 0. \qquad (47)$$

By differentiating the first of equations (46) with respect to α and taking account of the second of these equations, we obtain the equation

$$u d^2x + v d^2y + w d^2z = 0.$$

Hence, unless all the two-rowed minors that can be formed from the second and third rows of the determinant in the left member of (47) vanish, (47) reduces to

$$u(X-x) + v(Y-y) + w(Z-z) = 0.$$

But

$$ux + vy + wz + \theta = 0$$

and therefore

$$uX + vY + wZ + \theta = 0$$

is the osculating plane. But this is the same as (43). Hence

THEOREM 3. *The osculating plane of Γ at a given point is that one of the given planes that passes through this point.*

We leave it as an exercise for the reader to show that the locus of the tangents to the curve (32) is a developable surface.

192. Envelopes of two-parameter families of surfaces.

The equation

$$f(x, y, z, \alpha, \beta) = 0, \qquad (48)$$

where α and β are independent parameters, represents a two-parameter family of surfaces. If there is a surface E that touches each surface of the family in one or more points, we shall call it the *envelope* of the family. For example, the sphere

$$x^2 + y^2 + z^2 = 4$$

is the envelope of the two-parameter family of its tangent planes. If there is an envelope, the coordinates x, y, and z of its point, or points, of contact with one of the given surfaces are functions of α and β. We assume that these functions are differentiable. When they are introduced into the left member of (48) for x, y, and z, the equation becomes an identity. Then

$$\frac{\partial f}{\partial x}dx + \frac{\partial f}{\partial y}dy + \frac{\partial f}{\partial z}dz + \frac{\partial f}{\partial \alpha}d\alpha + \frac{\partial f}{\partial \beta}d\beta = 0, \quad (49)$$

when we put

$$dx = \frac{\partial x}{\partial \alpha}d\alpha + \frac{\partial x}{\partial \beta}d\beta,$$

$$dy = \frac{\partial y}{\partial \alpha}d\alpha + \frac{\partial y}{\partial \beta}d\beta,$$

$$dz = \frac{\partial z}{\partial \alpha}d\alpha + \frac{\partial z}{\partial \beta}d\beta.$$

Now dx, dy, and dz are proportional to the direction cosines of a line through the point of contact and lying in the tangent plane to E, while $\frac{\partial f}{\partial x}$, $\frac{\partial f}{\partial y}$, and $\frac{\partial f}{\partial z}$ are proportional to the directional cosines of the normal of the given surface at this point. Hence, since E and this surface touch here,

$$\frac{\partial f}{\partial x}dx + \frac{\partial f}{\partial y}dy + \frac{\partial f}{\partial z}dz = 0.$$

It then follows from (49) that

$$\frac{\partial f}{\partial \alpha}d\alpha + \frac{\partial f}{\partial \beta}d\beta = 0.$$

But α and β are independent and therefore $d\alpha$ and $d\beta$ are. We see then that the equations

$$\frac{\partial f}{\partial \alpha} = 0 \quad \text{and} \quad \frac{\partial f}{\partial \beta} = 0 \quad (50)$$

are satisfied by the coordinates of the point of contact. If

§ 192]   APPLICATIONS TO GEOMETRY   347

then we can solve equations (48) and (50) for x, y, and z in terms of α and β, or if we can eliminate α and β from these equations, the result in either case will give us the envelope, if there is one. It may be that

$$\frac{\partial f}{\partial x} = \frac{\partial f}{\partial y} = \frac{\partial f}{\partial z} = 0$$

for the values of the coordinates that satisfy (48) and (50). These are the coordinates of the singular points of the given surfaces. The process described will then give us the envelope, if there is one, or the locus of the singular points of the given surfaces if there are any such points, or a combination of these two loci.

It is possible in many ways to pick out from the surfaces (48) one-parameter families. This can be done by making one of the parameters a function of the other one, say, $\beta = \varphi(\alpha)$. The resulting one-parameter family may, or may not, have an envelope.

EXERCISES

1. Show that the envelope of the one-parameter family of spheres $x^2 + (y - \beta)^2 + (z - \gamma)^2 = 1$, where $\beta^2 + \gamma^2 = 4$, is a surface of revolution.

2. Find the envelope of the one-parameter family of planes $z = ax + by$, where $a^2 + b^2 = 1$.

3. Find the envelope of the two-parameter family of planes $\alpha x + \beta y + \sqrt{1 - \alpha^2 - \beta^2}\, z = 1$.

4. Find the envelope of the two-parameter family of planes $\alpha x + \beta y + (\alpha^2 + \beta^2)z = 1$.

5. Show that if $x = f(y, z)$ is the equation of the spheres in Exercise 1, then $x^2 \left[\left(\dfrac{\partial f}{\partial y} \right)^2 + \left(\dfrac{\partial f}{\partial z} \right)^2 + 1 \right] = 1$.

MISCELLANEOUS EXERCISES

1. What kind of locus is represented by the pair of equations $r = a$, $F(\theta, \varphi) = 0$, r, θ, and φ being spherical coordinates?

2. What is the nature of the locus of the equation $F(\theta, \varphi) = 0$? Of the locus of $F(r, \theta) = 0$? Of the locus of $F(r, \varphi) = 0$?

3. Find the radius of curvature of the cissoid $y^2 = \dfrac{x^3}{2a - x}$.

4. Find the radius of curvature of the curve $\dfrac{x^3}{a^3} + \dfrac{y^3}{b^3} = 1$.

5. Show that if a segment is laid off on the normal to the catenary from the point of contact in a direction away from the center of curvature and equal in length to the radius of curvature, the locus of the end of the segment is a straight line. This line is called the *directrix* of the catenary.

6. Express the square of the element of arc of a space curve in terms of spherical coordinates.

Are the following three curves rectifiable in the interval $(0, 1)$:

7. $y = \sin \dfrac{1}{x} \; (x \neq 0), \; y = 0 \; (x = 0)$.

8. $y = x \sin \dfrac{1}{x} \; (x \neq 0), \; y = 0 \; (x = 0)$.

9. $y = x^2 \sin \dfrac{1}{x} \; (x \neq 0), \; y = 0 \; (x = 0)$.

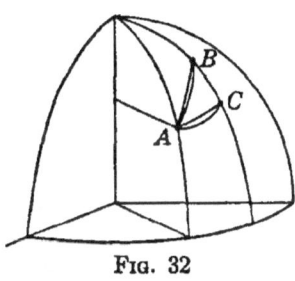

FIG. 32

10. Find the equations of a curve on the surface of a sphere that cuts all the meridians at the same angle.

Let the center of the sphere be at the origin and take the planes through the z-axis for the meridian planes. Denote by A and B two points on the curve and by C the intersection of the meridian through B with the parallel of latitude through A. If λ_1 is the angle formed by the chords $\overline{AB}$ and $\overline{AC}$, and μ_1 is the angle formed by the chords $\overline{CB}$ and $\overline{BA}$, then $\dfrac{\overline{CB}}{\overline{AC}} = \dfrac{\sin \lambda_1}{\sin \mu_1}$. But if r, θ, and φ are the spherical coordinates of A, and $r, \theta + \Delta\theta$, and $\varphi + \Delta\varphi$ those of B, then $\widetilde{CB} = r\Delta\varphi$ and $\widetilde{AC} = r \sin \varphi \cdot \Delta\theta$. As B approaches A along the curve

$$\lim \frac{\overline{CB}}{\overline{AC}} = \lim \frac{\widetilde{CB}}{\widetilde{AC}} = \lim \frac{r\Delta\varphi}{r \sin \varphi \cdot \Delta\theta} = \frac{1}{\sin \varphi} \frac{d\varphi}{d\theta}.$$

Moreover $\lim \lambda_1 = \lambda$, the complement of the given angle between the curve and the meridians, and $\lim \mu_1 = \dfrac{\pi}{2} - \lambda$. Hence

$$\frac{1}{\sin \varphi} \frac{d\varphi}{d\theta} = \tan \lambda,$$

and $\log \tan \dfrac{\varphi}{2} = \tan \lambda \cdot \theta + C$.

APPLICATIONS TO GEOMETRY

11. Show that the curve in Exercise 10 winds around the pole an infinite number of times, but is of finite length.

12. Find the equations of a curve on the cone $x^2 + y^2 - c^2z^2 = 0$ that cuts all the elements of the cone at the same angle.

13. Show that between a given point on the curve of Exercise 12 and the origin the curve winds around the axis of the cone an infinite number of times, but is of finite length.

14. Show that the orthogonal projection of the curve in Exercise 12 on the xy-plane is an equiangular spiral.

15. Find the angle between the curve $x = t \cos t$, $y = t \sin t$, $z = \dfrac{t}{c}$ and the element of the cone at the point $t = t_1$.

CHAPTER XIV

CALCULUS OF VARIATIONS

193. The simplest problem. In Chapter IV we considered the problem of finding the values of the independent variable x for which the function $f(x)$ is a maximum or a minimum and the corresponding problem for functions of two variables. In this chapter we shall consider another problem concerning maxima and minima; namely, having given a function $F(x, y, y')$ subject to certain conditions that will be described in detail later and two points $A = (a, c)$ and $B = (b, d)$ in the (x, y)-plane, to find the curve $y = f(x)$ passing through the points A and B and such that the integral

$$J = \int_a^b F(x, y, y')dx \tag{1}$$

shall be a maximum or a minimum.

This is the simplest problem in what is known as the *Calculus of Variations*. The thing to be determined here is not a point, as in the cases referred to, but a curve.

194. Examples from geometry and mechanics.

(a) Let $y = f(x)$ be a given curve passing through the points A and B on the same side of the x-axis. We shall assume that $a < b$. The area of the surface generated by revolving the arc AB completely around the x-axis is given by the formula

$$J = 2\pi \int_a^b y\sqrt{1 + y'^2}dx. \tag{2}$$

If we take another curve through A and B, the resulting value of J would in general be different. The problem is to see if there is one curve that will make this area a minimum, and if there is such a curve to determine its equation.

Why do we not ask if there is a curve that makes this area a maximum?

This problem is connected with the following problem that arises in mechanics:[1] Let a circular ring be dipped into a soap solution and then withdrawn. In this way a disc of film surrounded by the ring will be formed. If now a smaller circular ring concentric with the first one be pushed through the film in a direction perpendicular to the plane of the film, care being taken to keep the two planes parallel, the two rings will be connected by a film which is a surface of revolution whose axis is the line connecting the centers of the two circles. If we take this line for the x-axis, what is the curve $y = f(x)$ that generates this surface?

If we assume, as is proved in mechanics, that the film assumes the shape of a minimum surface of revolution, we have an answer to the first question raised; namely, as to the existence of a curve that will give a minimum surface of revolution. But the problem of determining the equation of this curve is one whose solution is to be found only in the theory of the Calculus of Variations.

(b) Let the (x, y)-plane be vertical with the positive direction of the y-axis downwards. We consider two points A and B in this plane, but not in the same vertical line, connected by a wire on which a bead is strung. Our problem is to

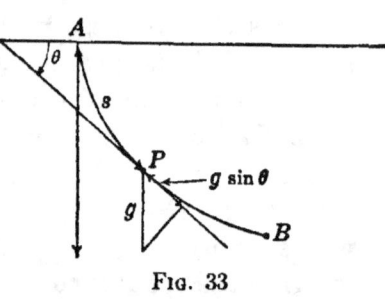

Fig. 33

determine the shape of the wire in order that the bead, starting from rest and moving without friction, shall fall from A to B under the influence of gravity in the shortest possible time.

It might seem that the answer is very simple, inasmuch as the straight line connecting the points is the shortest

[1] See Bliss, *The Calculus of Variations*, 1925, p. 7.

path between them. But a moment's reflection shows that the problem is not so simple as this because a longer path might be steeper in places and the increased speed of the bead along these parts of the path might more than make up for the greater distance to be traveled. Galileo surmised that the path of quickest descent was the arc of circle. We shall see later whether he was right or not.

It is not difficult to set up the integral which must be minimized. If θ is the inclination of the tangent to the path at any point P, we have

$$\frac{d^2s}{dt^2} = g \sin \theta.$$

But $\sin \theta = \frac{dy}{ds}$ (§ 38). Then

$$\frac{d^2s}{dt^2} = g \frac{dy}{ds}. \tag{3}$$

If we multiply each side of this equation by $2 \frac{ds}{dt}$, the result is

$$2 \frac{ds}{dt} \frac{d^2s}{dt^2} = 2g \frac{dy}{ds} \frac{ds}{dt} = 2g \frac{dy}{dt}.$$

Then integration gives us

$$\left(\frac{ds}{dt}\right)^2 = 2gy + c.$$

We shall assume that the origin has been taken at the point A. Then $y = 0$ and $\frac{ds}{dt} = v = 0$ when $t = 0$, since the bead starts from rest. Hence $c = 0$. We have then

$$\left(\frac{ds}{dt}\right)^2 = 2gy.$$

The time T for the fall is given by the formula

$$T = \int_0^l \frac{ds}{v} = \frac{1}{\sqrt{2g}} \int_0^l \frac{ds}{\sqrt{y}} = \frac{1}{\sqrt{2g}} \int_0^b \sqrt{\frac{1 + y'^2}{y}} \, dx, \tag{4}$$

where l is the length of the wire and b is the abscissa of B. It is this integral that is to be minimized. It is of the form

(1) with $J = \sqrt{\dfrac{1+y'^2}{y}}$. The factor $\dfrac{1}{\sqrt{2g}}$ can be disregarded since the two integrals $\dfrac{1}{\sqrt{2g}}\displaystyle\int_0^b \sqrt{\dfrac{1+y'^2}{y}}\,dx$ and $\displaystyle\int_0^b \sqrt{\dfrac{1+y'^2}{y}}\,dx$ are minimized by the same curve.

195. Euler's equation. In § 193 we referred to certain restrictions to be placed on $F(x, y, y')$. These are that F and its partial derivatives of the first and second orders shall be continuous when the point (x, y) lies in a given region S and y' has any value. We further assume that the curve C whose equation is $y = f(x)$ is interior to S and that $f(x)$ and $f'(x)$ are continuous in the closed interval (a, b), a and b being the abscissae of two points in S.

In our discussion it is unnecessary to consider in detail both maximum and minimum values of J, inasmuch as a function y that minimizes J also maximizes $-J$. We consider now in detail what it means for y to minimize J. Let $\eta = \varphi(x)$ be any function of x which together with its derivative η' is continuous in the closed interval (a, b) and which vanishes at $x = a$ and $x = b$. Since C lies within S, it is possible to keep η sufficiently small in absolute value (say, $|\eta| < \epsilon$) to assure us that $y + \alpha\eta$ also lies within S when $-1 \leq \alpha \leq 1$. When we say that y minimizes J we mean that $J_0 \leq J$, where

$$J_0 = \int_a^b F(x, y, y')\,dx, \qquad J = \int_a^b F(x, Y, Y')\,dx,$$

with $Y = y + \alpha\eta$. The integral J is a function of α whose derivative is given by the formula (§ 68)

$$J'(\alpha) = \int_a^b [\eta F_y(x, y + \alpha\eta, y' + \alpha\eta') \\ + \eta' F_{y'}(x, y + \alpha\eta, y' + \alpha\eta')]\,dx.$$

But $J(\alpha)$ has a minimum for $\alpha = 0$. Hence

$$J'(0) = \int_a^b [\eta F_y(x, y, y') + \eta' F_{y'}(x, y, y')] dx = 0.$$

We find by integration by parts that

$$\int_a^b \eta' F_{y'} dx = \eta F_{y'} \Big|_a^b - \int_a^b \eta \frac{dF_{y'}}{dx} dx.$$

But $\eta(a) = \eta(b) = 0$. Hence

$$J'(0) = \int_a^b \eta \left(F_y - \frac{dF_{y'}}{dx} \right) dx = 0. \tag{5}$$

The fact that equation (5) holds for any η subject to the conditions described enables us to conclude that

$$F_y - \frac{dF_{y'}}{dx} = 0. \tag{6}$$

For, if $F_y - \dfrac{dF_{y'}}{dx}$ were different from zero for a value c of x in the interval (a, b), it would be different from zero and of the same sign throughout a sufficiently small neighborhood (x_0, x_1) of c. We can in many ways select an η

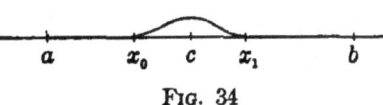

Fig. 34

that is continuous in (a, b), together with its derivative, positive within the interval (x_0, x_1), and zero everywhere else in (a, b). The function that is equal to $(x - x_0)^2(x - x_1)^2$ when $x_0 < x < x_1$ and equal to zero for every other value of x in (a, b) is a simple example of such a function. For such an η the integral $\int_a^b \eta \left(F_y - \dfrac{dF_{y'}}{dx} \right) dx$ would not be zero. Hence equation (6) is satisfied throughout (a, b). This is a condition that a function y must satisfy in order to minimize (or maximize) J. It can be written as follows:

$$F_y - F_{y'x} - F_{y'y} \cdot y' - F_{y'y'} \cdot y'' = 0. \tag{7}$$

This is known as *Euler's equation*. It is an ordinary differential equation of the second order, and its general solution therefore contains two arbitrary constants. These will be determined in general by the condition that the curve pass through the given points A and B.

Any solution of Euler's equation is called an *extremal*.

In case $F(x, y, y')$ does not contain the independent variable x explicitly, we can deduce a first integral of Euler's equation immediately from the fact that

$$\frac{d}{dx}(F - y'F_{y'}) = F_y y' + F_{y'} y'' - F_{y'} y'' - y' \frac{d}{dx} F_{y'}$$
$$= y'\left(F_y - \frac{d}{dx} F_{y'}\right) = 0.$$

This first integral is therefore

$$F - y'F_{y'} = c. \tag{8}$$

First and second variations. If we expand $J(\alpha)$ in powers of α, we obtain the series

$$J(\alpha) = J_0 + \alpha J'(0) + \frac{\alpha^2}{2!} J''(0) + \cdots.$$

The term of this expression that is of the first degree in α is called the *first variation* of J and is denoted by the symbol δJ. Thus

$$\delta J = \alpha J'(0).$$

The *second variation* $\delta^2 J$ is by definition $\alpha^2 J''(0)$. Euler's equation is equivalent to the equation $\delta J = 0$.

196. Minimum surface of revolution. We can apply (8) to the Euler equation connected with the problem of finding a minimum surface of revolution, since in this case $F(x, y, y') = y\sqrt{1 + y'^2}$. In this way we get the first integral

$$y\sqrt{1 + y'^2} - \frac{yy'^2}{\sqrt{1 + y'^2}} = \beta,$$

or

$$y - \beta\sqrt{1 + y'^2} = 0. \tag{9}$$

Then
$$y' = \pm \sqrt{\frac{y^2}{\beta^2} - 1},$$
$$dx = \pm \frac{\beta dy}{\sqrt{y^2 - \beta^2}},$$
$$x - \alpha = \pm \beta \log \frac{y + \sqrt{y^2 - \beta^2}}{\beta},$$
$$y + \sqrt{y^2 - \beta^2} = \beta e^{\pm(x-\alpha)/\beta},$$
$$y - \sqrt{y^2 - \beta^2} = \beta e^{\mp(x-\alpha)/\beta}.$$

Hence
$$y = \frac{\beta}{2}\left(e^{(x-\alpha)/\beta} + e^{-(x-\alpha)/\beta}\right). \tag{10}$$

This is the equation of a catenary whose directrix is the x-axis (see Exercise 5, p. 348).

The next step in the determination of the minimum surface of revolution is to determine if possible the constants α and β in such a way as to make this catenary pass through the points A and B. The mechanical theory of the soap film to which reference was made in § 194 shows that this is possible under certain circumstances, and furthermore that in this case we obtain in fact a minimum surface of revolution. What these circumstances are cannot be discussed here; we can merely state the general conclusion.

If from A we draw the line AC which meets the x-axis at C at a certain angle whose approximate value is 123° 32′ and from C the line CD making an angle with the x-axis equal to the supplement of this angle, B must be above the line CD in order that one of the curves (10) shall pass through the points A and B. In other words, if B is not above CD there is no minimum surface of revolution. But if one of these curves does pass through A and B, it does not follow that it generates a minimum surface of revolution. For this it is necessary and sufficient that the tangents to the catenary at A and B shall intersect above the x-axis. Even if there is no minimum surface, there is a lower limit to the areas of the possible surfaces, since all these areas are positive. This lower limit is the sum of the areas of the two circular discs whose radii are the ordinates of A and B.[1]

197. The brachistochrone. In § 194 we referred to the problem of determining the curve of quickest descent. A first integral of Euler's equation for this problem is given by (8). It is
$$1 + y'^2 = \frac{2c}{y}$$
or
$$\frac{dy}{dx} = \sqrt{\frac{2c-y}{y}}. \tag{11}$$

A second integration gives us
$$x = \int \frac{y\,dy}{\sqrt{2cy - y^2}}$$
$$= -\sqrt{2cy - y^2} + c \arccos \frac{c-y}{c} + C. \tag{12}$$

Since the point A is at the origin, $C = 0$. We leave it to the reader to verify that with $C = 0$ equation (12) represents the cycloid
$$x = c(\theta - \sin \theta),$$
$$y = c(1 - \cos \theta). \tag{13}$$

In order to determine whether c can be so determined that the curve (13) shall pass through B, we observe that the slope of the line from the origin to the point on the curve corresponding to a given value of θ is independent of the parameter c. If then we draw the curve
$$x = \theta - \sin \theta,$$
$$y = 1 - \cos \theta,$$
and denote by B' the point where the line OB (prolonged, if necessary) cuts it, the curve (13) with $c = \dfrac{OB}{OB'}$ will pass

---

[1] For details see Moigno-Lindelöf, *Leçons de calcul differential et de calcul intégral*, Vol. IV, pp. 205–211.

through B. Clearly not more than one of these curves passes through B. But is this unique curve through B a curve of quickest descent? It can be shown that it is, although the proof is too long to be given here. Because it is the curve of quickest descent it is known as the *brachistochrone*.[1]

EXERCISES

1. Show that the time taken for a particle to slide down the arc of the cycloid $x = a(\theta - \sin \theta)$, $y = a(1 - \cos \theta)$ from the vertex, the y-axis being directed downward, is proportional to the value of θ corresponding to the terminal point.

2. Find the time taken for a particle to slide down the chord from the vertex of the cycloid to the lowest point. How does this compare with the time taken to slide down the cycloid to its lowest point?

3. Two cycloids are generated by circles of radii $2a$ and a. If a particle starts to slide down the larger of these from the vertex at the same time another particle starts down the other cycloid from the vertex, where will the first particle be when the second one is at the lowest point of its cycloid?

4. Find the area of the surface generated by revolving the arc of the catenary $y = e^{(x-1)/2} + e^{-(x-1)/2}$ between the points $(0, e^{1/2} + e^{-1/2})$ and $(1, 2)$ around the x-axis.

5. Find the area of the surface generated by revolving the straight line connecting these points around the x-axis. Compare this result with the result obtained in Exercise 4.

6. Find a first integral of the Euler equation in case $F(x, y, y') = \dfrac{\sqrt{1 + y'^2}}{y}$, $y > 0$.

7. Show that the extremals in Exercise 6 are semi-circles with centers on the x-axis.

8. Find a first integral of the Euler equation and solve in case $F(x, y, y') = +\sqrt{y}\sqrt{1 - y'^2}$, $y > 0$.

9. Show that if $F(x, y, y')$ does not contain x or y and $F_{y'y'} \neq 0$ the extremals are straight lines.

10. Show that if $F(x, y, y') = y'^2(1 + y')^2$ and the slope of the straight line connecting A and B is either less than -1 or greater than 0, then the extremal connecting these points makes J a minimum.

Let J_0 be the value of J taken along the straight line $y = y(x)$ from A to B, and $\bar{J}$ its value along any other permissible path from A to B. This path can be represented by an equation of the form $y = y(x)$

---

[1] From the Greek, βραχιστος = *shortest*, and χρονος = *time*.

$+ \omega(x)$, where $\omega(a) = \omega(b) = 0$ and ω' is continuous in this interval. Then show that $J - J_0 > 0$.

11. Does the extremal from A to B for the same $F(x, y, y')$ as in Exercise 10 make J a minimum if the slope of AB is -1?

12. Find the Euler equation in case $F(x, y, y') = ay'^2 - 4byy'^3 + 2bxy'^4$.

13. Show that if $F(x, y, y') = 1 - y'^2$ the extremal from A to B maximizes J regardless of the relative positions of these points if only $a < b$.

14. Show that if $F(x, y, y')$ is a quadratic in y' with constant coefficients the extremal connecting A and B maximizes or minimizes J according as the coefficient of y'^2 is < 0 or > 0.

198. Two independent variables. In the preceding paragraphs of this chapter we have discussed briefly what is known as the simplest problem of the calculus of variations. We shall now consider a somewhat more complicated, but closely related, problem.

Let $F(x, y, z, p, q)$ be a function of the five arguments x, y, z, p, and q, which together with its partial derivatives of the first two orders is continuous when the point (x, y, z) lies within a given region V of space and p and q have any values. Let $z = f(x, y)$ be the equation of a surface and consider a part of this surface that lies in V and is bounded by a closed curve C. We suppose that $f(x, y)$ is continuous, together with its derivatives of the first two orders, when (x, y) is in the projection S of this part of the surface on the (x, y)-plane. We shall refer to such functions $f(x, y)$ as *admissible functions*. Our problem is to determine the minimum (or maximum) values of the integral

$$J = \iint_S F(x, y, z, p, q) dS \qquad (14)$$

when z is replaced by an admissible function $f(x, y)$ such that the surface $z = f(x, y)$ passes through C, and p and q stand for $\dfrac{\partial z}{\partial x}$ and $\dfrac{\partial z}{\partial y}$ respectively, provided there is such a minimum (or maximum).

Suppose that $f(x, y)$ is an admissible function that gives to J a minimum value. Take any admissible function $\xi(x, y)$ that vanishes on the boundary C_1 of S and that is sufficiently small in absolute value to bring the point (x, y, z) within the region V when $z = f(x, y) + \alpha\xi(x, y)$, where $|\alpha| \leq 1$. Then $z + \alpha\xi$ is an admissible function for which the corresponding J is a function of α:

$$J(\alpha) = \iint_S F(x, y, z + \alpha\xi, p + \alpha\xi_x, q + \alpha\xi_y)dS. \quad (15)$$

This function of α has by hypothesis a minimum for $\alpha = 0$ and a derivative at this point which is given by the formula

$$J'(\alpha) = \iint_S (\xi F_z + \xi_x F_p + \xi_y F_q)dS. \quad (16)$$

The arguments in the functions F_z, F_p, and F_q are x, y, $z + \alpha\xi$, $p + \alpha\xi_x$, and $q + \alpha\xi_y$. Now

$$J'(0) = 0,$$

and therefore

$$\iint_S (\xi F_z + \xi_x F_p + \xi_y F_q)dS = 0, \quad (17)$$

where the arguments are now x, y, z, p, and q. But

$$\iint_S \xi_x F_p dS = \int_{C_1} \xi F_p dy - \iint_S \xi \frac{\partial F_p}{\partial x} dS$$

by virtue of the formulae

$$\frac{\partial}{\partial x}(\xi F_p) = \xi_x F_p + \xi \frac{\partial F_p}{\partial x}$$

and

$$\iint_S \frac{\partial}{\partial x}(\xi F_p)dS = \int_{C_1} \xi F_p dy. \quad (\S 99.)$$

Similarly

$$\iint_S \xi_y F_q dS = -\int_{C_1} \xi F_q dx - \iint_S \xi \frac{\partial F_q}{\partial y} dS.$$

And $\xi = 0$ along the boundary C_1. Hence (17) can be written in the following form:

$$\iint_S \xi \left(F_z - \frac{\partial F_p}{\partial x} - \frac{\partial F_q}{\partial y} \right) dS = 0. \tag{18}$$

We have to keep in mind that ξ is to a large extent arbitrary. If we take advantage of this fact we can see, as in the parallel case for one variable (§ 195), that we must have

$$F_z - \frac{\partial F_p}{\partial x} - \frac{\partial F_q}{\partial y} = 0. \tag{19}$$

This is *Euler's equation* for double integrals. It gives a necessary condition for the existence of a minimum (or a maximum). In it the arguments upon which the functions depend are $x, y, z, \frac{\partial z}{\partial x}$, and $\frac{\partial z}{\partial y}$, where $z = f(x, y)$ is the function that makes J a minimum (or a maximum). Any solution of Euler's equation is called an *extremal*.

We leave it as an exercise for the reader to verify that Euler's equation for the integral

$$J = \iint_S (p^2 + q^2) dS$$

is

$$\frac{\partial^2 z}{\partial x^2} + \frac{\partial^2 z}{\partial y^2} = 0. \tag{20}$$

It is easy to see that any solution z of (20) that satisfies the required boundary conditions makes J an absolute minimum. For if v is any other admissible function put

$z - v = \omega$, or $v = z - \omega$. Now $\omega = 0$ on C_1. Moreover

$$\iint_S \left[\left(\frac{\partial v}{\partial x}\right)^2 + \left(\frac{\partial v}{\partial y}\right)^2 \right] dS$$

$$= \iint_S \left[\left(\frac{\partial z}{\partial x} - \frac{\partial \omega}{\partial x}\right)^2 + \left(\frac{\partial z}{\partial y} - \frac{\partial \omega}{\partial y}\right)^2 \right] dS$$

$$= \iint_S \left[\left(\frac{\partial z}{\partial x}\right)^2 + \left(\frac{\partial z}{\partial y}\right)^2 \right] dS$$

$$+ \iint_S \left[\left(\frac{\partial \omega}{\partial x}\right)^2 + \left(\frac{\partial \omega}{\partial y}\right)^2 \right] dS$$

$$- 2 \iint_S \left(\frac{\partial z}{\partial x} \frac{\partial \omega}{\partial x} + \frac{\partial z}{\partial y} \frac{\partial \omega}{\partial y} \right) dS.$$

If we put $P = \omega \dfrac{\partial z}{\partial y}$ and $Q = -\omega \dfrac{\partial z}{\partial x}$, we have

$$\iint_S \left(\frac{\partial P}{\partial y} - \frac{\partial Q}{\partial x} \right) dS = \iint_S \left(\frac{\partial z}{\partial x} \frac{\partial \omega}{\partial x} + \frac{\partial z}{\partial y} \frac{\partial \omega}{\partial y} \right) dS$$

$$+ \iint_S \omega \left(\frac{\partial^2 z}{\partial x^2} + \frac{\partial^2 z}{\partial y^2} \right) dS.$$

But $\dfrac{\partial^2 z}{\partial x^2} + \dfrac{\partial^2 z}{\partial y^2} = 0$, since z is a solution of (20). Hence the last integral vanishes. Then by Green's theorem

$$\iint_S \left(\frac{\partial z}{\partial x} \frac{\partial \omega}{\partial x} + \frac{\partial z}{\partial y} \frac{\partial \omega}{\partial y} \right) dS = -\int_{C_1} \omega \frac{\partial z}{\partial y} dx - \omega \frac{\partial z}{\partial x} dy = 0,$$

since $\omega = 0$ on C_1. It follows that

$$\iint_S \left[\left(\frac{\partial v}{\partial x}\right)^2 + \left(\frac{\partial v}{\partial y}\right)^2 \right] dS$$

is greater than $\displaystyle\iint_S \left[\left(\frac{\partial z}{\partial x}\right)^2 + \left(\frac{\partial z}{\partial y}\right)^2 \right] dS.$

199. Variable end points.

In §§ 193–197 we assumed that the end points of the path of integration were fixed. Problems present themselves however in which one or both of these points are subject merely to the condition that they lie on given curves. We might, for example, have a point A and a curve Γ in the

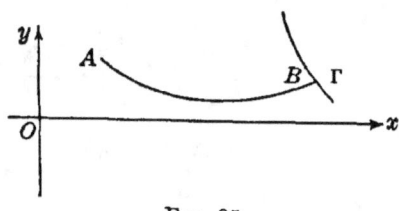

Fig. 35

upper half of the xy-plane and wish to determine the arc from A to Γ that generates a surface of smaller area than any other similar arc when revolved around the x-axis.

Suppose more generally that we wish to find the arc from A to Γ that minimizes (or maximizes) the integral

$$J = \int_a^b F(x, y, y')dx,$$

where $F(x, y, y')$ is subject to the same conditions as in § 195. If $y = f(x)$ is a curve passing through A and intersecting Γ in B_0 that minimizes J, it is clear that $f(x)$ must be a solution of Euler's equation, since of all admissible curves that connect the fixed points A and B_0 it must give a minimum value to J. This defines $f(x)$ in the interval (a, b_0), where b_0 is the abscissa of B_0. We can extend this definition to the interval (a, b'), where b' is slightly greater than b_0, in such a way as to make $f(x)$ continuous with a continuous derivative in the latter interval. We then form the curve $C : \bar{y} = y + k\eta$, where $\eta = \eta(x)$ is continuous, together with its first derivative, in (a, b') and $\eta(a) = 0$, $\eta(b_0) \neq 0$, and we define k by the condition that C shall go through $B = [b, \varphi(b)]$, where $b_0 \leqq b < b'$ and $y = \varphi(x)$ is the equation of Γ. This gives us

$$f(b) + k\eta(b) = \varphi(b),$$

or

$$k = \frac{\varphi(b_0 + \epsilon) - f(b_0 + \epsilon)}{\eta(b_0 + \epsilon)},$$

where $b = b_0 + \epsilon$. Then k is a function $k(\epsilon)$ of ϵ, and $k(0) = 0$, since $f(b_0) = \varphi(b_0)$. Now

$$k'(\epsilon) = \frac{\eta(b)[\varphi'(b) - f'(b)] - \eta'(b)[\varphi(b) - f(b)]}{\eta(b)^2}.$$

Hence

$$k'(0) = \frac{\varphi'(b_0) - f'(b_0)}{\eta(b_0)}.$$

But $\varphi'(b_0) = \tan \theta$, where θ is the inclination of the tangent to Γ at B_0. Then

$$k'(0) = \frac{\tan \theta - y'}{\eta(b_0)}.$$

For the curve y the integral J is a function of ϵ which has a minimum (or maximum) when $\epsilon = 0$.

$$J(\epsilon) = \int_a^b F(x, y + k\eta, y' + k\eta') dx$$

and

$$\left(\frac{dJ(\epsilon)}{d\epsilon}\right)_{\epsilon=0} = 0.$$

In getting the derivative of $J(\epsilon)$ with respect to ϵ, we must keep in mind that both the integrand and the upper limit b are functions of ϵ. Then (see § 68)

$$\left(\frac{dJ}{d\epsilon}\right)_{\epsilon=0} = \int_a^{b_0} [\eta F_y \cdot k'(0) + \eta' F_{y'} \cdot k'(0)] dx + F \Big|_{x=b_0}.$$

As in § 195

$$\int_a^{b_0} \eta' F_{y'} dx = \eta F_{y'} \Big|_a^{b_0} - \int_a^{b_0} \eta \frac{dF_{y'}}{dx} dx.$$

Then

$$\left(\frac{dJ}{d\epsilon}\right)_{\epsilon=0} = k'(0) \int_a^{b_0} \eta \left(F_y - \frac{dF_{y'}}{dx}\right) dx + [k'(0) \cdot \eta F_{y'} + F]_{x=b_0} = 0.$$

The integral in the second member of this equation vanishes,

since y is an extremal. It follows that
$$(\tan \theta - y')F_{y'} + F = 0$$
or
$$F_{y'} \sin \theta + (F - y'F_{y'}) \cos \theta = 0, \qquad (21)$$
where the arguments are $x = b_0$, $y = f(b_0)$, and $y' = f'(b_0)$. Equation (21) is a further necessary condition for a minimum (or a maximum). If we apply this general result to the problem of finding a minimum surface of revolution, we have $F(x, y, y') = y\sqrt{1 + y'^2}$ and therefore by (21)
$$\tan \theta = \frac{y'F_{y'} - F}{F_{y'}} = -\frac{1}{y'}.$$
Hence the minimizing curve must be perpendicular to Γ. The reader can readily verify that this condition of orthogonality holds also in the case of the brachistochrone.

200. Parametric equations. If the path of integration is given by equations in parametric form
$$\left.\begin{array}{l} x = x(t), \\ y = y(t), \end{array}\right\} \qquad (22)$$
we have to deal with the integral
$$J = \int_{t_0}^{t_1} F[x(t), y(t), x'(t), y'(t)]dt.$$
We assume that J takes on a minimum (or a maximum) value when the path of integration is a curve that lies in the region S and is represented by equations of the form (22), where $x(t)$, $y(t)$, $x'(t)$, and $y'(t)$ are continuous in the interval (t_0, t_1) and $x'(t)$ and $y'(t)$ do not vanish simultaneously. Then the equations
$$F_x - \frac{d}{dt}F_{x'} = 0, \qquad F_y - \frac{d}{dt}F_{y'} = 0 \qquad (23)$$
must be satisfied. These are the Euler equations for the problem in parametric representation. Their derivation involves some special points concerning parametric representation, and will be omitted.

If F depends on n functions $x_i(t)$ $(i = 1, 2, \cdots, n)$ and their first derivatives, a necessary condition for a maximum or a minimum is that these n functions satisfy the n equations

$$F_{x_i} - \frac{d}{dt} F_{x_i'} = 0. \qquad (24)$$

We shall see in the next paragraph an application of this condition to an important problem in theoretical mechanics.

201. Hamilton's principle. If the positions of a system of particles are determined by n parameters $q_1, q_2, \cdots, q_n$, the coordinates x_i, y_i, z_i of the separate particles are functions of the q's; and their derivatives $\dot{x}_i, \dot{y}_i, \dot{z}_i$ with respect to the time depend on the q's and their first derivatives with respect to the time.[1] Now the kinetic energy of the system is

$$T = \frac{1}{2} \sum m_i(\dot{x}_i^2 + \dot{y}_i^2 + \dot{z}_i^2).$$

Moreover

$$\dot{x}_i = \sum_{j=1}^{n} \frac{\partial x_i}{\partial q_j} \dot{q}_j, \qquad \dot{y}_i = \sum_{j=1}^{n} \frac{\partial y_i}{\partial q_j} \dot{q}_j, \qquad \dot{z}_i = \sum_{j=1}^{n} \frac{\partial z_i}{\partial q_j} \dot{q}_j.$$

It follows that the kinetic energy is a quadratic function of the $\dot{q}$'s whose coefficients are functions of the q's. The potential function U is a function of the coordinates of the separate particles and therefore a function of the q's alone.

As t changes from t_0 to t_1 the continuous function $T + U$ changes in a definite way that gives a definite value to the integral

$$J = \int_{t_0}^{t_1} (T + U) dt.$$

This integral is known as *Hamilton's integral*. The change in $T + U$ takes place in such a way as to give a minimum or a maximum value to J. There is no proof to be offered

---

[1] We here follow the convention in accordance with which differentiation with respect to the time is denoted by a dot placed above the function. Thus $\dot{x} = \dfrac{dx}{dt}$, $\ddot{x} = \dfrac{d^2x}{dt^2}$, etc.

for this statement. It is in fact an assumption which lies at the base of theoretical mechanics. When formulated somewhat differently it is known as *Hamilton's principle*.

HAMILTON'S PRINCIPLE. *The motion of a material system in the time from $t = t_0$ to $t = t_1$ is such as to make the first variation of Hamilton's integral vanish:*

$$\delta \int_{t_0}^{t_1} (T + U) dt = 0.$$

The Euler equations are

$$\frac{d}{dt} \frac{\partial T}{\partial \dot{q}_i} - \frac{\partial (T + U)}{\partial q_i} = 0 \quad (i = 1, 2, \cdots, n), \qquad (25)$$

since U does not depend on the $\dot{q}$'s. Equations (25) are *Lagrange's equations of motion*.

202. The spherical pendulum. A particle with mass that is constrained to move on the surface of a sphere under the influence of gravity is called a *spherical pendulum*. If the motion is in a vertical plane, we have a simple pendulum. If it is not in such a plane, we have a more complicated situation. We take the origin at the center of the sphere and use cylindrical coordinates r, θ, and z. The equation of the sphere is $r^2 + z^2 = a^2$ and $\overline{ds}^2 = \overline{dr}^2 + r^2\overline{d\theta}^2 + \overline{dz}^2$. But $r dr + z dz = 0$. Hence

$$\overline{ds}^2 = \frac{a^2}{a^2 - z^2} \overline{dz}^2 + (a^2 - z^2)\overline{d\theta}^2.$$

If the particle is of mass m,

$$T = \frac{1}{2} m v^2 = \frac{1}{2} m \left[\frac{a^2}{a^2 - z^2} \dot{z}^2 + (a^2 - z^2) \dot{\theta}^2 \right].$$

Now U is that function of the coordinates of the particle whose derivative in a given direction is the component of the force in that direction. In this case therefore $U = -mgz$. Then

$$T + U = \frac{1}{2} m \left[\frac{a^2}{a^2 - z^2} \dot{z}^2 + (a^2 - z^2) \dot{\theta}^2 \right] - mgz.$$

We now use Lagrange's equations of motion (25), which were shown in § 201 to be consequences of Hamilton's principle. The position of the particle is determined by z and θ. We take them therefore for the q_1 and q_2 respectively of (25). Then

$$\frac{a^2 z \dot{z}^2}{(a^2-z^2)^2} - z\dot{\theta}^2 - g - \frac{d}{dt}\frac{a^2 \dot{z}}{a^2-z^2} = 0. \tag{26}$$

and

$$\frac{d}{dt}(a^2-z^2)\dot{\theta} = 0. \tag{27}$$

From (27) we have immediately

$$(a^2-z^2)\dot{\theta} = C,$$

or

$$\dot{\theta} = \frac{C}{a^2-z^2}. \tag{28}$$

We substitute this value for $\dot{\theta}$ in (26):

$$\frac{a^2 z \dot{z}^2}{(a^2-z^2)^2} - \frac{C^2 z}{(a^2-z^2)^2} - g - a^2\frac{(a^2-z^2)\ddot{z} + 2z\dot{z}^2}{(a^2-z^2)^2} = 0.$$

Or

$$\frac{a^2 \ddot{z}}{a^2-z^2} + \frac{a^2 z \dot{z}^2}{(a^2-z^2)^2} + \frac{C^2 z}{(a^2-z^2)^2} + g = 0.$$

Multiplying through by $2\dot{z}$:

$$\frac{2a^2 \dot{z}\ddot{z}}{a^2-z^2} + \frac{2a^2 z \dot{z}^3}{(a^2-z^2)^2} + \frac{2C^2 z \dot{z}}{(a^2-z^2)^2} = -2g\dot{z}.$$

It is easy to verify that the left member of this equation is the derivative of

$$\frac{a^2 \dot{z}^2}{a^2-z^2} + \frac{C^2}{a^2-z^2}.$$

Hence

$$a^2 \dot{z}^2 + C^2 = (a^2-z^2)(-2gz + C_1),$$

$$dt = \frac{a\,dz}{\sqrt{(a^2-z^2)(-2gz+C_1) - C^2}}. \tag{29}$$

We have here an elliptic integral for the determination of the time taken by the particle in descending from a height z_0 to a height z_1; namely,

$$t = a \int_{z_0}^{z_1} \frac{dz}{\sqrt{(a^2 - z^2)(-2gz + C_1) - C^2}}.$$

The cubic under the radical in (29) has three real roots. For it is positive when z has a large positive value, negative when $z = a$, positive when $z = z_0 (-a < z_0 < a)$, since the initial velocity is real, and negative when $z = -a$. If we denote these roots in the order of increasing magnitude by z_1, z_2, and z_3, we can write (29) as follows:

$$dt = \frac{A\,dz}{\sqrt{(z - z_1)(z - z_2)(z - z_3)}},$$

where $A = \dfrac{a}{\sqrt{2g}}$.

203. An application in economics. We consider here an application of the preceding theory to a question connected with the economics of an exhaustible mine.

Let p be the net price per unit of product received by the owner of the mine at the time t. This depends upon the current rate of production, which we represent by q, and upon the past production, as well as upon t. Now $q = \dfrac{dx}{dt}$, where x denotes the amount already removed from the mine at the time t. Then $p = p(t, x, q)$. The present value of a unit of profit to be realized at the end of the time t is $e^{-\gamma t}$, where γ is a constant called the *force of interest*. The discounted profit at the present time—that is, the present value of the mine— is given by the integral

$$J_0 = \int_0^T F(t, x, q)\,dt,$$

where

$$F(t, x, q) = p(t, x, q)q e^{-\gamma t},$$

if it is assumed that the mine will be exhausted in the time

$t = T$. At this time we shall have $x = a$, the amount originally in the mine.

The problem is so to determine the rate of production as to make the mine as profitable as possible to its owner. To do this we have to determine x as a function of t in such a way as to maximize J_0. This will determine the rate of production q. Before we can do anything toward the solution of this problem we must know the form of the function $p(t, x, q)$. If we assume that this is linear in its arguments —that is, that $p = \alpha - \beta q - cx + gt$, where α, β, c, and g are constants—the conditions laid down in the general discussion are satisfied and the first step in the solution is to set up Euler's equation.

We must remember that the second end point is not fixed, but is restricted to lying on the line $x = a$. In other words, T is not known in advance. We know from (21) that under these circumstances we must select an extremal subject to the condition

$$F - qF_q = 0.$$

For the particular form of p that we are assuming this requires that $\beta q^2 = 0$, or $q = 0$, if we assume that $\beta \neq 0$. This result is what we should expect, since when the mine is exhausted the rate of production must be zero.

If instead of considering the course of exploitation that would be most profitable to the owner we look for that course that would be best socially, we are led to seek to maximize the integral

$$J_S = \int_0^T F(t, x, q) dt,$$

where

$$F(t, x, q) = \left(\alpha q - \frac{\beta q^2}{2} - cxq + gtq \right) e^{-\gamma t}.$$

This is on the assumption that, as before,[1] $p = \alpha - \beta q - cx + gt$.

---

[1] See Hotelling, "The Economics of Exhaustible Resources," *Journal of Political Economy*, Vol. 39 (1931), pp. 137–175.

CALCULUS OF VARIATIONS

EXERCISES

1. It is shown in § 199 that if $F(x, y, y') = y\sqrt{1 + y'^2}$ and the second end point is required to be on the curve $y = \bar{y}(x)$, an extremum must be perpendicular to this curve. Show that the same conclusion holds when $F(x, y, y') = G(x, y)\sqrt{1 + y'^2}$.

2. Find the Euler equation in case $F(x, y, z, p, q) = \sqrt{1 + p^2 + q^2}$.

3. Find the Euler equation in case $F(x, y, z, p, q) = p^2 - q^2$.

4. Set up Euler's equation for J_0 in case $p = \alpha - \beta q - cx + gt$.

5. Set up Euler's equation for J_* under the same circumstances.

CHAPTER XV

FUNCTIONS OF A COMPLEX VARIABLE

204. The geometrical representation of complex numbers. To every complex number $\alpha = x + iy$ we associate the point in the (x, y)-plane whose coordinates are x and y; and conversely, to every point (x, y) of the plane we associate the complex number $\alpha = x + iy$. This one-to-one association of the complex numbers with the points of the plane is such that the real numbers are associated with the points of the x-axis, and the pure imaginary numbers with the points of the y-axis. For this reason the x-axis is referred to as the *real axis* (or axis of reals) and the y-axis as the *imaginary axis* (or axis of imaginaries). The real part of any complex number is the x-coordinate of the associated point, and the coefficient of i is the y-coordinate. We shall speak of the plane in which these points lie as the *complex plane*.

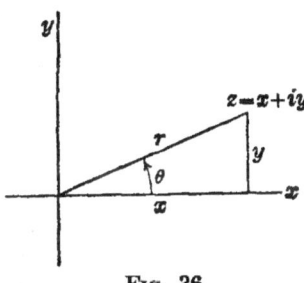

FIG. 36

This association of complex numbers with the points of the plane makes it possible to interpret a complex number as a two-dimensional vector, and carries with it a simple geometric interpretation of the operations of addition and multiplication, together with the inverse operations of subtraction and division. We can speak either of the number $\alpha = x + iy$ or the point $\alpha = (x, y)$.

205. Geometric interpretation of addition and multiplication.

(a) *Addition.* In § 11 we defined the sum of $\alpha_1 = x_1 + iy_1$ and $\alpha_2 = x_2 + iy_2$ as the number $\alpha_1 + \alpha_2 = (x_1 + x_2) + i(y_1 + y_2)$. The point $\alpha_1 + \alpha_2$ is then the fourth vertex of the parallelogram one of whose sides is the line from the origin to the point α_1, and another one the line from the origin to α_2. The figure shows both the sum $\alpha_1 + \alpha_2$ and the difference $\alpha_1 - \alpha_2$.

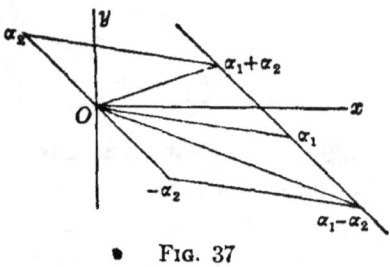

Fig. 37

(b) *Multiplication.* If P is the point (x, y) the distance OP is called the *modulus* of $\alpha = x + iy$ and is denoted by the symbol $|\alpha|$. It is always taken as positive (except when P is at the origin). We also represent it by r and accordingly have $r = \sqrt{x^2 + y^2}$. The angle measured from the positive end of the x-axis to OP is called the *argument* of α. Then $x = r \cos \theta$, $y = r \sin \theta$ and

$$\alpha = r(\cos \theta + i \sin \theta).$$

This gives us another way of writing a complex number.

We are now in a position to give a geometric interpretation to the product $\alpha_1 \cdot \alpha_2$ of the numbers α_1 and α_2. We have

$$\begin{aligned}
\alpha_1 \alpha_2 &= r_1 r_2 (\cos \theta_1 + i \sin \theta_1)(\cos \theta_2 + i \sin \theta_2) \\
&= r_1 r_2 [(\cos \theta_1 \cos \theta_2 - \sin \theta_1 \sin \theta_2) \\
&\qquad + i(\sin \theta_1 \cos \theta_2 + \cos \theta_1 \sin \theta_2)] \\
&= r_1 r_2 (\cos \overline{\theta_1 + \theta_2} + i \sin \overline{\theta_1 + \theta_2}).
\end{aligned}$$

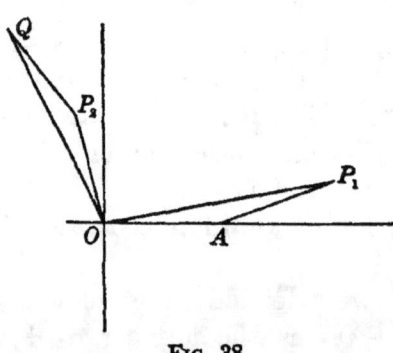

Fig. 38

That is, *the modulus of the product of two numbers is equal to the product of their moduli, and the argument of the product is equal to the sum of their arguments.*

In order to plot the point $\alpha_1 \alpha_2$ we lay off the unit segment OA from the origin on the positive end of the real axis. We then construct the triangle OP_2Q similar to OAP_1. The point Q corresponds to the number $\alpha_1 \cdot \alpha_2$.

206. Division. From the formula

$$\frac{\alpha_1}{\alpha_2} = \frac{r_1(\cos\theta_1 + i\sin\theta_1)}{r_2(\cos\theta_2 + i\sin\theta_2)} = \frac{r_1}{r_2}(\cos\overline{\theta_1 - \theta_2} + i\sin\overline{\theta_1 - \theta_2})$$

we see that the modulus of the quotient of two numbers is equal to the modulus of the dividend divided by the modulus of the divisor, and that the argument of the quotient is equal to the argument of the dividend minus the argument of the divisor.

It is now easy to give a geometric interpretation of division. We leave the details to the reader.

207. The roots of a number. The relation between a complex number and its various nth roots (n a positive integer) admits of a simple geometric exposition. If $\alpha = r(\cos\theta + i\sin\theta)$, then

$$\sqrt[n]{\alpha} = \sqrt[n]{r}\left(\cos\frac{\theta}{n} + i\sin\frac{\theta}{n}\right).$$

The modulus of the nth root is the positive nth root of the modulus of α. The argument of α has many values differing by multiples of 2π. The different values of the argument of $\sqrt[n]{\alpha}$ differ therefore by multiples of $\dfrac{2\pi}{n}$. If θ is one value of the argument of α, then $\dfrac{\theta}{n}$, $\dfrac{\theta + 2\pi}{n}$, $\cdots$, $\dfrac{\theta + 2(i-1)\pi}{n}$, $\cdots$ are values of the argument of $\sqrt[n]{\alpha}$. Any one of these differs from one of the first n of them by a multiple of 2π, while no two of the first n have such a difference. There are therefore exactly n different values of $\sqrt[n]{\alpha}$. These are equally spaced on the circle with center at the origin and radius equal to the modulus of $\sqrt[n]{\alpha}$. The figure shows the fifth roots of 1.

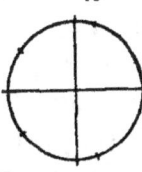

Fig. 39

208. Some elementary functions. When we are dealing with complex numbers it is customary to represent the independent variable by z and the dependent variable by w.

(a) $w^2 = z$. We know from § 207 that in this case there are two values of w that correspond to any value of z, except zero. These are

$$w_1 = \sqrt{r}\left(\cos\frac{\theta}{2} + i\sin\frac{\theta}{2}\right)$$

and

$$w_2 = \sqrt{r}\left(\cos\frac{\theta + 2\pi}{2} + i\sin\frac{\theta + 2\pi}{2}\right)$$
$$= -\sqrt{r}\left(\cos\frac{\theta}{2} + i\sin\frac{\theta}{2}\right) = -w_1.$$

If we fix our attention on a particular value of z, say $z_0 \neq 0$, there are two associated values of w. We choose one of them and call it w_1:

$$w_1 = +\sqrt{r_0}\left(\cos\frac{\theta_0}{2} + i\sin\frac{\theta_0}{2}\right).$$

As z_0 moves along a curve w_1 will also move, since at least one of the coordinates of z_0 will change. If z_0 comes back to its initial position, will w_1 return to its initial position? We can answer this

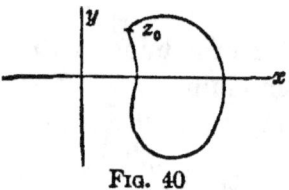

Fig. 40

question by observing the changes in θ_0, since the final

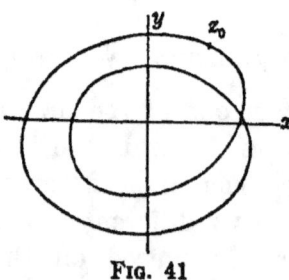

Fig. 41

value of r_0 must be the same as its initial value. As to θ_0 it is clear that it will come back to its initial value if the path described by z_0 does not include the origin. But if z_0 goes once around the origin before it comes back to its original position, the final value of θ_0 will equal its initial value increased or diminished by 2π according to the direction of the movement. Then w_1 will not come back to its original position since

$$\sqrt{r_0}\left(\cos\frac{\theta_0 \pm 2\pi}{2} + i\sin\frac{\theta_0 \pm 2\pi}{2}\right) = -w_1 = w_2.$$

An even number of circuits around the origin in the same direction will leave w unchanged, while an odd number will change its sign.

(b) *The exponential function.* Consider the infinite series

$$1 + \frac{z}{1} + \frac{z^2}{2!} + \cdots + \frac{z^n}{n!} + \cdots. \tag{1}$$

This series is suggested by the expansion of e^x in powers of x when x is real. If we write z^n in the form

$$z^n = r^n(\cos n\theta + i\sin n\theta)$$

the sum of the first $n + 1$ terms of this series is seen to be equal to

$$\left(1 + r\cos\theta + \cdots + \frac{r^n}{n!}\cos n\theta\right)$$
$$+ i\left(r\sin\theta + \cdots + \frac{r^n}{n!}\sin n\theta\right). \tag{2}$$

Now the series $1 + r + \cdots + \frac{r^n}{n!} + \cdots$ converges for all real values of r (§ 142) and $\left|\frac{r^n}{n!}\cos n\theta\right| \leq \frac{r^n}{n!}$. Hence the series in the first parentheses converges for all real values of r and θ as $n \to \infty$. A similar argument shows that the coefficient of i also converges for all real values of r and θ. Hence (1) converges for all values of z. We represent the sum by the symbol e^z. This symbol is chosen because series (1) converges to e^x when x is real. By definition then

$$e^z = 1 + \frac{z}{1} + \frac{z^2}{2!} + \cdots + \frac{z^n}{n!} + \cdots. \tag{3}$$

§208] FUNCTIONS OF A COMPLEX VARIABLE

This is called the *exponential function*. It has the following fundamental properties:

$$e^0 = 1, \qquad e^{z_1} \cdot e^{z_2} = e^{z_1+z_2}. \tag{4}$$

The first of these is immediately evident. The second can be proved by forming the Cauchy product of the two series that define e^{z_1} and e^{z_2} (see § 141).

(c) *The trigonometric functions.* We leave it to the reader to apply the argument used for (1) to show that the two series

$$z - \frac{z^3}{3!} + \cdots + (-1)^n \frac{z^{2n+1}}{(2n+1)!} + \cdots \tag{5}$$

and

$$1 - \frac{z^2}{2!} + \cdots + (-1)^n \frac{z^{2n}}{(2n)!} + \cdots \tag{6}$$

converge for all values of z. If z is equal to the real number x these are the expansions in powers of x of $\sin x$ and $\cos x$ respectively. We therefore define $\sin z$ and $\cos z$ for all values of z by the respective formulae:

$$\sin z = z - \frac{z^3}{3!} + \cdots + (-1)^n \frac{z^{2n+1}}{(2n+1)!} + \cdots \tag{7}$$

and

$$\cos z = 1 - \frac{z^2}{2!} + \cdots + (-1)^n \frac{z^{2n}}{(2n)!} + \cdots. \tag{8}$$

It follows from (8) that

$$\cos 2i = 1 + \frac{4}{2!} + \cdots + \frac{2^{2n}}{(2n)!} + \cdots > 3.$$

This may cause surprise to the reader who has learned that the sine and cosine never exceed 1 in absolute value. The explanation is that this limitation holds only for real values of the argument. For proper choices of the complex number z we can have $\sin z$ or $\cos z$ equal to any preassigned number, real or complex.

By comparing (3), (7), and (8) we find that
$$e^{iz} = \cos z + i \sin z \qquad (9)$$
and
$$e^{-iz} = \cos z - i \sin z. \qquad (10)$$
Hence
$$e^z = e^{x+iy} = e^x e^{iy} = e^x(\cos y + i \sin y). \qquad (11)$$
It follows from (9) and (10) that
$$\sin z = \frac{e^{iz} - e^{-iz}}{2i} \quad \text{and} \quad \cos z = \frac{e^{iz} + e^{-iz}}{2}. \qquad (12)$$

EXERCISE. Show that
$$\sin(z_1 \pm z_2) = \sin z_1 \cos z_2 \pm \cos z_1 \sin z_2,$$
$$\cos(z_1 \pm z_2) = \cos z_1 \cos z_2 \mp \sin z_1 \sin z_2.$$

From (9) we have $e^{2\pi i} = \cos 2\pi + i \sin 2\pi = 1$. Moreover $e^{z+2\pi i} = e^z e^{2\pi i}$. Hence
$$e^{z+2\pi i} = e^z.$$
That is, the function e^z has the period $2\pi i$.

It follows from (12) that
$$\sin(z + 2\pi) = \frac{e^{i(z+2\pi)} - e^{-i(z+2\pi)}}{2i} = \frac{e^{iz} - e^{-iz}}{2i} = \sin z$$
and
$$\cos(z + 2\pi) = \frac{e^{i(z+2\pi)} + e^{-i(z+2\pi)}}{2} = \frac{e^{iz} + e^{-iz}}{2} = \cos z.$$

Hence $\sin z$ and $\cos z$ have the period 2π.

We define the other trigonometric functions for complex values of the argument by the formulae that hold for real values of the argument. Thus:
$$\tan z = \frac{\sin z}{\cos z},$$
$$\cot z = \frac{\cos z}{\sin z},$$
$$\sec z = \frac{1}{\cos z},$$
$$\csc z = \frac{1}{\sin z}.$$

§ 208] FUNCTIONS OF A COMPLEX VARIABLE 379

(d) *Hyperbolic functions.* The functions

$$\frac{e^z + e^{-z}}{2} \quad \text{and} \quad \frac{e^z - e^{-z}}{2}$$

are important for complex values of the argument, as well as for real values. The first one is called the *hyperbolic cosine* of z(cosh z) and the second one, the *hyperbolic sine* of z(sinh z). Thus:

$$\cosh z = \frac{e^z + e^{-z}}{2}, \quad \sinh z = \frac{e^z - e^{-z}}{2}.$$

(e) *The logarithm.* If z and w are connected by the relation

$$z = e^w,$$

we say that w is the *logarithm of z:*

$$w = \log z. \tag{13}$$

The following properties of the logarithm can be deduced immediately from the definition:

(1) $\qquad \log 1 = 0,$

(2) $\qquad \log z^n = n \log z,$

(3) $\qquad \log (z_1 z_2) = \log z_1 + \log z_2,$

(4) $\qquad \log \dfrac{z_1}{z_2} = \log z_1 - \log z_2.$

Since $e^{2\pi i} = 1$ it follows that if $w = \log z$ then also $w + 2\pi i = \log z$; or, more generally, $w + 2k\pi i = \log z$, where k is any integer, positive or negative. In other words, $\log z$ is like $\sqrt{z}$ in that for any value of z ($\neq 0$) there is more than one corresponding value of $\log z$. This function is a many-valued function. But it differs radically from $\sqrt{z}$ in that it has an infinite number of values for every value of z ($\neq 0$), whereas $\sqrt{z}$ has only two.

EXERCISES

Find the real part, the imaginary part, the modulus, and the argument of each of the following numbers:

1. $(2 - \sqrt{-5})(3 + \sqrt{-1})$.   2. $(1 - 2i)^2$.   3. $\dfrac{1+i}{1-i}$.

4. $\dfrac{2}{3+4i}$.   5. $\dfrac{1+\sqrt{3}i}{1-\sqrt{3}i}$.   6. z^3 and z^4 when $z = x + iy$.

7. $\dfrac{1}{c+di}$.   8. $\dfrac{a+bi}{c+di}$, where a, b, c, and d are real.

9. $\dfrac{az+b}{cz+d}$.   10. Find and plot the cube roots of 1 and -1.

11. Do the same for the fourth and fifth roots of 1 and -1.

12. Plot all the fifth roots of $1 + i$ and show that their sum is zero.

13. Show that the sum of all the nth roots of any number a is zero.

14. Show that if $a + ib$ is a root of the equation $P(x) = 0$, where $P(x)$ is a polynomial in x with real coefficients, then $a - ib$ is also a root.

15. Give all the values of log 12.   16. Of log $(1 + 2i)$.

17. What does the relation $|z - \alpha| = r$ imply as to the relation connecting x, y, a, b, and r, if $z = x + iy$, $\alpha = a + ib$?

18. Show that $|z - \alpha| = k|z - \beta|$ is the equation of a circle if $\beta = c + id$.

19. Show how to locate $\dfrac{1}{z}$ graphically in case $0 < |z| < 1$.

209. Analytic functions. As in the case of real variables, we say that w is a function of z in a given region if to every value of z in this region there is associated one or more definite values of w. We indicate that w is a function of z by the equation $w = f(z)$.

If $w = f(z)$ and there is associated to every positive number ϵ a positive number $\delta(\epsilon)$ such that

$$|f(z) - A| < \epsilon$$

when $0 < |z - a| < \delta(\epsilon)$, we say that

$$\lim_{z \to a} f(z) = A.$$

If in addition to this condition we have

$$f(a) = A,$$

we say that $f(z)$ is *continuous* at the point $z = a$.

Derivative of a function. If the function $f(z)$ is such that

$$\lim_{h \to 0} \frac{f(a+h) - f(a)}{h}$$

exists and is independent of the way in which h approaches zero, the limit is called the *derivative of $f(z)$ at $z = a$*, and $f(z)$ is said to be *differentiable* at this point. We denote the derivative of $f(z)$, as in the case of functions of a real variable, by $f'(z)$. A function that is differentiable at every point of a region is said to be *analytic* in this region. Thus the functions defined in § 208 are analytic in any finite region within which they are defined. But not every function of z is differentiable. For example, if $w = x + 2iy$ its value is determined by the values of x and y, which in turn are determined by the value of z. Then w is a function $f(z)$ of z. If h is real

$$\lim_{h \to 0} \frac{f(z+h) - f(z)}{h}$$

$$= \lim_{h \to 0} \frac{x + h + 2iy - (x + 2iy)}{h} = 1.$$

But if h is a pure imaginary, $h_1 i$

$$\lim_{h \to 0} \frac{f(z+h) - f(z)}{h}$$

$$= \lim_{h_1 \to 0} \frac{x + 2i(y + h_1) - (x + 2iy)}{ih_1} = 2.$$

Since these two limits are not the same, w is not a differentiable function of z.

210. Condition for differentiability. In view of the fact that not every function is differentiable, we look for necessary and sufficient conditions for differentiability. We observe in the first place that if $f(z)$ is differentiable at $z = z_0$ it is continuous at this point. For

$$\frac{f(z_0 + h) - f(z_0)}{h} - f'(z_0) = \eta,$$

where $\eta \to 0$ along with h. Hence

$$|f(z_0 + h) - f(z_0)| = |hf'(z_0) + h\eta| \leq |hf'(z_0)| + |h\eta|.$$

And the last expression approaches zero with h.

We denote the real part of w by $u(x, y)$ and the imaginary part by $iv(x, y)$ and write w in the form

$$w = u(x, y) + iv(x, y).$$

If we give to x the increment Δx and let Δw be the resulting increment of w, then, since $\Delta z = \Delta x$,

$$\lim_{\Delta z \to 0} \frac{\Delta w}{\Delta z} = \lim_{\Delta x \to 0} \frac{\Delta u + i \Delta v}{\Delta x} = \frac{\partial u}{\partial x} + i \frac{\partial v}{\partial x}.$$

If on the other hand we give to y the increment Δy, $\Delta z = i \Delta y$, and we have

$$\lim_{\Delta z \to 0} \frac{\Delta w}{\Delta z} = \lim_{\Delta y \to 0} \frac{\Delta u + i \Delta v}{i \Delta y} = -i \frac{\partial u}{\partial y} + \frac{\partial v}{\partial y}.$$

But if the function is differentiable these two limits are the same:

$$\frac{\partial u}{\partial x} + i \frac{\partial v}{\partial x} = \frac{\partial v}{\partial y} - i \frac{\partial u}{\partial y}.$$

Hence

$$\left. \begin{aligned} \frac{\partial u}{\partial x} &= \frac{\partial v}{\partial y}, \\ \frac{\partial v}{\partial x} &= -\frac{\partial u}{\partial y}. \end{aligned} \right\} \qquad (14)$$

These conditions are necessary for differentiability; and if $u(x, y)$ and $v(x, y)$ together with their first partial derivatives are continuous, they are also sufficient to insure differentiability. For

$$\Delta u = \frac{\partial u}{\partial x} \Delta x + \frac{\partial u}{\partial y} \Delta y + \eta_1 \Delta x + \eta_2 \Delta y,$$

$$\Delta v = \frac{\partial v}{\partial x} \Delta x + \frac{\partial v}{\partial y} \Delta y + \eta_3 \Delta x + \eta_4 \Delta y,$$

§ 210] FUNCTIONS OF A COMPLEX VARIABLE 383

where the η's approach zero as $\Delta x \to 0$ and $\Delta y \to 0$ (see § 33). Now

$$\frac{\Delta w}{\Delta z} = \frac{\Delta u + i \Delta v}{\Delta x + i \Delta y}$$

$$= \frac{\left(\frac{\partial u}{\partial x} + i \frac{\partial v}{\partial x}\right)\Delta x + \left(\frac{\partial u}{\partial y} + i \frac{\partial v}{\partial y}\right)\Delta y + \epsilon_1 \Delta x + \epsilon_2 \Delta y}{\Delta x + i \Delta y},$$

where $\epsilon_1 = \eta_1 + i\eta_3$ and $\epsilon_2 = \eta_2 + i\eta_4$. But by virtue of (14)

$$\frac{\partial u}{\partial y} + i \frac{\partial v}{\partial y} = -\frac{\partial v}{\partial x} + i\frac{\partial u}{\partial x} = i\left(\frac{\partial u}{\partial x} + i\frac{\partial v}{\partial x}\right).$$

Hence

$$\frac{\Delta w}{\Delta z} = \frac{\partial u}{\partial x} + i\frac{\partial v}{\partial x} + \frac{(\eta_1 + i\eta_3)\Delta x}{\Delta x + i\Delta y} + \frac{(\eta_2 + i\eta_4)\Delta y}{\Delta x + i\Delta y}.$$

Moreover $|\Delta x| \leq |\Delta x + i\Delta y|$ and $|\Delta y| \leq |\Delta x + i\Delta y|$, and therefore

$$\frac{(\eta_1 + i\eta_3)\Delta x}{\Delta x + i\Delta y} \to 0 \quad \text{and} \quad \frac{(\eta_2 + i\eta_3)\Delta y}{\Delta x + i\Delta y} \to 0$$

as $\Delta x \to 0$ and $\Delta y \to 0$. But $\Delta x \to 0$ and $\Delta y \to 0$ as $\Delta z \to 0$. Hence

$$\lim_{\Delta z \to 0} \frac{\Delta w}{\Delta z} = \frac{\partial u}{\partial x} + i\frac{\partial v}{\partial x}.$$

This is a unique limit and w is therefore differentiable.

The restriction that equations (14) hold is in reality a severe restriction, as we shall see later. These equations are known as the *Cauchy-Riemann differential equations*.

If w is an analytic function of z, it can be shown that the second partial derivatives of u and v exist. Then from (14) we find that

$$\frac{\partial^2 u}{\partial x^2} = \frac{\partial^2 v}{\partial x \partial y},$$

$$\frac{\partial^2 u}{\partial y^2} = -\frac{\partial^2 v}{\partial x \partial y}.$$

Hence
$$\frac{\partial^2 u}{\partial x^2} + \frac{\partial^2 u}{\partial y^2} = 0.$$

Similarly
$$\frac{\partial^2 v}{\partial x^2} + \frac{\partial^2 v}{\partial y^2} = 0.$$

That is, if $w = u(x, y) + iv(x, y)$ is differentiable, $u(x, y)$ and $v(x, y)$ satisfy Laplace's equation

$$\frac{\partial^2 \omega}{\partial x^2} + \frac{\partial^2 \omega}{\partial y^2} = 0. \tag{15}$$

Moreover, if we take for u any real solution of Laplace's equation (15) and then determine v by the condition

$$v = \int_{(a,b)}^{(x,y)} -\frac{\partial u}{\partial y} dx + \frac{\partial u}{\partial x} dy$$

the function $w = u(x, y) + iv(x, y)$ is analytic. In the first place, since u is a solution of (15), $v(x, y)$ is a definite function of x and y (Chap. VIII, Theorem 2) and, in the second place,

$$\frac{\partial v}{\partial x} = -\frac{\partial u}{\partial y}, \qquad \frac{\partial v}{\partial y} = \frac{\partial u}{\partial x}.$$

But these are the Cauchy-Riemann equations.

EXERCISES

Which of the following functions satisfy the Cauchy-Riemann differential equations?

1. $x^2 + y^2 + 2ixy$.   2. $x^2 - y^2 + 2ixy$.   3. $x + y + i(x^2 - y^2)$.
4. $x - iy$.   5. e^{x+iy}.   6. $e^{x^2-y^2+2ixy}$.   7. $\log \sqrt{x^2 + y^2} + i \arctan \frac{y}{x}$.

8. Express the Cauchy-Riemann differential equations in polar coordinates.

211. Integration. Let z_0 and Z be two points in a region S and let $y = \varphi(x)$ be the equation of a curve C connecting these two points and lying wholly in S. We assume that $\varphi(x)$ is a continuous function of x for every x corresponding to points on the arc from z_0 to Z and that $f(z) = u(x, y) + iv(x, y)$ is continuous in S. Then we divide the arc into

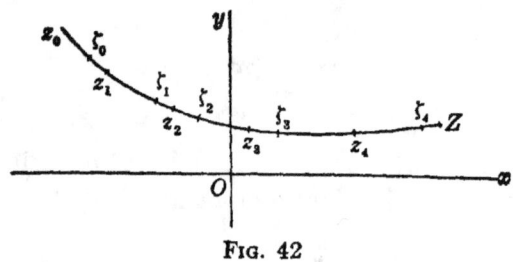

Fig. 42

n parts by the points $z_1, z_2, \cdots, z_{n-1}$ and denote by ζ_k any value of z on the arc $z_k z_{k+1}$. If the sum

$$\sum_{k=0}^{n-1} f(\zeta_k) \Delta z_k,$$

where $\Delta z_k = z_{k+1} - z_k$, approaches a unique limit as the maximum length of all the arcs approaches zero, we call this limit the *integral of $f(z)$ over the curve C from z_0 to Z* and represent it by the symbol

$$\int_C f(z) dz.$$

We have now to show that this limit exists under the given hypothesis concerning $f(z)$; namely, that $f(z)$ be continuous. If we put

$$f(\zeta_k) = u_k + iv_k,$$

then

$$f(\zeta_k)\Delta z_k = (u_k + iv_k)(\Delta x_k + i\Delta y_k)$$
$$= u_k \Delta x_k - v_k \Delta y_k + i(v_k \Delta x_k + u_k \Delta y_k),$$

and

$$\lim_{n \to 0} \sum f(\zeta_k)\Delta z_k = \lim \sum (u_k \Delta x_k - v_k \Delta y_k)$$
$$+ i \lim \sum (v_k \Delta x_k + u_k \Delta y_k)$$
$$= \int_C u dx - v dy + i \int_C v dx + u dy \quad (\S\ 99). \quad (16)$$

CAUCHY'S INTEGRAL THEOREM. *If C is a closed curve and $f(z)$ is analytic throughout the region bounded by C and continuous on C, then $\int_C f(z)dx = 0$.*

The proof of this theorem follows immediately from (16) since each of the line integrals in the right member is of the form

$$\int P dx + Q dy,$$

with $\dfrac{\partial P}{\partial y} = \dfrac{\partial Q}{\partial x}$. This last equation is merely a restatement of one of the Cauchy-Riemann differential equations which $u(x, y)$ and $v(x, y)$ satisfy.

212. Cauchy's integral formula. Let C be a closed curve lying wholly in a simply connected region S within which $f(z)$ is analytic. We assume that C is composed of one or more curves, or parts of curves, whose equations are of the form $y = \varphi_i(x)$, where $\varphi_i(x)$ is continuous within the interval determined by the corresponding arc. Select any point a within the region enclosed by C and draw the circle C' within this region with center at a. The function $\dfrac{f(z)}{z-a}$ is analytic within the region bounded by C and C'.

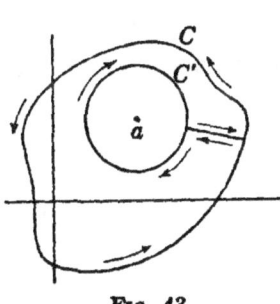

Fig. 43

In order to have a continuous boundary for this region we draw a cross cut from a point on C' to a point on C and make this line a part of the boundary. We know from the integral theorem that the integral of $\dfrac{f(z)}{z-a}$ taken entirely around this continuous boundary is zero. In this integration the cross cut is integrated over twice, but in opposite directions, and these two integrations offset each other. Hence

$$\int_C \frac{f(z)}{z-a} dz + \int_{C'} \frac{f(z)}{z-a} dz = 0.$$

The arrows indicate the direction of integration. We have then

$$\int_{C} \frac{f(z)}{z-a} dz = \int_{C'} \frac{f(z)}{z-a} dz. \tag{17}$$

Now

$$\int_{C'} \frac{f(z)}{z-a} dz = \int_{C'} \frac{f(a)}{z-a} dz + \int_{C'} \frac{f(z)-f(a)}{z-a} dz. \tag{18}$$

Since $f(a)$ is a constant,

$$\int_{C'} \frac{f(a)}{z-a} dz = f(a) \int_{C'} \frac{dz}{z-a}.$$

For points on C' we have $z - a = \rho e^{i\theta}$ where ρ is constant, and $dz = i\rho e^{i\theta} d\theta$. Then

$$\int_{C'} \frac{dz}{z-a} = \int_0^{2\pi} \frac{i\rho e^{i\theta} d\theta}{\rho e^{i\theta}} = 2\pi i$$

and

$$\int_{C'} \frac{f(a) dz}{z-a} = 2\pi i f(a). \tag{19}$$

We denote the last integral in (18) by J:

$$J = \int_{C'} \frac{f(z)-f(a)}{z-a} dz.$$

Since $f(z)$ is analytic, and therefore uniformly continuous in S, we can take ρ sufficiently small to satisfy the inequality

$$|f(z) - f(a)| < \epsilon$$

for any pre-assigned positive ϵ, when z is any point on C'. For such a choice of ρ

$$|J| < \frac{\epsilon}{\rho} \cdot 2\pi\rho = 2\pi\epsilon.$$

Then J approaches zero when ϵ does. But the other integral in the right member of (18) has a value that is

independent of ρ. Hence
$$\int_C \frac{f(z)}{z-a} dz = \int_{C'} \frac{f(a)}{z-a} dz = 2\pi i f(a),$$
or
$$f(a) = \frac{1}{2\pi i} \int_C \frac{f(z)}{z-a} dz. \tag{20}$$

This is *Cauchy's integral formula*. It bears out the statement made in § 210 to the effect that analytic functions are a severely restricted class of functions. For it shows that if $f(z)$ is analytic in S its value at any point of S is wholly determined by its values along any closed curve of the kind described that lies wholly in S and encloses the given point.

If we differentiate each side of (20) with respect to a we get the formula
$$f'(a) = \frac{1}{2\pi i} \int_C \frac{f(z) dz}{(z-a)^2}.$$

The validity of the differentiation under the integral sign can be seen as follows:

$f(a + \Delta a) - f(a)$
$$= \frac{1}{2\pi i} \int_C f(z) \left[\frac{1}{z - a - \Delta a} - \frac{1}{z-a} \right] dz$$
$$= \frac{\Delta a}{2\pi i} \int_C \frac{f(z) dz}{(z - a - \Delta a)(z-a)}.$$

Hence
$$\frac{f(a + \Delta a) - f(a)}{\Delta a} - \frac{1}{2\pi i} \int_C \frac{f(z) dz}{(z-a)^2}$$
$$= \frac{1}{2\pi i} \int_C f(z) \left[\frac{1}{(z - a - \Delta a)(z-a)} - \frac{1}{(z-a)^2} \right] dz$$
$$= \frac{\Delta a}{2\pi i} \int_C \frac{f(z) dz}{(z - a - \Delta a)(z-a)^2}.$$

But this last integrand is uniformly bounded along C as

§ 213] FUNCTIONS OF A COMPLEX VARIABLE 389

$\Delta a \to 0$. We conclude therefore that the integral is bounded and that

$$f'(a) = \frac{1}{2\pi i} \int_C \frac{f(z)dz}{(z-a)^2}.$$

We can see in a similar way that

$$f^{(n)}(a) = \frac{n!}{2\pi i} \int_C \frac{f(z)dz}{(z-a)^{n+1}}. \tag{21}$$

213. Taylor's theorem. Let $f(z)$ be a function that is analytic within a region S and continuous on its boundary C. If a is any point within S we draw the largest circle possible with a as center that contains no point of C in its interior. For any fixed point z in the interior of the circle and a variable point ζ on C we have $|z - a| < r$ and $|\zeta - a| \geq r$, where r is the radius of the circle. Now

$$\frac{1}{\zeta - z} = \frac{1}{\zeta - a - (z-a)} = \frac{1}{\zeta - a} \cdot \frac{1}{1 - \dfrac{z-a}{\zeta - a}}.$$

But

$$\frac{1}{1 - \dfrac{z-a}{\zeta - a}} = 1 + \frac{z-a}{\zeta - a} + \cdots + \frac{(z-a)^{n-1}}{(\zeta - a)^{n-1}}$$

$$+ \frac{(z-a)^n}{(\zeta - z)(\zeta - a)^{n-1}}.$$

Hence

$$\frac{1}{\zeta - z} = \frac{1}{\zeta - a} + \frac{z-a}{(\zeta - a)^2} + \cdots + \frac{(z-a)^{n-1}}{(\zeta - a)^n}$$

$$+ \frac{(z-a)^n}{(\zeta - z)(\zeta - a)^n},$$

and

$$\frac{1}{2\pi i} \int_C \frac{f(\zeta)d\zeta}{\zeta - z} = \sum_{k=0}^{n-1} \frac{(z-a)^k}{2\pi i} \int_C \frac{f(\zeta)d\zeta}{(\zeta - a)^{k+1}}$$

$$+ \frac{1}{2\pi i} \int_C \frac{(z-a)^n f(\zeta)d\zeta}{(\zeta - z)(\zeta - a)^n}.$$

In the light of (20) and (21) we see that this is equivalent to

$$f(z) = \sum_{k=0}^{n-1} \frac{(z-a)^k}{k!} f^{(k)}(a) + \frac{1}{2\pi i} \int_C \frac{(z-a)^n f(\zeta) d\zeta}{(\zeta - z)(\zeta - a)^n}.$$

We have now to consider what happens to this last integral as we increase n without limit.

In the first place $\left|\dfrac{z-a}{\zeta-a}\right| \leq \left|\dfrac{z-a}{r}\right| = b$, where b is a positive constant less than 1. Now $|\zeta - z| \geq r - |z| = c$, a positive constant. Since the function $f(\zeta)$ is bounded on C there is a number M such that for every point on C we have $|f(\zeta)| < M$. Then, in case C is rectifiable,

$$\left| \int_C \frac{(z-a)^n f(\zeta) d\zeta}{(\zeta - z)(\zeta - a)^n} \right| < \frac{M b^n l}{c},$$

where l is the length of C. It follows from this that the absolute value of the integral approaches zero as n increases without limit. Hence the series $\sum_{k=0}^{\infty} \dfrac{(z-a)^k}{k!} f^{(k)}(a)$ converges to $f(z)$ for every value of z within the circle. That is,

$$f(z) = \sum_{k=0}^{\infty} \frac{(z-a)^k}{k!} f^{(k)}(a). \tag{22}$$

This discussion shows that if $f(z)$ is analytic in the neighborhood of a, the expansion (22) is valid, provided that z is sufficiently near to a. If no such expansion is valid—that is, if $f(z)$ is not analytic in the neighborhood of a—the point a is called a *singular point* of $f(z)$. We can say that the series (22) converges to $f(z)$ in the circle whose center is at a and whose radius is equal to the distance from a to the nearest singular point of $f(z)$. For if b is such a singular point and z is any point within the circle through b with center at a, we can select an r such that $|z - a| < r < |b - a|$, and then take for C the circle of radius r and center a. It can be shown that the series diverges for every z outside this larger circle. It may converge or diverge for a value of z on the circumference.

Formula (22) is of great importance in the theory of functions of a complex variable. We have introduced it here however primarily for the light it throws on certain questions concerning the behavior of real functions of a real variable. Consider, for example, the expansion of $\tan x$ in powers of x. It is practically impossible to determine directly the radius of convergence of this series because we cannot write the general term of the series. But if we assume that the only singular points of the function $\tan z$ are $(2k+1)\frac{\pi}{2}$, we see immediately from the preceding discussion that this radius is $\frac{\pi}{2}$.

EXERCISES

Get the first three non-vanishing terms in the expansion of each of the following functions, and find the radius of convergence in each case:

1. $\sec x$ in powers of x. 2. $\cot x$ in powers of $x - \frac{\pi}{2}$.

3. $\dfrac{x}{1+x^2}$ in powers of $x - 3$. 4. $\dfrac{2x+1}{x^2 - 3x + 2}$ in powers of $x + 1$.

5. $\dfrac{1}{x^2 + 2x + 3}$ in powers of x.

214. Real integrals. In this paragraph we shall apply Cauchy's integral theorem to the evaluation of certain definite integrals in the real domain.

(a) $\int_0^\infty \dfrac{\sin x}{x} dx$. We have already seen that this improper integral is convergent (§ 83). In order to evaluate it we consider the

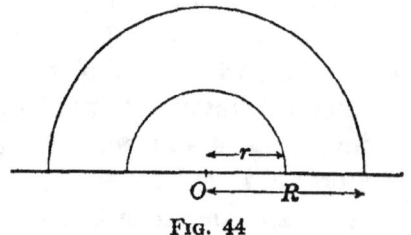

Fig. 44

integral $\int \dfrac{e^{iz}}{z} dz$ taken in the positive direction over the path indicated in the figure. The integrand is analytic in

the region bounded by the path and is continuous along it.
Hence by the theorem

$$\int_C \frac{e^{iz}}{z} dz = \int_r^R \frac{e^{ix}}{x} dx + i\int_0^\pi e^{-R \sin\theta + iR \cos\theta} d\theta$$
$$+ \int_{-R}^{-r} \frac{e^{ix}}{x} dx + i\int_\pi^0 e^{-r \sin\theta + i r \cos\theta} d\theta = 0.$$

If we put $x = -t$,

$$\int_{-R}^{-r} \frac{e^{ix}}{x} dx = -\int_r^R \frac{e^{-it}}{t} dt = -\int_r^R \frac{e^{-ix}}{x} dx.$$

Hence

$$\int_r^R \frac{e^{ix}}{x} dx + \int_{-R}^{-r} \frac{e^{ix}}{x} dx = \int_r^R \frac{e^{ix} - e^{-ix}}{x} dx = 2i \int_r^R \frac{\sin x}{x} dx,$$

and therefore

$$2\int_r^R \frac{\sin x}{x} dx + \int_0^\pi e^{-R \sin\theta + iR \cos\theta} d\theta$$
$$+ \int_\pi^0 e^{-r \sin\theta + i r \cos\theta} d\theta = 0.$$

If we determine the limit of the second integral in the left member of this equation as $R \to \infty$ and the limit of the third integral as $r \to 0$, we shall have the value of the integral $\int_0^\infty \frac{\sin x}{x} dx$. Now

$$\left| \int_0^\pi e^{-R \sin\theta + iR \cos\theta} d\theta \right| \leq \int_0^\pi e^{-R \sin\theta} d\theta = 2\int_0^{\pi/2} e^{-R \sin\theta} d\theta,$$

since $|e^{-R \sin\theta + iR \cos\theta}| = e^{-R \sin\theta}$. Let $f(\theta)$ be defined as follows:

$$f(\theta) = \frac{\sin\theta}{\theta} \quad \left(0 < \theta \leq \frac{\pi}{2}\right),$$
$$f(\theta) = 1 \quad (\theta = 0).$$

Then in the interval $0 < \theta < \frac{\pi}{2}$

$$f'(\theta) = \frac{\theta \cos\theta - \sin\theta}{\theta^2} < 0.$$

[§ 214] FUNCTIONS OF A COMPLEX VARIABLE

This inequality follows from the fact that for the values of θ in the given interval $\tan \theta > \theta$ and that therefore $\theta \cos \theta - \sin \theta = \cos \theta (\theta - \tan \theta) < 0$. For $\theta = \frac{\pi}{2}$, $\theta \cos \theta - \sin \theta = -1$. Hence as θ increases from 0 to $\frac{\pi}{2}$ the function $f(\theta)$ decreases from 1 to $f\left(\frac{\pi}{2}\right) = \frac{2}{\pi}$. In other words,

$$\sin \theta \geq \frac{2\theta}{\pi} \tag{23}$$

when $0 \leq \theta \leq \frac{\pi}{2}$. From this we conclude that

$$e^{-R \sin \theta} \leq e^{-2R\theta/\pi}.$$

Then

$$\int_0^{\pi/2} e^{-R \sin \theta} d\theta \leq \int_0^{\pi/2} e^{-2R\theta/\pi} d\theta = \frac{\pi}{2R}(1 - e^{-R}) \underset{R \to \infty}{\longrightarrow} 0.$$

The integral $\int_\pi^0 e^{-r \sin \theta + ir \cos \theta} d\theta$ is a function of r, say $F(r)$.

$$F(r + \Delta r) - F(r)$$
$$= \int_\pi^0 e^{r(-\sin \theta + i \cos \theta)} \left[e^{\Delta r(-\sin \theta + i \cos \theta)} - 1 \right] d\theta.$$

We restrict r to the interval $(0, h)$, where $h > 0$. Now
$$|F(r + \Delta r) - F(r)|$$
$$\leq \int_\pi^0 \left| e^{r(-\sin \theta + i \cos \theta)} \left[e^{\Delta r(-\sin \theta + i \cos \theta)} - 1 \right] \right| d\theta.$$

This integrand is less than ϵ for all values of r and θ under consideration, provided that Δr is sufficiently small in absolute value. Hence $F(r)$ is a continuous function of r for $0 \leq r < h$. But $F(0) = -\pi$. Then $F(r) \underset{r \to 0}{\longrightarrow} -\pi$. We have therefore

$$\int_0^\infty \frac{\sin x}{x} dx = \frac{\pi}{2}.$$

(b) $\int_0^\infty e^{-x^2}dx$. This is an important integral in the theory of probability. It can be readily evaluated without resort to complex numbers. However we give such an evaluation here because of its bearing on the next integral we wish to discuss.

Consider the real double integral

$$I = \iint e^{-x^2-y^2}dxdy.$$

Let I_a be the value of this integral when taken over the square with center at the origin and sides parallel to the axes and equal to $2a$.

$$I_a = \left(\int_{-a}^a e^{-x^2}dx\right)^2.$$

If we take the integral over the circle with center at the origin and radius equal to R, we get, in terms of polar coordinates,

$$I_R = \int_0^R dr \int_0^{2\pi} re^{-r^2}d\theta = \pi(1 - e^{-R^2}) \quad \text{(see § 102)}.$$

The square is contained between the circle of radius $R = a$ and the one of radius $R = a\sqrt{2}$. Hence the integral over the square lies between the integrals over these two circles, since the integrand is positive everywhere.

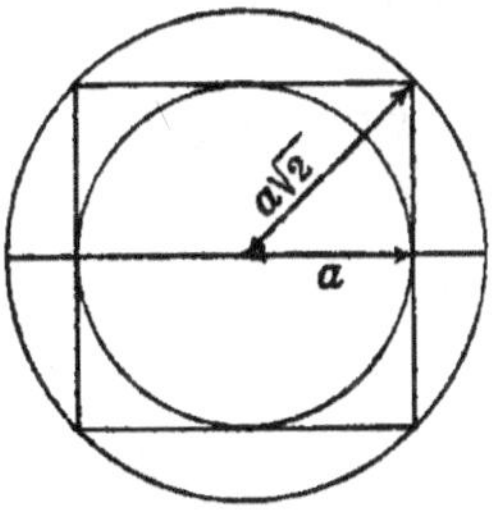

$$\pi(1 - e^{-a^2}) < \left(\int_{-a}^a e^{-x^2}dx\right)^2 < \pi(1 - e^{-2a^2}).$$

Fig. 45

If we let $a \to \infty$, we get

$$\left(\int_{-\infty}^\infty e^{-x^2}dx\right)^2 = \pi,$$

or

$$\int_0^\infty e^{-x^2}dx = \frac{\sqrt{\pi}}{2}.$$

(c) $\int_0^\infty \sin x^2 dx$, $\int_0^\infty \cos x^2 dx$. These integrals have the same value, as we shall see. They are the Fresnel integrals that occur in the theory of diffraction.

In order to evaluate them we start with the integral $\int e^{-z^2} dz$ taken around the boundary of the sector indicated in the figure. Since the integrand is analytic in every finite region, the value of this integral is zero. In integrating over the arc we put $z = Re^{i\theta} = R(\cos \theta + i \sin \theta)$.

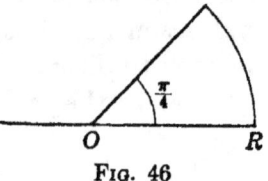

Fig. 46

$$\int_0^R e^{-x^2} dx + i \int_0^{\pi/4} e^{-R^2(\cos 2\theta + i \sin 2\theta)} Re^{i\theta} d\theta + \int_{Re^{i\pi/4}}^0 e^{-z^2} dz = 0.$$

Now

$$\left| \int_0^{\pi/4} e^{-R^2(\cos 2\theta + i \sin 2\theta)} Re^{i\theta} d\theta \right| \leq R \int_0^{\pi/4} e^{-R^2 \cos 2\theta} d\theta$$

$$= \frac{R}{2} \int_0^{\pi/2} e^{-R^2 \sin \varphi} d\varphi < \frac{\pi}{4R}(1 - e^{-R^2}) \underset{R \to \infty}{\to} 0,$$

where $2\theta = \frac{\pi}{2} - \varphi$. The last inequality follows from (23).

In the integral $\int_{Re^{i\pi/4}}^0 e^{-z^2} dz$ put $z = \frac{1+i}{\sqrt{2}} t$. When

$$z = R\left(\cos \frac{\pi}{4} + i \sin \frac{\pi}{4}\right) = \frac{R}{\sqrt{2}}(1+i),$$

$t = R$. Hence

$$\int_{Re^{i\pi/4}}^0 e^{-z^2} dz = -\frac{1+i}{\sqrt{2}} \int_0^R e^{-it^2} dt$$

$$= -\frac{1+i}{\sqrt{2}} \int_0^R (\cos t^2 - i \sin t^2) dt.$$

In the limit as $R \to \infty$

$$\int_0^\infty e^{-x^2} dx = \frac{1+i}{\sqrt{2}} \int_0^\infty \cos t^2 dt - \frac{i(1+i)}{\sqrt{2}} \int_0^\infty \sin t^2 dt.$$

If we equate the real and the imaginary parts on the two sides of this equation, we find that

$$\frac{\sqrt{\pi}}{2} = \frac{1}{\sqrt{2}}\int_0^\infty \cos t^2 dt + \frac{1}{\sqrt{2}}\int_0^\infty \sin t^2 dt,$$

$$0 = \frac{1}{\sqrt{2}}\int_0^\infty \cos t^2 dt - \frac{1}{\sqrt{2}}\int_0^\infty \sin t^2 dt.$$

From this we conclude that

$$\int_0^\infty \sin t^2 dt = \int_0^\infty \cos t^2 dt = \frac{1}{2}\sqrt{\frac{\pi}{2}}.$$

Suggestions For Further Reading

While the kind of course you've just experienced is a thing of the past at most universities, the books that evolved from that era have not vanished. In fact, many of the most popular and influential have been reissued inexpensively thanks to the venerable Dover Books. We're going to focus on texts that fit the "old school advanced calculus" criteria established by the description given in the Preface and by Fite as the example. Consequently, this will be a relatively short Recommended Reading section compared with the other books republished so far by Blue Collar Scholar.

Of all the classical texts on advanced calculus, to me, the most comprehensive and representative of the old-school AC course is the wonderful book by Louis Brand (3). Indeed, Brand is in the public domain and it was my first choice as the basis for this book. But Dover beat me to republishing it and did a fine job. So although it would be legal to reissue it myself, it would have been utterly stupid to do so. (Especially when a fine second choice like Fite was available.) Brand covers all of the same topics as Fite except that the chapter on the calculus of variations is replaced with a chapter on vector algebra, which is used extensively in the sections on multivariable calculus. Moreover, it is more modern and advanced in presentation, has many more examples and is somewhat more detailed then Fite. It is an excellent text, in some ways superior to Fite. I would certainly recommend it at least as supplementary reading.

Dover has recently reissued another major advanced calculus text that's worth mentioning in some detail: the famous 1940 text by Phillip Franklin (4). Again, it's pretty striking how similar in overall subject content all these textbooks are, which demonstrates how stable the classical analysis course following calculus has been in America. The differences between the books lie in how detailed each author is in certain aspects depending on the expertise of the author and how this reflects what they perceive the role of the advanced calculus course is. Franklin states in his preface that he believes the role of the advanced calculus course is to supply all the rigorous theory of analysis-both real and complex-that mathematics and serious physical science students will need as a foundation for graduate study and research. As a result, the book is very rigorous in nature. It emphasizes analysis over geometry much more so then Fite or Brand-in fact, virtually to the exclusion of all else. Franklin's book is more sophisticated than either previous book. It focuses on careful definitions and proofs of all major analysis ideas, theorems and lemmas. It has entire chapters on sequences of functions, Fourier integrals, the Steiljes integral and special functions like the gamma function. It also has 2 chapters giving a careful-if old fashioned-introduction to complex analysis. Important theorems of analysis in a classical form are featured, such as complete proofs of the Heine-Borel and Bolzano-Wierstrass theorems on the real line and $\mathbf{R}^2$ Trigonometric, exponential, and logarithmic functions are developed from the functional equations they satisfy instead of geometric definitions. There are some applications, such as arc-length and surface area, but all of them are presented in analytic terms with careful derivations and proofs. There is virtually no discussion of vectors even in old-fashioned "arrow" algebra in $\mathbf{R}^2$ or $\mathbf{R}^3$ Most striking, there are deliberately no numerical or concrete examples in the book, all cases and theorems are presented and proven in general terms. Indeed, Franklin suggests in the preface students should keep their calculus book handy to supply examples! I know many older mathematicians who swear by this book and say that you can't get a better education in basic analysis then working through this book side by side with a standard calculus book. I don't know, but it certainly is a crisply written, authoritatively thorough and very beautiful book for its' focused goals, authored by one of the great American mathematicians of the first half of the 20th century. If you're fascinated by classical analysis and the old fashioned textbooks that present it, this is a great addition to your library-and now that it's available in Dover, there's no reason not to have it.

Another book by another famous mathematician that's also been reissued by Dover is David Widder's 1947 text (5). Widder was a longtime analyst at Harvard University and this book emerged from decades of his teaching the course. Unfortunately, Widder's take on AC is more 19th century then any of the books above. In fact, in some ways, this book is the opposite of the book by Franklin. There is almost no rigorous analysis. Basically, Widder's book is a careful and visual treatment of old school multivariable calculus from the old "arrows" point of view with topics in infinite series and special functions needed for applications, particularly to differential equations. Students that buy it hoping for a good treatment of classical analysis are going to be disappointed. That being said, the book is quite readable and careful. Widder fills the book with lots of nice exercises of a diverse nature. An instructor could teach a very nice and careful multivariable calculus course from the first half and a nice methods course for physicists or engineers from the rest of it. An old fashioned but inexpensive and versatile book by a master.

Yet another very solid book in this category saved from limbo for an inexpensive price by Dover is the excellent book by Avner Friedman (6). Again, the chosen topics in the book are quite standard and you won't see anything bizarre jump out at you from the Table of Contents. In fact, it doesn't cover any complex variables or variational analysis, so in a sense it covers less and is more focused then the previous books. The book is extremely well written and organized. Like Franklin, it has more of an analytical bent then the other texts discussed here. Unlike Franklin, it is far more modern in its treatment of functions of several variables-utilizing both vector algebra and the topology of the $\mathbf{R}^2$ and $\mathbf{R}^3$ -and has a host of wonderful examples and problems. Although not as complete as Brand's text, Friedman is a superior expositor. Also, there's a number of topics that aren't discussed in the other texts discussed here, such as semi-continuity, absolute convergence, Jordan content, harmonic functions and ordinary differential equations. In Friedman, we probably have the single most modern and readable of all the classical advanced calculus texts currently in print and it is well worth considering for study.

By the way, Friedman is a guy who's had a very odd and interesting career as a mathematician and I think it's worth noting in passing. After getting his PhD from the Hebrew University way back in 1959, he taught over the next 40 years at Northwestern University, Purdue University and The University of Minnesota where he did research in the theory of partial differential equations and taught a number of post-calculus level analysis courses, including the one that formed the basis for this book. While Friedman was well respected among analysts with a solid if not spectacular research record, he kept a pretty low profile and wasn't exactly a famous name. In 1987, after spending his entire teaching/research career as a pure mathematician and at an age when most mathematicians are considering retirement, he suddenly took an interest in the relatively new field of industrial mathematics. His late blooming second career as an applied mathematician took off. In 2001, he arrived at Ohio State University as another nascent field-mathematical bioscience-was beginning. He applied his mastery of partial differential equations to many problems in that field, including modeling human heart rhythms and disorders that disrupt it. He became the Director for the Mathematical Biosciences Institute there until he retired to become Distinguished University Professor. 2 lessons from his story: a) Hardy's old dictum about young men proving theorems and old men writing books is BS *because Friedman did the exact opposite in his career* and b) As long as you're in good health and willing to work, nothing in your chosen field is impossible at any age.

For historical interest, you may wish to look at the very first published advanced calculus books from the early 20th century, many of which have entered the public domain and therefore have been republished very cheaply by many small publishers. The texts by Osgood (7), Wilson (8) and Woods (9) are all very representative of the kinds of proto-advanced calculus courses taught at top universities in the early decades of the last century. Indeed, Wilson in particular has a preface which is most enlightening about the motivations behind his thinking. He describes the book as "a further course in calculus" and that the book's main thrust is "vigor rather than rigor". While these books aren't completely intuitive and do prove several definitions and results carefully, it's clear this is not the primary purpose of these courses. Their primary purpose is to build on the student's elementary calculus of one variable course which would be insufficient for those looking to achieve careers in either mathematics or the physical sciences. This means a course primarily in multivariable calculus, differential equations, infinite series and their associated applications. Woods specifically describes his course as being designed with mostly students of applied mathematics, physics and engineering in mind. What modern students may find most surprising in these three texts is how similar many of the chapters are to those in conventional modern textbooks on multivariable calculus and differential equations. Indeed, the old fashioned treatments contained therein will be superior in both insight and presentation to that in many standard recent books.

I'd be derelict in my duty if I didn't recommend 2 last textbooks for the student interested in classical advanced calculus to look at. And they're both by the same co-author. When Fite wrote his text in the late 1930's and the advanced calculus course was beginning to be revised into the form it had before it ended in the 1980's at most universities, most analysts were either using their own lecture notes or a famous textbook to teach the sequence: Richard Courant's calculus book (10). Yes, that's right. If you look carefully at the contents of Courant's classic-it more or less corresponds precisely to the post- WWII year-long AC course! Volume 1 gives a rigorous one variable calculus course. Volume II gives a late 19th century style multivariable calculus course utilizing vector calculus on $\mathbf{R}^2$ and $\mathbf{R}^3$, complete with a huge number of applications. Many universities did in fact use Courant when teaching the "new" AC course in America-both in its' original German and in the later English translation. If you're interested in classical analysis, you simply must look at this classic. When math programs became much stronger after WW II, Courant's book was considered too old fashioned to teach classical analysis from and the author himself recognized this. Therefore, Courant wrote an "Americanization" of his old classic with Fritz John for use with modern advanced calculus courses (11). To be honest, "Americianization" is really a misnomer-although Courant was referring to the un-rigorous manner in which elementary calculus was traditionally taught in the US. A more accurate way of describing the "new" book was as a *modernization* of Courant's old course using the concepts of linear algebra and topology. It preserves much of the same spirit of the original textbook-great precision, thorough proofs and many applications. It's really in the second volume on multivariable calculus that the differences between the 2 versions becomes much more profound, as linear transformations, matrices and differential forms all play a vital role in defining the structures of manifolds embedded in $\mathbf{R}^n$. I don't agree that differentiable manifolds should play a role in courses at this level and many of the exercises are far too difficult. But otherwise, it's a remarkable course. In many ways, (11) represents the apex of the advanced calculus sequence-it has all the best versions of the courses' virtues and all the vices of over ambition of the post-Bourbaki American mathematics curricula of the 1970's that ultimately spelled it's doom. In any event, it's a wonderful book by 2 masters and well worth checking out. I would strongly recommend considering (11) if you want to experience the best of the advanced calculus courses. Be warned, you *will* have to supplement the course extensively with many simpler exercises-which fortunately are easy to obtain today on the Web. Unfortunately, the new edition by Springer-Verlag is quite expensive and hard to get. See if you can get a second-hand copy of the 1965 and 1972 editions of Volumes 1 and 2 respectively.

All the above texts can and should serve as excellent supplements to either an honors course, self study-or just flat out for people that love great mathematics and know that the good old days didn't get everything wrong.

References

1) Kline, Morris, *Mathematical Thought From Ancient To Modern Times*, 2nd ed, Oxford University Press, 1990, Volumes 1-3

2) Tucker, Alan, "The History of the Undergraduate Program in Mathematics in the United States", *The American Mathematical Monthly*, *120*(**8**), October 2013

3) Brand, Louis, *Advanced Calculus: An Introduction To Classical Analysis*, Dover Publications, 2006

4) Franklin, Phillip, *A Treatise On Advanced Calculus*, Dover Publications, 2016

5) Widder, David, *Advanced Calculus*, Dover Publications, 2nd ed, 1989

6) Friedman, Aver, *Advanced Calculus*, Dover Publications, 2007

7) Osgood, William Fogg, *Advanced Calculus*, Macmillan Company, 1933

8) Wilson, Edwin Bidewell, *Advanced Calculus: A Text Upon Select Parts of Differential Calculus, Differential Equations, Integral Calculus, Theory of Functions, With Numerous Exercises,* Forgotten Books, 2017

9) Woods, F., *Advanced Calculus*, Ginn, Boston, 1926.

10) Courant, Richard; McShane, Edward James, *Differential And Integral Calculus Volumes 1 and 2,* Ishi Press, 2010

11) Courant, Richard; John, Fritz, *Introduction to Calculus and Analysis*, Vol. 1, Springer-Verlag, part 1, 1998, Volume 2, part 2 1999 ; Volume 2 part 1, 2000, Volume 2, 2001

About The Author

William Benjamin Fite (1869-1932) got his PhD in mathematics at Cornell University in 1901. His advisor was George Abram Miller (1863-1951) and the title of the dissertation was "On Metabelian Groups". He was appointed Professor of Mathematics in Columbia University, New York, in 1911. He was best known as the Treasurer of the American Mathematical Society during the period 1921-1929. Besides writing a research-level treatise on metabelian groups, which was a popular research subject in the 1930's, he wrote high school and university level textbooks such as *College Algebra, First course in algebra, Second course in algebra* and several others. He also wrote several very significant papers on group theory and differential equations. *Advanced Calculus*, published in 1938, was his most sophisticated textbook, although not his best known. It is hoped this republication with new material will give it the significance it lacked during his lifetime.

About the Publisher/ Editor:

Karo Maestro aka The Mathemagician is the *nome-de-plume* of a former graduate student in mathematics. His true identity will remain a secret for now, but one day soon will be revealed to all. Some things this online enigma can reveal: He was a distinguished undergraduate student as a double major in mathematics and biochemistry whose poor health prevented completing graduate studies. Unbowed and undaunted, he plans to return to ultimately obtain a PhD before dying. Partially to that end, he is building a publishing company-Blue Collar Scholar-committed to making high quality sources of mathematics-both original works and reprints-available widely and inexpensively to students of all backgrounds. The volume you hold is the second published. He also hopes to fall in love with and marry one of the most beautiful women on Earth before he dies, but frankly the doctorate is far more likely. He has 3 bachelors' degrees, in philosophy, physical chemistry and mathematics as well as minors in biochemistry and psychology. He is also a past reviewer of textbooks for the Mathematical Association Of America. Among his more recent achievements are the website, TULOOMATH (www.tuloomath.com) , which is designed to be a one stop hub for free downloadable lecture notes and online textbooks in university mathematics from high school algebra to PhD level topics. He's also the author of *Tables,Chairs And Beermugs* (https://tableschairsandbeermugsmathemagician.blogspot.com/) , the associated blog for the website where he reviews textbooks and vents on matters mathematical and academic. He'll soon be beginning 2 more blogs on progressive politics and pop culture. He's painfully blunt, opinionated and has a comment on just about everything in creation-from math textbooks to progressive politics to how to make a perfect cup of organic green tea. He loves JRR Tolkien, science fiction, comic books and movies of all kinds, the comedy of the late great George Carlin, playing with his wonderful nieces, real barbeque, creative burgers and fresh cut French fries, tall curvaceous women and popular music, particularly Bon Jovi and U2. Among his current musical favorites are Alysia Cara, Adele, Imagine Dragons and Ed Sheeran. He is currently working on 2 original books on how to earn a PhD in pure mathematics by self-study alone.

(No, he's not insane-at least he doesn't think so.)

And he still has no idea what people see in Ariana Grande. Still no clue.

INDEX

Abel's lemma, 224
  test for uniform convergence, 225
Addition, complex, 13, 372
  real, 7
Area of curved surface, 178

Barrier, 174
Bilateral surface, 187
Bliss, 351
Bôcher, 118
Borel, 240
Bound, upper and lower, 6
Brachistochrone, 357

Carslaw, 281
Cauchy form of remainder, 65
  integral formula, 386
  integral theorem, 386
  product, 246
  root criterion, 212
Cauchy-Riemann differential equations, 383
Center of mass, 109
Change of variables in a double integral, 174
  in a triple integral, 185
Characteristic curve, 342
Continuity, definition of, 18
  uniform, 23, 40, 42
Convergence, Abel's test for uniform, 225
  absolute, 213
  absolute, for double series, 243
  conditional, 213
  d'Alembert's test for, 212
  quasi-uniform, 239
  uniform, 218
  uniform, of improper integrals, 144
  uniform, of infinite integrals, 143
  Weierstrass' test for uniform, 223
Contour lines, 53
Covariant, 312
Cross cut, 174
Curvature, 328

Darboux's theorem, 90
De la Vallée-Poussin, 61, 306

Derivative, definition of, 26, 381
  directional, 51
  of a composite function, 28
  of a power series, 250
  partial, 43
Developable surface, 344
Differential, 35
  exact, 54
  of arc, 338
  of higher order, 36
  total, 44, 200
Differentiation of infinite integrals, 227
  of improper integrals, 229
  under the integral sign, 101, 142
Directrix of a catenary, 348
Discontinuity, point of, 19
  first kind, 19
Division, by power series, 257
  complex, 374
  real, 10
Du Bois Reymond, 164

Economics, application in, 369
Edge of regression, 344
Envelope of curves, 321
  of surfaces, 342, 345
Euler's constant, 69
  equation, 353, 361
  theorem, 59
Evolute, 331

Force of interest, 369
Fourier constants, 267
  series, 267
Functions, admissible, 359
  analytic, 380
  Beta, 148
  continuous without a derivative, 237
  differentiable, 28, 44
  dominant, 252
  elementary, 375–380
  Gamma, 146
  integrable, 91, 157, 182
  orthogonal, 291

sectionally continuous, 19
sectionally smooth, 19

Goursat-Hedrick, 179
Gradient of a function, 53
Greatest limit, 205, 213
Green's theorem for the plane, 166
   for space, 187

Hamilton's integral, 366
   principle, 367
Hankel, 16
Hessian, 311
Hobson, 164
Hotelling, 370

Indeterminant forms, 72
Infinite, 34
Infinitesimal, definition of, 34
   order of, 34
   principal part of, 35
Inflection, point of, 76
Integrals, definite, 91
   Dirichlet, 285
   double, 157
   elliptic, 122
   Fourier, 288
   improper, 133
   indefinite, 103
   infinite, 139
   line, 166
   real, 391
   repeated double, 160
   repeated triple, 182
   surface, 186
   triple, 181
Integration, approximate, 105
   by parts, 142
   complex, 384
   field of, 182
   of rational functions, 117
   under the integral sign, 142, 235
Interval, closed and open, 20
Intrinsic equation, 333
Inversion, 306

Jacobian, 308

Lagrange multiplier, 81
   remainder, 65
Landen's transformation, 127
Legendre polynomials, 295
Lemniscate, 121

Length of arc, 336
Lower sum, 89

MacLaurin's series, 67
Maximum and minimum values of
   functions of one variable, 76
   of functions of two variables, 79
Mean value, 112
Mean value theorem for derivatives, 32
   for double integrals, 159
   for integrals, 97, 98
Moigno-Lindelof, 357
Moment, 110
   of inertia, 111
Monotonic, 13
Multiplication, complex, 13, 372
   of series, 245
   real, 8
Multiply connected regions, 172

Normal plane, 30
Numbers, complex, 13
      argument of, 373
      modulus of, 373
   hypercomplex, 16
   irrational, 3
   rational, 1
   real, 3

Oscillation of a function, 24
Osculating circle, 329
   plane, 339
Osgood, 327

Parametric equations, 319, 365
Pendulum, simple, 124
   spherical, 367
Pierpont, 297
Potential, 195, 296
Power series, simple, 247
   double, 259

Radius of convergence, 248
   of curvature, 329
Reciprocal of a number, 9
Rolle's theorem, 31
Roots of a number, 374

Schwarz's inequality, 264
Section, 2
Sequence, 205
   convergent, 205

INDEX

divergent, 205
  greatest limit of, 205
Series, infinite, 208
  double, 240
  trigonometric, 264
Serret-Scheffers, 153, 154, 232, 327
Simply connected regions, 172
Simpson's rule, 106
Singular points, 51, 306, 347, 390
Solid of revolution, 96
Stirling's formula, 150
  series, 153
Stokes' theorem, 188
Subtraction, complex, 13
  real, 8
Surface of revolution, minimum, 355

Tangent line, 30
  plane, 48
  to a skew curve, 307
Taylor's series, 67
  generalized, 77
  theorem, 389

Unicursal curves, 121
Uniform approach, 93
Unilateral surface, 186, 187
Upper sum, 89

Variations, 355

Weber and Wellstein, 16
Weierstrass' theorem, 290
Work, 198

www.ingramcontent.com/pod-product-compliance
Lightning Source LLC
Chambersburg PA
CBHW052235220526
45471CB00001B/52